Locating, Classifying and Countering Agile Land Vehicles

David D. Sworder • John E. Boyd

Locating, Classifying and Countering Agile Land Vehicles

With Applications to Command Architectures

 Springer

David D. Sworder
Department of ECE, UCSD
University of California, San Diego
La Jolla, CA, USA

John E. Boyd
Cubic Defense Applications
San Diego, CA, USA

ISBN 978-3-319-19430-1   ISBN 978-3-319-19431-8   (eBook)
DOI 10.1007/978-3-319-19431-8

Library of Congress Control Number: 2015944259

Springer Cham Heidelberg New York Dordrecht London
© Springer International Publishing Switzerland 2016

Printed on acid-free paper

Springer International Publishing AG Switzerland is part of Springer Science+Business Media (www. springer.com)

Preface

In many modern tracking applications, target detection and placement are performed to support some specific decision or action. Estimates of target position and motion will be used for some command purpose, perhaps beginning with determining target intent and then selecting an appropriate course of action like intercepting or otherwise countering the target. The challenge is to construct an architecture that uses noisy data sequences of indicated location not only to derive estimated position, velocity, and attitude of the target but also to provide the information required to support effective command functions. This book focuses on command architectures—estimation and information structures that can effectively inform useful command decisions. The target engagements we have in view almost always demand that a command architecture function in real time. Frequently, there are strict limits on the interrogation and countermeasure resources expended. Command architectures considerably expand the character and complexity of target trackers, whether relatively simple, like alpha-beta algorithms, or more sophisticated, like extended Kalman filter and multiple-model estimators.

We contrast the performance of command architectures based upon a single-model representation of motion with those based on multiple-model approaches to the same problem. The latter start with the same raw measurement data sets as the former but use more sophisticated situation assessment tools to facilitate the simultaneous tasks of locating, classifying, and countering. Our purpose is to highlight the advantages a hybrid problem formulation provides for the decision-making command algorithms.

We explore several prototypical engagements. In each, we determine the quality of a single-model algorithm and explore its sensitivities when the target is uncooperative and given to abrupt changes in kinematic state. We contrast the estimator's actual performance with that predicted on the basis of the model. We then turn to a hybrid algorithm that integrates an element of situation awareness with the same kinematic data set found deficient in the single-model algorithm. Again the sensitivities are presented. The comparison is done with a view to the requirements of a command architecture and the effectiveness of indicated decisions.

The book is intended for engineers who are tasked with developing command algorithms for applications that must operate in highly variable and ambiguous environments. The cases presented and the architectures considered will also serve to illustrate the crucial dependence of effective command on model complexity, model realism, and estimation quality for real-world plants and systems.

We have included, for the interested reader, detailed analytical developments in an appendix. The appendix also provides an example of a step-by-step implementation of the hybrid estimation algorithm. To function in a wildly changing environment, the hybrid algorithm is necessarily quite intricate. For the architectures in this book, we have chosen to work in 2D (motion in the plane) and, to make the engagements concrete, have referred to the target as a ground object. These constraints can be easily relaxed.

The authors' earlier book, *Estimation Problems in Hybrid Systems* (Cambridge, 1999), explored the mathematical structure of one form of hybrid estimation. This book utilizes that approach in an assortment of different engagement applications. The strict mathematical basis of tracking algorithms is often violated in applications. This book explains and illustrates the engineering judgments that must be employed when the developmental hypotheses are violated.

This work contrasts the performance of command architectures in several particular engagements. In each case, the algorithms are presented the same kinematic data stream. The idiosyncratic engagements are chosen to illustrate the distinctive sensitivities of alternative modeling approaches rather than to provide a detailed performance comparison for a specific application.

The authors wish to thank Cubic Defense Applications for their support in publishing this book and, in particular, for the graphics and editing support of their technical publications department.

La Jolla, CA, USA   David D. Sworder
San Diego, CA, USA   John E. Boyd
February 2015

Contents

Symbols and Notational Conventions

Precise description, the defining and comparison of models, estimators, and command architectures requires rather complex notation. The authors have attempted to enhance accessibility of this book by following particular notational conventions. This section briefly discusses these principles and spells out the usage or meaning of many of the symbols used in the text. Some pronunciation hints are provided.

The work, and the definitions given here, assume familiarity with mathematical analysis of probability and stochastic processes and with estimation of dynamical systems. A search of the Internet may yield useful reviews and quick definitions where the authors' assumptions are optimistic.

Notational Conventions

- Font styles are used to distinguish related concepts. For example, x is used for the kinematic state 4-vector of a target moving in the plane. The Greek letter chi, χ, is used to denote the location coordinate 2-vector of such a target.
- Math calligraphic style is used for collections, especially evolving time series collections, of information such as filtrations.
- Diacritical and accent marks name a quantity derived from and related to a variable. For example, "hat" of x, $\hat{x}$, indicates the estimated value (mean value) of the variable x.

Symbology

Operators

$\hat{x}$ A carat set above the symbol for a variable, e.g., $\hat{x}$, is used to denote the estimated or mean value of the variable, in this case x. ["x hat"]

$\tilde{x}$ A tilde set above the symbol for a variable, e.g., $\tilde{x}$, is used to denote the error in an estimate of the variable. In this case $\tilde{x} = x - \hat{x}$. ["x tilde"]

$\vee$ The filtration generated by the union of two filtrations, e.g., $\mathscr{Y}[k] \vee \mathscr{Z}[k]$. ["vee"]

E Mathematical expectation.

$|$ In an expression for a random variable, the vertical bar, $|$, denotes a conditional. For example, $\mathbb{P}(x|y)$ is the probability of x given the condition y. [read as "given"]

diag The main diagonal of a square matrix or, if D is an n-vector, diag(D) is the n by n diagonal matrix of D.

cov Covariance.

$\triangleq$ Defined. For example, $\mathrm{cov}(x) \triangleq E[\tilde{x}\tilde{x}']$ means $\mathrm{cov}(x)$ is defined as the covariance of the error in x.

$\mathbf{H}^+$ The mapping by which the tracker's kinematic state estimate is adjusted to accommodate the regime for the next time step.

$\mathbf{C}^{\mathscr{M}}$ The mapping by which a kinematic state estimate is returned to the closest point on the map. Putting the target position onto the nearest road.

Estimators

KF Kalman filter

EKF Extended Kalman filter. Any version of the Kalman optimal linear estimator that is applied to a problem that does not meet the linearity, independence, and Gaussian assumptions required by the Kalman filter. A common example in this work arises from the conversion of range-bearing measurements of target position to cartesian coordinates using as the conversion reference point the estimated target position, violating the independence assumptions required by the KF.

GWE Gaussian Wavelet Estimator.

Filtrations and Information Sets

$\mathscr{F}$ A filtration, an increasing sequence of σ-algebras. ["F"]

$\mathscr{Y}$ The filtration generated by a time sequence of kinematic measurements, i.e., position and velocity. ["Y"]

$\mathscr{Z}$ The filtration generated by a time sequence of modal, or regime, observations. ["Z"]

$\mathscr{G}$ The composite filtration of $\mathscr{Y}$ and $\mathscr{Z}$: $\mathscr{G} \overset{\Delta}{=} \mathscr{Y} \vee \mathscr{Z}$. ["G"]

$\mathscr{O}[k]$ Filtration generated by the exogenous process of system modes or regimes. ["O"]

$\mathscr{M}$ The set of all information contained in a map: road locations, nominal local speeds, sensor obstructions, intersections, and the like. $\mathscr{M}$ contains all the information known to the estimator but is not a filtration. ["M"]

Symbols

1 Vector of 1s.

I The identity matrix.

$\boldsymbol{\alpha}$, $\boldsymbol{\alpha}_t$ Often used in a Gaussian sum as the normalized or unnormalized modal distribution.

D Discernibility matrix: the inverse of a covariance matrix.

$\mathbf{e}_i$ The ith unit vector.

$\mathbf{E_i}$ The ith unit matrix. If $i \leq n$, $\mathbf{E_i} = \mathrm{diag}(\mathbf{e}_i)$.

ι A regime sequence of length L.

ι^+ A regime sequence of length $L + 1$.

κ The modal set. The set of possible regime sequences of length L.

$\boldsymbol{\phi}$, $\boldsymbol{\phi}_t$, $\boldsymbol{\phi}[k]$ Our convention is to use ϕ to represent the mode or regime of a hybrid process.

$\mathbb{P}$ Probability. Used for the probability distribution function of a random variable.

$\mathbb{P}^\sim$ The projected distribution. A Gaussian approximation to $\mathbb{P}$ matching $\mathbb{P}$'s first two moments. ["P tilde"]

P_{xx} Covariance of the kinematic state, x.

R_x Covariance of the sensor noise process.

ϱ The distance between two points. For example, the distance between an expected position measurement and its actual value. Also used for Mahalanobis distance, the distance of a point from a distribution, typically measured in units of standard deviation. ["rho"]

S The regime index. The total number of regimes is S.

v The velocity components of a kinematic state.

x_t The value at time t of the continuous-time kinematic state process x.

$x[k]$ The value at time kT of the discrete time kinematic state process x at $t = kT$.

χ The location coordinates of a target in a 2D plane. Thus the target kinematic state x is the "stack" of χ and v: $x_t = \begin{bmatrix} \chi_t \\ v_t \end{bmatrix}$.

Chapter 1
A Model for Tracking and Classification

Abstract In modern system architectures, the command algorithm must do more than merely locate a point equivalent of an agile target. The algorithm must place the target, classify it, and, if need be, counter any threat. The command algorithm designer must produce a causal transformation that not only tracks the motion but also places an assurance window about the target for interrogation and perhaps neutralization. When the target is uncooperative, this requires a sophisticated multi-model approach. This chapter frames the problem and introduces the notation used in what follows.

1.1 Introduction to the Problem

System integration requires an understanding of the system's basic capabilities along with a sense of the interrelationships of the system states and the influences of the operational environment. The system architect begins with an analytical representation of the internal and external excitations and infers the system response based upon these processes. In a practical problem, the system is required to accomplish some specific task with limits on available resources. For example, the system must direct countermeasures from a limited inventory toward an unwelcome object passing toward a secured location. In another example, the system must locate and visually classify an object of indeterminate intent. In both examples, the command system must locate a dynamic object and then make a decision based on the quality of its location estimate. This book presents a study of command architectures for estimation and resource allocation. In the context of several rather unadorned engagements, we will discuss the sensitivities of command algorithms to simplified environmental models.

Over at least the past century, model-based approaches have proven useful in system integration. Such approaches broadly require:

- An analytical model of the system that includes all interfaces with the outside world. This system model relates objective inputs and outputs to an internal group of dynamic states that produce the distinctive system behaviors. In the target tracking problems we consider in this book, these behaviors are primarily the target kinematics. In this context, *state* has a specific meaning: a set of internal

© Springer International Publishing Switzerland 2016
D.D. Sworder, J.E. Boyd, *Locating, Classifying and Countering Agile
Land Vehicles*, DOI 10.1007/978-3-319-19431-8_1

variables that allow the external valuables to be extrapolated forward in time. For simplicity in architectural design, however, the model is a reductionist abstraction that creates a notional future that may differ significantly from the actualized future.

- An observational model that delineates the situational measurements available to the command system. This model not only lists the elements of this data stream, but it also provides the timing and the quality of the individual components. Some architectures are able to avail themselves of more information than others. We want to quantify the utility of auxiliary measurements to insure the cost of acquiring them is reflected in their utility.
- A causal algorithm selected to map the observations into an estimate of the relevant states of the system. This, along with a high quality measure of the reliability of the estimate, permits rational command decisions to be made. Based upon the algorithm, countermeasures can be deployed autonomously, action agents can be enabled.

Command architectures require that we must first locate the object of inquiry (the target). This part of the problem is frequently referred to as target tracking. A vehicle is moving within some fixed motion space and we must find it. The commander may want to classify the target as friendly or hostile, predict the future position of the target, counter the target at an opportune time, etc. For example, the target vehicle may be a car following a road path toward a designated area, and the engineer seeks an algorithm that provides current position and classifies the vehicle as friend or foe. In another example, the vehicle is an airplane moving toward a protected area, and the engineer seeks an algorithm that classifies the airplane's intent and selects appropriate countermeasures. In both cases, we wish to determine where the vehicle is, and where it is likely to go. In the target tracking problem, the system is the moving vehicle, its model is a mathematical representation of how its observable outputs (position, velocity, perhaps orientation or current maneuver mode) derive from its inputs like operator control (if known), random influences, and constraints (e.g., roads). We assume the system outputs are measured with quantifiable accuracy and frequency. We require an implementable algorithm (the *estimator* or *tracker*) that maps a measurement sequence into an estimate of the current (or future) state of the vehicle.

For the tracking sub-architecture (the system estimator), recognition that not all target classes move and maneuver in the same way can greatly improve tracker accuracy and command effectiveness. There may be different kinds of targets in view, and any one target might have multiple maneuver modes. Trucks and motorcycles turn and accelerate differently. An aircraft may spend much of its time in constant velocity flight, but in turns its motion is better modeled as constant lateral acceleration. A tracker that can account for these differences is greatly aided by a system model that acknowledges the different classes and modes. In a comprehensive model of the engagement, the *kinematic* states of the system are augmented by *regime* states that correspond to target type and motion mode.

Fig. 1.1 A tank is moving on an irregular terrain

A carefully constructed tracker will take the regime process into account, and a classifier, if available, can greatly aid estimation in this hybrid environment.

Tracking is enhanced by the fact that target motion occurs within a context. For example, consider a tank moving in the open area as shown in Fig. 1.1. The basic information set available to the tracker is a sequence of such images. From the camera gimbal angle, the bearing to a point object is measured. But there is far more information implicit in the image process. For example, we might classify the target by comparing the measured image with a list of stored images of like type; e.g., Fig. 1.2. A fit metric quantifies a class distribution. Further processing could then yield the longitudinal orientation of the vehicle. Range could be inferred from the size of the image. All of these target properties flow from careful image processing. From these rather ambiguous image interpretations, the command algorithm must generate a region within which the target will be found and an action appropriate to the encounter. Commonly, we split this problem into sub-problems: identify the target type and intent, neutralize the target if appropriate.

Figure 1.1 is an optimistic case in which the target is contrasted with an undifferentiated backdrop. The target bearing can be measured. Its range can be estimated from its size. Direction can be estimated from its aspect angle. Of course, all of these attributes are subject to uncertainty.

Even the existence of a target in the field of view may not be a sure thing. Figure 1.3 shows a situational image in which there are no targets of interest. But simple edge or contrast detection algorithms might well identify many targets. Hence, in a complex engagement, the classifier must be aware of the possibility of false detections.

Fig. 1.2 A single template in an archive of possible targets

Fig. 1.3 An image in which there are no targets of interest

Figure 1.1 provides a clear visual image of the target. But other sensors are not nearly as sharp. Figure 1.4 is a FLIR-image of a tank. The engine is the differentiated sub-region of the figure A sequence of FLIR images is more difficult to interpret because of the low contrast.

The kinds of measurements used in tracking and prediction are of various sorts: range-only (GPS), range-bearing (radar), optical target image (TV), infrared target image (FLIR), etc. The measurements may yield information on a spatially distributed object (an optical silhouette), or they may be reduced to information about a point-equivalent target (range to the center-of-reflection).

Fig. 1.4 A FLIR image of a tank. The engine in the main region of contrast. At some aspect angles, the engine is obscured

Other relevant tracking information is not explicitly temporal: a topographic map may display regions of possible motion along with prohibited regions. Or a road profile could be accessible by the tracker with the intent of correlating the measured motion with the profile [11]. Broadly, we will refer to such non-temporal resources as an inventory of para-measurements. Para-measurements will be integrated with the conventional kinematic measurements to produce the final actions in an engagement.

To illustrate some of the algorithm development issues, consider ballistic missile defense. A tracker/classifier identifies an approaching ballistic target (BT) within a specified category list. Impact point prediction (IPP) is accomplished by extrapolating the estimated kinematic state of the BT forward to impact. Accurate IPP requires rapid target classification, a high quality motion model, and good missile state estimates during flight.

A sophisticated tracker will do more than simply provide IPP. It will compute complementary performance metrics including bias (in this example, the mean deviation of impact point from a missile aim-point), and the *circular error probable* (CEP). The CEP is a circle of such radius that it captures 50 % of the impacts. The use of CEP assumes that the actual impact points are distributed over a roughly circular area. Thus, a comprehensive figure of merit for a BT tracker would involve both the magnitude of the point-placement bias and the size of the dispersal region (the CEP).

In the applications we will investigate, the uncertainty regions in tracking and prediction are not well modeled as circular. The target may have preferred directions as it is constrained by internal and external factors. For this reason, we will generalize CEP, and talk about *elliptical error probable* (EEP). The EEP ellipse is such that it captures the target with probability 50 %.

In a command architecture, the estimator's inferred accuracy influences the actions that follow. We may deploy countermeasures to intercept the target with a probability of, say, 90 %. A flawed calculation of EEP may lead either to a too little cover area and a failed intercept or to a too big cover area with a waste of limited resources.

For reasons developed above, the self-reported estimate of tracker accuracy, whether CEP, EEP, error-covariance matrix, or other metric, assumes considerable importance in the command architectures. Even in Kalman filters and their kin, the update of target location estimate uses a filter gain (a multiplier of the difference between the extrapolated observation and that measured) to adjust target placement. The gain depends upon the measurement noise statistics and the uncertainty in the filter's extrapolated state. If the latter is overly large, the filter gain is large and the location estimate volatile. If the latter is overly small, the filter gain is small and the location estimate has not the power to follow rapid changes in direction.

In the engagements we will explore in the sequel, the system uncertainties cannot be expressed in terms of the second order statistics of the tracking error. We will encounter multi-mode location densities that will test our engineering judgements. We will use the EEP as a broad confidence measure. But precise asset allocation will require a more nuanced view of situational awareness.

Model-based trackers integrate the measurement aggregate (including para-measurements) into an estimate of the state of a target using an engagement model as an intermediary. The first part of the model, the motion model, describes the temporal evolution of the target's kinematic state.

Within the motion model, the engineer distinguishes the idiosyncrasies of various target types: nominal speed and turn rate, location constraints, number and coordination of vehicles, etc. For example, the target shown in Fig. 1.5 is distinguishable from the tank in Fig. 1.1 both in its maneuverability and terrain capability. A kinematic model for the former would not be appropriate for the latter, and if used, estimator accuracy is likely to be degraded.

The second part of the model—the measurement model—describes the type and quality of the measurements. In addition to the temporal measurements; e.g., GPS pseudo-range, this sub-model contains a description of the para-measurement aggregate.

The tracker architecture depends on the cooperation, or lack thereof, of the target itself. A cooperative vehicle will declare its type and intent to the tracker; e.g., an aircraft in an air-traffic-control application is cooperative insofar as it gives the tracker its identification number and flight plan. Type and intent information is useful for reducing the uncertainties in the kinematic state estimate. For example, the tracker is cued to look for particular flight patterns when intent is signaled. The cooperative target is the most commonly studied because classificational ambiguities are avoided in the engagement model.

Alternatively, tracking an uncooperative target is more difficult because important global identifiers of the engagement are missing. If such qualities are required,

Fig. 1.5 Shape analysis allows the tracker to identify the specific target type and match the kinematic model with the mobility of the target

they must be inferred concurrently with state estimation. Indeed, some uncooperative targets make location prediction difficult by deliberately masking relevant identifiers; e.g., the hostile target masks intent. In this circumstance, the tracker/classifier must predict position by inferring intent from the motion patterns of the target.

In this book, we will look at the problem of tracking and classifying an agile ground target. We will see that the cooperative–uncooperative dichotomy oversimplifies many engagements; e.g., a cooperative target may wish to signal intent to the tracker but the data link is not adequate for the task. We will, therefore, consider tracking algorithms in a mixed environment in which the classificational attributes of the target may be only partially known. The objective is to present high quality, implementable (recursive) estimators of both the kinematic state and also such category variables as are relevant. The form that the estimator takes depends upon the sensor architecture used and the para-measurements available.

The performance of the proposed algorithms will usually be illustrated with an example of planar motion. A 2D motion space is rich enough to display important issues in tracker synthesis without becoming so complex as to become tedious. A space of higher dimension has the same broad character as 2D motion, but the trackers become more complex in appearance. Unfortunately, even in 2D, the notation required to describe the tracking environment is more convoluted than is typical in estimation theory. The next sections introduce some conventions used in the sequel and explain their need.

1.2   A Hybrid State Model of a Maneuvering Target

The concept of the target state conveys the notion of predictability. For example, the target state at time t_0 is the information required to uniquely determine the output at $t > t_0$ given the forward excitation [2]. The extrapolation requires an analytical model that relates the relevant processes. We will call this the kinematic model.

In this book we will consider vehicles that move in an n-dimensional Euclidean state space with kinematic state vector x_t. The state has position components, χ_t, and velocity components, v_t along with such other dynamic variables as are appropriate. If the target is moving in 2D, χ_t is two-dimensional as is the velocity. Necessarily, n is at least 4. But n will be larger than 4 when the target model moves in a higher dimensional space or when the target has other relevant states like acceleration, actuator dynamics, and so on. The totality of these vector components is called the *global kinematic state* of the target. For convenience, we will stack position over velocity at the top of x_t.

Extrapolation is based upon a model of the dynamic properties of the target. But such properties depend upon the type of target we are following. A large truck has not the agility of a motorcycle even though both may traverse the same road course. Either could be described by a simple Newtonian state model, but the coefficients in the equations would be different. So when we develop a command architecture, we must first settle on the class of targets we can expect to see in a particular engagement.

To impose a structure on the exercises that follow, we will need a notational convention flexible enough to delineate uncertainty in both situational and kinematic conditions. The former we call the regime state; the latter, the kinematic state.

1.2.1   The Modal State

1.2.1.1   The Operating Regime

We will have more to say about x_t in the next section. But for now we observe that sophisticated engagement models acknowledge that the target vehicle (or vehicles) will confront different macro-conditions as the engagement evolves. We use a discrete variable, the regime state, ϕ_t, to point to the current situational status of the engagement [19].

For example, suppose the engagement involves tracking a moving vehicle. The target could be an M1 Abrams tank or it could be a motorcycle. The tracker must distinguish between two macro-conditions: regime #1 ($\phi_t = 1$) if the target is an M1 Abrams tank; regime #2 ($\phi_t = 2$) if the target is a motorcycle. (We will frequently use a right arrow to indicate such mappings, and later, transitions and implications. Here, $\phi_t = 1 \mapsto$ the target is a tank.) The agility and directional restrictions on these two vehicle classes are very different, and they must be distinguished in the kinematic model.

In contrast to the kinematic state vector, the regime index ϕ_t is a pointer. We call ϕ_t the regime state (though it may not have the predictive properties we want from a state variable), and assume it has a finite range. As an example, there might be a list of possible targets, each distinguished by its own kinematic model. ϕ_t points to the particular kind of target in view. More generally, for any kind of regime, we can list the S alternatives. Let $\mathbf{S}$ denote the sequence of the first S natural numbers: $\mathbf{S} = \{1, 2, \ldots, S\}$. Then the ith regime is identified by $\phi_t = i \in \mathbf{S}$. The regime state is identified with a number in a counting system with radix S. In this framework, $\mathbf{S}$ is called the regime alphabet, and it represents an ordered list of possible regimes. In the (tank, motorcycle) scenario, we make the identification: $\phi_t = 1$ means the target is a tank; $\phi_t = 2$ means the target is a motorcycle. The kinematic extrapolation equation utilizes $\phi_t = i$ where i is chosen appropriately.

The regime structure can be described in another way. Let $\mathbf{e}_i$ be the ith canonical unit vector in $\mathbb{R}^S$; e.g., $\mathbf{e}_1 = (1, 0, \ldots, 0)'$. We can associate the ith unit vector with the regime, the target is of type i: $\mathbf{e}_i \mapsto$ target is type i. The regime being $\mathbf{e}_1$ means the target is a tank. For expositional simplicity, we will use the vector and numerical designations interchangeably and without comment: $\phi_t = \mathbf{e}_1$ and $\phi_t = 1$ are alternative labels for the statement that the target is a tank. The modal primitive is classificational and has none of the common analog properties of the conventional kinematic state; e.g., $\mathbf{e}_1 + \mathbf{e}_2$ is not a regime because it is not a unit vector. So the regime space is not a vector space in the usual mathematical sense.

When the regime variable is target type, the interpretation of ϕ_t is unambiguous; the target is a tank or it isn't. In many applications, however, the separation of local- and macro-conditions is not so clear cut. Suppose the target is a motorcycle constrained to follow a rectangular road grid aligned to the four cardinal directions. With the target type known, the regime space covers travel north (N or $\phi_t=1$), travel south (S or $\phi_t=2$), travel east (E or $\phi_t=3$), travel west (W or $\phi_t=4$) with perhaps *stop* appended. So if $\phi_t = \mathbf{e}_1$, the north-specific motion model should be used with strong longitudinal accelerations and weak lateral accelerations. In this example, ϕ_t is a pointer to a velocity within four directional bins. But *north* is an analog variable in that north motion includes a range of velocities that are only close to true north. The ambiguity concerning direction of motion within the $\phi_t = \mathbf{e}_1$ bin must be transferred to the kinematic state which contains velocity as a component. In this case, we see that the vagary of direction is masked in our simplified definition of the regime.

The regime can point to composite events. For example, $\{\mathbf{e}_1, \ldots, \mathbf{e}_4\}$ may correspond to the tank moving on the (north, south, east, west) grid; e.g., $\mathbf{e}_1$ implies a tank is headed north. Suppose $\{\mathbf{e}_5, \ldots, \mathbf{e}_8\}$ corresponds to a motorcycle moving on the same motion space. In this case, the regime space with two target types and four directions is partitioned into eight bins.

If the regime is target type, ϕ_t is unchanging. In other applications the macro-conditions change during the engagement; e.g., the northbound motorcycle may come to a road junction and turn east. The range of the regime state is still $\mathbf{S}$, but ϕ_t may vary in time: $\phi_t = \mathbf{e}_5$ transitions to $\phi_t = \mathbf{e}_7$ at the turn. We will suppose that the regime process, $\{\phi_t\}$, is constant with isolated discontinuities. When $\{\phi_t\}$ changes from the jth regime state to the ith state, $\Delta\phi_t = \phi_t - \phi_{t-} = \mathbf{e}_i - \mathbf{e}_j$. We will say that ϕ_t experienced a transition event at time t. The regime process is a right-continuous temporal process with range space $\mathbf{S}$.

The global state of the target evolves on some time interval $[0, \mathsf{T}]$. But in the work that follows, the tracker/classifier will be designed on the basis of a time-discrete approximation to the state process. Let T be a sample interval. The time-continuous regime process is reduced to a sequence, $\{\phi[k] = \phi_{kT}; k \in \{0, 1, 2, \ldots\}\}$. If the time interval of the engagement is not specified, the complete regime process becomes a string of arbitrary length—or a radix S number with an arbitrary number of digits.

We will call $\{\phi[k]\}$ the regime event process. If $\phi[k + 1] \neq \phi[k]$, there has been a regime transition in the interval $(kT, (k + 1)T]$; if $\phi[k + 1] = \phi[k]$, the regime continues unchanged—at least at the sample times.

There are some important modeling ambiguities that arise when we approximate a time-continuous process with a time-discrete representation. The discontinuities in the regime process can occur at any time in $[0, \mathsf{T}]$. When we place these discontinuities at the sample times, we ignore the influence on system of the precise time of a change in regime. We also ignore multiple transition events within an interval; e.g., $\phi[k] = \phi[k + 1]$ does not imply $\phi_t \equiv \phi[k] \in (kT, (k + 1)T]$. In what follows, the consequences of these temporal uncertainties is implicit, but T will be assumed to be small enough that we can aggregate all of these effects within the basket of exogenous uncertainties.

An issue of more subtlety is the choice of the principal modal condition over a sample interval. If there is a regime transition in the interval, when did it occur? Using our discrete time process model and notation, the time-continuous process $\{(x_t, \phi_t)\}$ is mapped to $\{(x[k], \phi[k])\}$. The vector $(x[k], \phi[k])$ appears as an initial condition in a kinematic extrapolation formula for the forward step. However, the actual mapping $x[k] \mapsto x[k + 1]$ is an explicit function of the regime (or regimes) of operation over $t \in [kT, (k+1)T)$. The initial value $\phi[k]$ does not capture the portion of the kth-time interval during which the regime is actually $\phi[k]$. Within this interval, perhaps $\phi_t = \phi[k]$ for a long time. Or perhaps $\phi_t = \phi[k + 1]$ is dominant—or, in the case of multiple modal transitions in a single interval, even a regime different from either. The temporal partition precludes a nuanced description of the intra-interval variation in the regime. With this is mind, we will suppose that multiple intra-sample transitions occur infrequently, and further, that the initial regime will serve as a proxy over the whole of the interval at least as concerning the kinematic extrapolation to $x[k + 1]$. Of course, this adds to our basket of uncertainty.

1.2.1.2 Regime Sequences and Languages

Extrapolation in the command architecture depends upon both where you are and where you have been—the future depends upon the past and not just the present. Our regime model requires concepts and notation for the past–present–future of the regime state. Focusing on the kth intra-sample interval, the regime proxy is $\phi_{kT} = \phi[k] = \mathbf{e}_i$. The kinematic extrapolation uses the ith primitive over the full interval, $[kT, (k+1)T)$. Label the principal regime in the preceding interval $\mathbf{e}_j$, and the successor regime by $\mathbf{e}_p$. Then $[\mathbf{e}_p, \mathbf{e}_i, \mathbf{e}_j]$, also represented as pij, is a regime string centered on $[kT, (k+1)T)$, representing a segment of state history in reverse time. We can view pij as a word with alphabet $\mathbf{S}$, or a three-digit number with radix S. In this context, we will use "word," "number," and "string" interchangeably. If the target is a motorcycle moving on the usual road grid, the event 144 points to the motorcycle making a north turn from the west at $t = (k+1)T$.

Our notation differs from that used to describe languages in automata [2]. In that convention, the state sequence is listed from left to right with the oldest event to the left; e.g., in the above illustration, the motorcycle motion would be labeled 441 or (west, west, turn north). We wish to emphasize the current and future regimes, and we will do so by placing them at the left of the regime sequence. Our strings are retrograde when contrasted with the usual automata sequence labels.

We will have frequent occasion to consider these sequences of regimes. Let us define three sets of interest in this application. Let L be an integer memory length in the application. Let κ_L (which we shall usually abbreviate as κ) be the set of all S-radix, L-digit numbers with elements ι. Define $\kappa^+ = \kappa_{(L+1)}$ to be the set of all S-radix, $(L+1)$-digit numbers with elements ι^+, and define $\kappa^- = \kappa_{L-1}$ to be the set of all S-radix, $(L-1)$-digit numbers with elements $\iota^- \in \kappa^-$. For example if $L=3$, $\iota = 144 \in \kappa$ and $\iota^+ = 1144 \in \kappa^+$ (west, west, turn north, north), reading right to left as is our convention. In this case $\iota^- = 44 \in \kappa^-$. Here ι is a word in κ, ι^- is a word in κ^-, and ι^+ is a word in κ^+.

As we model the evolving system behavior, we will refocus on the next sample time, $t = (K+1)T$, and we will consider a sequence of regimes of length $L+1$ moving back from $t = (k+2)T^-$. We can indicate this string by $\iota^+ = [\mathbf{e}_p, \mathbf{e}_i, \mathbf{e}_j, \ldots, \mathbf{e}_r, \mathbf{e}_l] = pij \ldots rl \in \kappa^+$ where κ^+ is the set of all words of length $L+1$ from the alphabet $\mathbf{S}$. Again, ι^+ is a word in κ^+.

The augmented regime strings can be designated with numbers or unit vector arrays. If ι^+ is an $S \times (L+1)$ array of unit vectors, it can also be represented as an $(L+1)$-digit, S-radix number, or a word of length $(L+1)$, or a string of regimes of length $(L+1)$. The set κ^+ is the collection of all such regime strings however they are indicated. For example, if $L = 4$, then $\iota^+ = [\mathbf{e}_1, \mathbf{e}_1, \mathbf{e}_1, \mathbf{e}_1, \mathbf{e}_3] = (11113)$ identifies the event sequence $[\phi[k+1] = \mathbf{e}_1, \phi[k] = \mathbf{e}_1, \phi[k-1] = \mathbf{e}_1, \phi[k-2] = \mathbf{e}_1, \phi[k-3] = \mathbf{e}_3]$. In the illustration in which the regimes are direction events, $\iota^+ = [\mathbf{e}_1, \mathbf{e}_1, \mathbf{e}_1, \mathbf{e}_1, \mathbf{e}_3]$ describes a north-bound motion coming from the east: the vehicle was E-bound four samples ago but then turned and continued north. In this illustration, κ^+ contains all possible directional sequences over an interval of length $5T$.

To limit the complexity of the command algorithms we shall consider, we limit the memory length L. At each sample time, before proceeding, we will reduce the augmented regime sequence by "forgetting" the oldest regime. To illustrate, suppose that at the current time the vehicle is moving north after a eastbound interval. With an $L = 4$ memory, $\iota = [\mathbf{e}_1, \mathbf{e}_1, \mathbf{e}_1, \mathbf{e}_3]$. Now suppose the vehicle continues north: $\iota^+ = [\mathbf{e}_1, \mathbf{e}_1, \mathbf{e}_1, \mathbf{e}_1, \mathbf{e}_3]$. As we move forward in time, our event memory must truncate ι^+. We do so by dropping the oldest (rightmost) regime from ι^+ to obtain the updated $L = 4$ regime history, $[\mathbf{e}_1, \mathbf{e}_1, \mathbf{e}_1, \mathbf{e}_1] \mapsto \iota$. In this example, command algorithm is no longer aware of the previous east motion.

1.2.1.3 Regime Dynamics

Change in system mode or regime is unpredictable. All of the random processes[1] that follow are defined on a probability space $(\Omega, \mathscr{F}, \mathbb{P})$ and time interval $t \in [0, \mathsf{T}]$. These random processes are adapted to a common right-continuous filtration $\{\mathscr{F}_t; 0 \le t \le \mathsf{T}\}$. Specifically, the regime process, $\{\phi_t\}$, is constant in the main, but isolated events cause abrupt changes.

To first order, the regime state will be thought to be exogenous—though much more will be said of this later. Specifically, we will suppose $\{\phi_t\}$ is a right continuous, piecewise constant, $\mathscr{F}_t$-adapted Markov process taking on values in the set of canonical unit vectors in $\mathbb{R}^S$ ($\phi_t \in \{\mathbf{e}_1, \ldots, \mathbf{e}_S\}$).

We will represent the incremental behavior of the regime state with an $S \times S$-matrix Q: if $i \ne p$, $\mathbb{P}(\phi_{t+dt} = \mathbf{e}_p \mid \phi_t = \mathbf{e}_i) = Q_{ip}\,dt$ with $Q_{ii} = -\sum_{p \ne i} Q_{ip} > 0$. It is known that the mean sojourn time in state $\phi_t = \mathbf{e}_i$ is $-Q_{ii}^{-1}$, and if $\phi_t = \mathbf{e}_i$, the probability that the next modal transition will be $\mathbf{e}_i \mapsto \mathbf{e}_p$ is $-Q_{ip}/Q_{ii}$. Consequently, Q can be evaluated from observations of the regime process.

The regime state process is described by the stochastic equation

$$d\phi_t = Q'\phi_t dt + dm_t \tag{1.1}$$

with initial condition ϕ_0. The second term in (1.1) is a purely discontinuous $\mathscr{F}_t$-martingale increment: $E\{dm_t \mid \mathscr{F}_t\} = 0$ [4].

For now, we will suppose that the transition rates of the regime process are constant. This restriction will be considerably modified in the context of specific applications. But more of that later. The model of the time-discrete regime state process derives from (1.1). If we sample $\{\phi_t\}$ every T seconds, we have

$$\phi[k + 1] = \Pi\phi[k] + m[k + 1], \tag{1.2}$$

where $\mathscr{F}[k] = \mathscr{F}_{kT}$, $\{m[k]\}$ is a time-discrete, $\mathscr{F}[k]$-martingale difference sequence, $(E\{m[k + 1] \mid \mathscr{F}[k]\} = 0)$, and $\Pi_{pi} = \mathbb{P}(\phi[k + 1] = \mathbf{e}_p \mid \phi[k] = \mathbf{e}_i)$ is the modal transition matrix. With a fixed sample interval T, $\Pi = \exp(Q'T)$. The time-discrete regime model may allow events prohibited in (1.1). For example, a transition from $\mathbf{e}_1$ to $\mathbf{e}_2$ may be prohibited—a sudden U-turn from north-to-south is not allowed

---

[1]Readers seeking an introduction to stochastic processes and terms like martingale, adapted, and filtration will find an abundance of material in the published literature. Introductions and overviews abound on the worldwide web. See, for example, Wikipedia articles on stochastic processes and probability space.

and $Q_{12} = 0$. However a transition from $\mathbf{e}_1$ to $\mathbf{e}_2$ through $\mathbf{e}_4$ may be permissible. Since a north-west-south segment is allowed over a time interval of length T, the north-south transition possibility will appear in Π, $(\Pi_{21} > 0)$, albeit with small probability if T is small.

Equation (1.2) gives the time-discrete regime dynamics. The evolution of a modal state $\iota \in \kappa$ derives from this. The regime sequence, $\{\phi[k]\}$, is a random process with initial value $\phi[0]$. The $\{\phi[k]\}$ process is adapted to the $\{\mathscr{F}[k]\}$-filtration. Clearly, the realization of ι is adapted to $\mathscr{F}[k]$, and ι^+ is adapted to $\mathscr{F}[k+1]$ though neither gives the full regime history. To refer to this latter, we will use $\iota^\oplus \in \mathscr{F}[k]$. Then ι is a prefix of $\iota^\oplus$—the most recent part of the modal string.

We have used the notation $\phi[k] = \mathbf{e}_i$ to express the fact that the current regime is the ith. Let us extend this notation to say that $\phi[k] = \mathbf{e}_\iota$ if the L-prefix of $\iota^\oplus$ is ι. Or more broadly, ϕ_ι points to the L-string ι: $\phi_\iota[k] = 1$ if ι is true and $\phi_\iota[k] = 0$ otherwise. Indeed, $\phi_{\text{string}} = 1$ if the string condition is satisfied and zero otherwise. Or we might even say $\phi[k] = \mathbf{e}_{\text{string}}$. This notation is useful in sums; e.g., $\sum_{\iota^- \in \kappa_{(i,\iota^-)}} (M_\iota)\phi_\iota$ is a function of $i \in \mathbf{S}$.

1.2.2   Kinematic State

1.2.2.1   Fundamental Models

At first order, we have supposed that the modal process is exogenous. But the kinematic state, $\{X_t\}$, is tightly linked to the modal state. The range space of the kinematic state is $\mathbb{R}^n$. The evolution of the kinematic state depends upon the operating regime, past regime intervals, endogenous commands, and an agglomeration of exogenous disturbances.

At some inclusive level, target motion would naturally be represented with a non-linear stochastic differential equation. During intervals in which the motion mode is constant, $\phi_t \equiv \mathbf{e}_i$, the evolution of the kinematic state equation would be delineated by

$$dX_t = \mathbf{f}(X_t, u_t, \mathbf{e}_i)\, dt + \mathbf{g}(X_t, \mathbf{e}_i)\, dw_t, \tag{1.3}$$

where X_t is this global kinematic state, $\{u_t\}$ is an endogenous command, and the random process $\{w_t\}$ represents the unstructured and unpredictable exogenous influences on the system.

The kinematic model is an intermediary for synthesizing an implementable estimator. Except in special cases, (1.3) is far too complex to be used directly. Such equations are difficult to use to create an implementable tracker/classifier.

In some applications, there is an $(n \times S)$ array of stasis conditions, one for each regime: $X_t \approx X_{:i}^S$ when $\phi_t = \mathbf{e}_i$. The local base-state is the deviation of the global state from the current reference, $x_t = X_t - X^S \phi_t$, and the ith local model provides the forward dynamics. If the global kinematic state process is continuous and if $\mathbf{e}_i \mapsto \mathbf{e}_p$

at time t, the reference level changes from $X^S_{.i}$ to $X^S_{.p}$, and $x_t = x_{t-} + X^S_{.i} - X^S_{.p}$. In the time-discrete case, $x[k + 1]^- \mapsto x[k + 1]^- + X^S_{.i} - X^S_{.p}$. Thus, there are several simultaneous values of the deviation variables for the same global kinematic state.

On the other hand, if the common reference state is the origin in the state space, the global state and the local state are identical. In algorithm development, we will suppose that $\{X_t\}$ evolves in a regime-constrained mode of operation in which the nonlinear kinematic model can be replaced locally by a linear dynamic equation. Each such localization is indexed by a regime state (or more broadly, the mode). The global state, X_t, is replaced by the pair (x_t, ϕ_t) where x_t is deviation from the nominal kinematic state associated with index ϕ_t. The deviation variable, x_t, is called the base or kinematic state and carries the same labels as did X_t; e.g., the first component of x_t is horizontal position referenced to a nominal condition associated with ϕ_t, the second component is vertical position, and so on. This will be the common situation in what follows.

To illustrate this, consider a target moving in the (east, north)-plane. The motion equation in the X-direction is

$$\frac{d^2}{dt^2}X = a_X, \tag{1.4}$$

where $\{(a_X)_t\}$ combines the endogenous acceleration with exogenous disturbances in the X-direction. Motion in the Y-direction is described similarly.

The planar model integrates both motions:

$$\frac{d}{dt}\begin{bmatrix} \chi \\ v \end{bmatrix} = \begin{bmatrix} 0_{2\times2} & I_{2\times2} \\ 0_{2\times2} & 0_{2\times2} \end{bmatrix}\begin{bmatrix} \chi \\ v \end{bmatrix} + \begin{bmatrix} 0_{2\times2} \\ I_{2\times2} \end{bmatrix}\begin{bmatrix} a_X \\ a_Y \end{bmatrix}, \tag{1.5}$$

where χ_t is position with respect to the (X, Y) coordinate system and $v_t = (V_X, V_Y)'$ the velocity. Equation (1.5) is a global description of motion in the plane.

1.2.2.2 Constant Velocity and Constant Turn Models: Continuous and Discrete

In many tracking applications, the endogenous acceleration is integrated into the reference path. The residual acceleration is a wide-band random process which creates the ensemble of possible deviations about the nominal. Commonly, this residual is represented with a vector white-noise process.

Complicated motions are achieved by selecting the endogenous acceleration in keeping with some objective; e.g., evasion [14]. The endogenous accelerations create primal motion templates, and a composite motion is generated by joining these templates at random times. For example, suppose an agile target is capable of a coordinated turn with angular frequency $\omega_i; i \in \mathbf{S}$. An endogenous acceleration can be chosen to execute this motion. Indeed, the acceleration is a linear function of the kinematic state, and (1.5) can be adjusted to yield a linear differential equation that delineates a *constant turn* (CT) motion:

$$\frac{d}{dt}\begin{bmatrix} X \\ Y \\ V_X \\ V_Y \end{bmatrix} = \begin{bmatrix} 0 & 0 & 1 & 0 \\ 0 & 0 & 0 & 1 \\ 0 & 0 & 0 & -\omega_i \\ 0 & 0 & \omega_i & 0 \end{bmatrix} \begin{bmatrix} X \\ Y \\ V_X \\ V_Y \end{bmatrix}. \tag{1.6}$$

When the turn rate is zero, we have a *constant velocity* (CV) motion.

A *jinking* motion is created by mixing periods of turning with nearly CV flight. For example, the target can intersperse intervals in which the endogenous acceleration is zero with intervals in which the turn rate is $\pm\omega_1$. In this instance, the kinematic states in the local turn-specific models are all referenced to the same point in motion space, $\chi = 0$; $\nu = 0$. But the different motion regimes have different kinematic descriptions.

The process model given in (1.6) describes a point target moving without exogenous forcing. The target is actually an extended object. Further, the vehicle is subject to various exogenous accelerations that will be viewed as wide-band in the context of the tracking problem. A more useful motion model would be

$$\frac{d}{dt}\begin{bmatrix} X \\ Y \\ V_X \\ V_Y \end{bmatrix} = \begin{bmatrix} 0 & 0 & 1 & 0 \\ 0 & 0 & 0 & 1 \\ 0 & 0 & 0 & -\omega_i \\ 0 & 0 & \omega_i & 0 \end{bmatrix} \begin{bmatrix} X \\ Y \\ V_X \\ V_Y \end{bmatrix} + \begin{bmatrix} 0 & 0 \\ 0 & 0 \\ \sqrt{P_x^w} & 0 \\ 0 & \sqrt{P_y^w} \end{bmatrix} \begin{bmatrix} \dot{w}_X \\ \dot{w}_Y \end{bmatrix}, \tag{1.7}$$

where $\dot{w}_X$ is Gaussian white noise and similarly $\dot{w}_Y$. Equation (1.7) aggregates all of the disturbances and modeling errors into the white-noise process $\{\dot{w}\}$.

In what follows, when operating in the ith regime, we will use the local model

$$dx_t = A_i x_t \, dt + \sqrt{P_i^w} \, dw_t, \tag{1.8}$$

where $\{w_t\}$ is an $\mathscr{F}_t$-unit Brownian motion (an $\mathscr{F}_t$-martingale) independent of other exogenous excitations.

In each regime, there is a specific local model what manifests the idiosyncrasies of the kinematic state in that operating condition. We will assume that the state labels are the same for each regime; e.g., the first components of x_t are χ. But the deviation variables are referenced to the common nominal; i.e., the kinematic state is *centered* at the origin in the state space. In this way, the single nonlinear motion model in (1.3) is replaced with the family of S localizations in (1.8).

Equation (1.8) is a plausible approximation when regime state is constant. However, the kinematic state may be discontinuous at regime transition times: $\Delta\phi_t \neq 0$ might imply $\Delta x_t \neq 0$. If the reference point for a kinematic state changes, the perturbation state will experience a corresponding increment. For example, when a target moves from CV-motion to a coordinated turn, the position and velocity are continuous across the transition, but the jerk is not. If jerk is a component of the kinematic state, the kinematic state is discontinuous at a turn event.

Again it is simpler to express motion on a discrete time grid. As was done with the regime process, the kinematic state will be sampled with period T, and the $\{x_t\}$ process replaced with $\{x[k]\}$. If it were true that $\phi_t \equiv e_i$, we would simply use the orthodox continuous-to-discrete mapping of the model given in (1.8) to create the local time-discrete model

$$x[k+1] = A_i x[k] + \sqrt{P_i^w}\, w[k+1], \tag{1.9}$$

where $\{w[k]\}$ is an $\mathscr{F}[k]$-unit Gaussian-white sequence (an $\mathscr{F}[k]$-martingale increment). In what follows, P_i^w is always nonnegative. Equation (1.9) uses the evocative notation $([A_i, P_i^w]; i \in S)$ from (1.8) even though (A_i, P_i^w) are different matrices in the two equations; e.g., $\exp(A_i T) \mapsto A_i$. Equation (1.9) is a linear-Gauss–Markov (LGM) motion model: A_i is called the dynamic matrix and $\sqrt{P_i^w}$ is called the action matrix. Since i ranges over the set S, (1.9) is actually a family of local LGM-models.

The dynamic matrix is a discrete approximation to the classical point-equivalent kinematics of the target. The action matrix weights an aggregate of many influences, both exogenous and endogenous.

Let us look at the influence of $w_X[k+1]$ in (1.9) on an interval of length T. The velocity increment due to this acceleration is Gaussian with mean zero and variance $P_X^w T$. The standard deviation of the increment is $\sqrt{P_X^w T}$. We will say that the strength of the acceleration is $\sqrt{P_X^w / T}$.

For example, suppose the target is moving mainly north on an irregular terrain. If $\sqrt{P_i^w}\, w[k+1]$ represents the size of target acceleration, then the longitudinal (north-south) uncertainty would be greater than the lateral acceleration (east-west). So P_i^w is be used to shape the strength of the acceleration to fit the nominal motion. A conservative model might increase the size of all components of P_i^w to make the tracker less dependent on past events. This modification of the model is called adding pseudo-noise. The choice of P_i^w is usually based upon engineering judgement, and it is often used to reduce the sensitivities in the reductionist model.

Equation (1.9) would suffice for tracker synthesis if $\{\phi_t\}$ were known and constant. Actually, we should better write (1.9) so as to display the interplay of the regime state and the kinematic state:

$$x[k+1] = \sum_i (A_i x[k] + \sqrt{P_i^w}\, w[k+1])\phi_i[k], \tag{1.10}$$

where $\phi_i[k]$ is the principal regime index over the kth interval. Unfortunately, $\{\phi_t\}$ may change during a sample interval, and (1.10) does not capture this variability.

1.2.2.3 State Discontinuities

As written, (1.10) suggests a temporal concatenation of single period extrapolations. The terminal value of the kinematic state over one interval becomes the initial value for the succeeding interval. The regime is assumed constant over the sample period and equal to the regime at the beginning of the period.

Equation (1.10) must be reformed when there are non-trivial displacements associated with a change in the modal primitive. Suppose ϕ_t makes an $i \mapsto p$ transition during the kth sample interval. Let us isolate the endpoints of a sample interval: $x[k]^- = x_{kT-}$ and $x[k]^+ = x_{kT+}$. Then the kinematic extrapolation formula would be written

$$x[k+1]^- = \sum_\iota (A_\iota x[k]^+ + \sqrt{P_\iota^w} w[k+1]) \phi_\iota[k] \tag{1.11}$$

$$x[k+1]^+ = x[k+1]^- + \Delta_{\iota+} x[k+1]. \tag{1.12}$$

Equation (1.11) is the intra-sample kinematic extrapolation. It carries the base-state from its post-sample initial condition, $x[k]^+$, to the pre-sample terminal condition, $x[k+1]^-$. The extrapolation is accomplished under $\phi[k]$. Between $[k+1]T^-$ and $[k+1]T^+$, the kinematic state is adjusted for the change in regime.

To illustrate the motion model, suppose a motorcycle is following an evasive path in which alternate periods of CV motion are separated by sudden turns; a classic jinking motion. In (1.7), let us identify a representative set of turn rates (including $\omega = 0$) that describes the engagement: $\omega = \omega_i; i \in S$. Distinguish the regime with a specific angular rate: $\omega = \omega_i \mapsto \phi_t = \mathbf{e}_i$. By temporal sampling, and we create a family of LGM models in which the regime state is the current turn rate. If position and velocity are continuous across regime changes, then $\Delta_{\iota+} x[k+1] = 0$ for all $\mathbf{e}_i \mapsto \mathbf{e}_p$.

More generally, the kinematic-state must be adjusted to accommodate the forward regime. The adjustment can occur at the end of an interval or at the beginning of the successor interval. Equation (1.12) is called the handoff transformation, $\mathbf{H}^+$: $x[k+1]^- \xrightarrow{\mathbf{H}^+} x[k+1]^+$. If we wish to distinguish (1.11) from (1.12), we will call the former the *interval* extrapolation and the latter the *handoff* extrapolation [19].

To illustrate, let us modify the jinking motion of (1.6), and suppose the speed of the motorcycle slows by 10% when beginning a turn with a return to the original speed when CV-motion resumes. If a CV$\mapsto$CT transition occurs, we would have the side condition

$$\begin{bmatrix} X Y \\ V_X \\ V_Y \end{bmatrix}[k+1]^+ = \begin{bmatrix} 1 & 0 & 0 & 0 \\ 0 & 1 & 0 & 0 \\ 0 & 0 & 0.9 & 0 \\ 0 & 0 & 0 & 0.9 \end{bmatrix} \begin{bmatrix} X Y \\ V_X \\ V_Y \end{bmatrix}[k+1]^-. \tag{1.13}$$

In (1.13), the inter-sample transformation depends on the regime proxies for the current and the succeeding intervals. The complete extrapolation model would be written

$$x[k + 1] = A_\iota x[k] + \sqrt{P_\iota^w}\, w[k + 1] + \Delta_\iota + x[k + 1].\tag{1.14}$$

In the above example of jinking motion, the dynamic matrix carried the structured uncertainty. The reference point for the kinematic state was the origin for all regimes.

In other applications, the handoff conditions are more complex. Suppose that at time $t = (k + 1)T$, $\mathbf{e}_i \mapsto \mathbf{e}_p$ and the handoff map includes translations and rotations. Handoff can be written

$$x[k + 1]^+ = A_{pi} x[k + 1]^- + B_{pi},\tag{1.15}$$

where for completeness $[A_{ii}, B_{ii}] = [\mathbf{I}, \mathbf{0}]$.

To illustrate handoff, suppose a planar target is moving on a rectangular road grid. The coordinate reference system is centered at the origin. The upper components of x_t are: $[\chi = $ east, north; $v = $ east velocity, north velocity]. The regime index is ordered (north, south, east, west). A west-bound vehicle ($\phi_t = \mathbf{e}_4$) is such that $v_2 \approx 0$ and $v_1 < 0$. If the $\mathbf{e}_4 \mapsto \mathbf{e}_1$ event occurs, the target turns north: $v_1 \approx 0$ and $v_2 > 0$.

We can represent this grid-constrained motion with an intra-sample CV model:

$$\frac{d}{dt}\begin{bmatrix} X\ Y \\ V_X \\ V_Y \end{bmatrix} = \begin{bmatrix} 0\ 0\ 1\ 0 \\ 0\ 0\ 0\ 1 \\ 0\ 0\ 0\ 0 \\ 0\ 0\ 0\ 0 \end{bmatrix}\begin{bmatrix} X\ Y \\ V_X \\ V_Y \end{bmatrix} + \begin{bmatrix} 0 & 0 \\ 0 & 0 \\ \sqrt{P_{x;\iota}^w} & 0 \\ 0 & \sqrt{P_{y;\iota}^w} \end{bmatrix}\begin{bmatrix} \dot{w}_X \\ \dot{w}_Y \end{bmatrix},\tag{1.16}$$

where again $\{w_t\}$ is a unit Brownian motion. In this model, the regime enters into the direction of uncertainty: the P^w matrix magnifies $\{w_t\}$ more in the longitudinal direction than in the lateral direction.

When a constant speed west $\mapsto$ north turn event occurs ($\mathbf{e}_4 \mapsto \mathbf{e}_1$), the position is unchanged, but the velocity must be adjusted. Suppose the nominal exit velocities from a turn in the cardinal directions are $\{v_i; i \in \mathbf{S}\}$. Then

$$\begin{bmatrix} X\ Y \\ V_X \\ V_Y \end{bmatrix}[k + 1]^+ = \begin{bmatrix} 1\ 0\ 0\ 0 \\ 0\ 1\ 0\ 0 \\ 0\ 0\ 0\ 0 \\ 0\ 0\ 0\ 0 \end{bmatrix}\begin{bmatrix} X\ Y \\ V_X \\ V_Y \end{bmatrix}[k + 1]^- + \begin{bmatrix} 0 \\ 0 \\ v_2 \\ -v_1 \end{bmatrix}.\tag{1.17}$$

So if $\phi[k + 1] = \mathbf{e}_4$, the north speed is ≈ 0 and initially the west speed is positive. The west $\mapsto$ north transition leaves the position unchanged but resets the velocity to nominal north, v_1. Of course, if the turn does not preserve speed, (1.17) would need to be adjusted.

Handoff could occur on either side of the $t = (k + 1)T$ separator. The principal regime in the kth interval is $\mathbf{e}_i$ and that in the successor interval is $\mathbf{e}_p$. For convenience, let us suppose handoff occurs at $t = (k + 1)T^+$ and that $\mathbf{H}^+$ depends only on the modal prefix (p, i). The kinematic extrapolation model is written

$$x[k + 1]^- = \sum_{\iota}(A_{\iota}(\mathsf{A}_{ij}x[k]^- + \mathsf{B}_{ij}) + \sqrt{P_{\iota}^w}w[k + 1])\phi_i[k + 1]\phi_j[k]. \qquad (1.18)$$

Equation (1.18) describes the evolution of the state process over the kth time interval. The initial base-state is $x[k]^-$ and the modal proxy for the preceding interval is $\mathbf{e}_j$. The $\mathbf{H}^+$-transformation maps $x[k]^- \mapsto x[k]$ and the regime proxy becomes $\mathbf{e}_i$. The specific extrapolation model is the ιth, and this model carries $x[k] \mapsto x[k + 1]^-$. The process is then repeated.

Even with this reductionist model several complications arise. The kinematic state depends upon the sequence of past regimes: the kinematic state at $t = kT$ is a stochastic mapping from $(x[0], \phi[0])$ to $(x[k]^-, \phi[k])$ under $\iota^\oplus \in \kappa_k$. The cardinality of κ_k grows so rapidly as to preclude an implementable tracking algorithm. We will limit our focus to the $(x[0], \phi[0]) \mapsto (x[k]^-, \phi[k])$ transformation under ι, a prefix in κ_k. In this way, we maintain only S^L distinguishable kinematic models.

1.3   The Measurement Process

1.3.1   Regime-State Measurement

The state vector of a hybrid tracker consists of a discrete part and a continuous part. In like manner, the measurement vector may include direct measurements of both elements of the state. Let us begin with the sequence of modal measurements. These we will call $\{z[k]\}$. The modal primitive is a category variable, and its lineal measurement set will often mimic the regime classifications. For example, if a target type is identified from a noisy silhouette, each measured image is placed in a classificational bin associated with a single target type; e.g., the image of a tank at distance is placed in the regime bin "tank" to contrast it with the regime bin "motorcycle."

If the engagement is cooperative, the target transmits its regime proxy to the tracker and $z[k] \equiv \phi[k]$. Unfortunately, if the target is in the far field, classificational errors will occur; i.e., the notional regime may be in error. Over time, the tracker accumulates both noisy regime measurements and noisy kinematic measurements. They are then fused to produce higher quality regime estimates than achieved by simply setting $\phi[k] = z[k]$.

When the regime state is an aggregate of a set of more diffuse kinematic conditions; e.g., the classification "north" as a direction, an analog state element is forced into a discrete regime bin. In this case, both the raw regime measurement and the regime itself have a weaker intrinsic identity. The association of the measurement with a regime statement is more fragile, and the noisy measurement of regime is not clearly distinguishable from those of its proximate peers. Still, clever processing of a sequence of low quality regime measurements can be useful when combined with kinematic data.

The number of measurement bins is arbitrary but finite. For illustration here, we will suppose that the range space of the regime measurement is that of the regime state: $z[k] = \mathbf{e}_i; i \in \mathbf{S}$. When the range of $z[k]$ is $\mathbf{S}$, the language of the regime observation is that of the regime sequence itself. The quality of the classification is represented by the $S \times S$ discernibility matrix $\mathbf{D} = [\mathbf{D}_{ij}]$ where $\mathbf{D}_{ij}$ is the probability that the regime sensor selects $\mathbf{e}_i$ if $\mathbf{e}_j$ is the true regime at time of measurement. Hence, $\mathbf{D}_j$ is a probability vector: the column sums of $\mathbf{D}$ are one, and to avoid singularities we will assume that $\mathbf{D}_{ij} > 0$ for all (i, j).

The ideal regime sensor is one where $\mathbf{D} = \mathbf{I}$—though this violates the positivity of $\mathbf{D}_{ij}$. This is the commonly studied situation in which a cooperative target transmits its regime state to the tracker. In a cooperative engagement, the kinematic model is stochastic, but the current regime is known to the tracker at all times. To the degree that $\mathbf{D} \neq \mathbf{I}$, the tracker must estimate both the kinematic state and the regime by fusing measurements of both.

There are many applications in which there is no regime sensor. In this event, we will say that the discernibility matrix is $\mathbf{D} = (1/S) * \text{ones}(S \times S)$; i.e., $\{z[k]\}$ is uninformative.

The precise timing of the state measurements is important in tracking. For kinematic extrapolation, the proxy states are those at the beginning of an intra-sample interval, $(\phi[k], x[k]^+)$. Temporal sampling introduces an error that is partially captured in the channel and the action matrices. However, sensor latency requires subtle adjustments into the tracker.

Let us look first at the regime measurement. The quality of $\{z[k]\}$ is measured by the discernibility matrix $\mathbf{D}$. But the actual quantity measured depends in part upon the complexity of the processing required to label the raw data set. If we ignore multiple transitions within a sample interval, $\phi[k + 1]^- = \mathbf{e}_i$ and $\phi[k + 1] = \mathbf{e}_p$. The regime measurement at $t = [k + 1]T$ could be associated with either. If the raw measurement is an image, a significant period of time is necessary to properly tag the image. In this event, the measurement at $t = [k + 1]T$ is of the regime that is dominant over the kth time interval; i.e., $\mathbf{e}_i$:

$$z[k + 1] = \mathbf{D}\phi[k] + \eta[k + 1], \tag{1.19}$$

where $\{\eta[k]\}$ represents the error in the regime measurement.

On the other hand, if the regime state were velocity direction, a doppler-radar would give a snapshot of conditions at $t = (k + 1)T$; i.e., $\mathbf{e}_p$:

$$z[k + 1] = \mathbf{D}\phi[k + 1] + \eta[k + 1]. \tag{1.20}$$

The observation model in (1.19) gives a measurement of the regime proxy over the interval preceding the observation. Alternatively, the model (1.20) looks forward to the next interval. In the former case, the tracker must wait a sample period to measure the dominant regime. In either case, we will denote the filtration generated by $\{z[k]\}$ by $\{\mathscr{Z}[k]\} \subset \mathscr{F}[k]$. The sequence $\{\eta[k]\}$ is an $\mathscr{F}[k]$-martingale difference. In tracking, (1.20) is to be preferred to (1.19) because the tracker avoids the processing delay. But the engineer is constrained by the application.

1.3.2   Kinematic-State Measurement

In addition to the regime measurement, there is a measurement of the local kinematic-state [13, 14]. The latter is assumed to be a linear, albeit mode-dependent, measurement of the base-state vector with perhaps a bias:

$$y[k] = \sum_{\iota} (H_{\iota} x[k] + \sqrt{P_{\iota}^{n}} n[k] + Y_{.\iota}^{M}) \phi_{\iota},$$

where Y^{M} is an array of observation offsets. In what follows, the measurement vector will be translated by the bias vector without comment to yield the centered measurement $y[k] - Y_{.\iota}^{M} \mapsto y[k]$:

$$y[k] = \sum_{\iota} (H_{\iota}[k] x[k] + \sqrt{P_{\iota}^{n}[k]} n[k]) \phi_{\iota}, \tag{1.21}$$

where the measurement noise $\{n[k]\}$ is an independent (temporally and spatially), unit $\mathscr{F}[k]$-Gaussian white sequence that is independent of $\{w[k]\}$. The pair (1.10) and (1.21) is called a (local) LGM model for the engagement.

The gain matrix, H_{ι}, commonly depends upon the geometry of the engagement. The positive channel noise matrices, P_{ι}^{n}, shape the measurement noise, and they too can be mode-dependent. We will suppose that $P_{\iota}^{n}[k] > 0$ for all (ι, k). The inverse of the channel noise matrix has a positive symmetric square root: $(P_{\iota}^{n})^{-1} = D_{\iota}^{n} = (F_{\iota}^{n})^{2} > 0$. We will label the data set (filtration) generated by the kinematic measurements $\{\mathscr{Y}[k]\}$.

Equation (1.21) is used as the comprehensive measurement model in common tracker synthesis procedures; e.g., in the extended Kalman filter (EKF). There are the usual caveats attached to (1.21). In applications, the raw kinematic measurements are often nonlinear functions of the kinematic state; e.g., range-bearing is a nonlinear function of position when expressed in a Cartesian coordinate system. Additionally, the linearization is based upon an uncertain target/sensor geometry. Calculating the gain and noise matrices in (1.21) requires replacing the current target location with a good estimate—and even replacing a sensor location with an estimate if it is not known precisely. The assumed geometry of the engagement is typically based upon estimates of the kinematic state: the coefficients $(H_{\iota}[k], P_{\iota}^{n}[k]; \iota \in \kappa)$ are themselves random and adapted to the measurement dataset. This creates a circular link in which the observation model depends explicitly upon kinematic estimates which are themselves based upon earlier observations. Not surprisingly, this feedback can lead to instability; e.g., if the the tracking error is large, the conversion of the raw measurements to $\{\mathscr{Y}[k]\}$ is faulty, and the tracking error increases evermore.

Equation (1.21) makes explicit the fact that the sensor coefficients depend upon a modal sequence, and not simply upon the current regime. The kinematic observation also depends on the precise time when handoff takes place. If, for example, handoff occurs at the end of the predecessor interval; i.e., when $\iota \mapsto \iota^+$, then $\mathbf{H}^+$ will include the $i \mapsto p$ condition: $x[k] = x[k]^- + \Delta_\iota x[k]$. We will assume for now that measurement takes place before handoff: $y[k] = y[k]^-$.

1.4   State Estimates Using the Gaussian-Sum Distribution

The tracker avails itself of whatever data sets are available. Suppose that at $t = kT$, the tracker has only the kinematic data set $\mathscr{Y}[k]$. We would call any statistic based exclusively upon this data set a $\mathscr{Y}[k]$-statistic; e.g., the conventional EKF-estimate of target location would be called the $\mathscr{Y}[k]$-mean location of the target.

If there are direct regime measurements, we have the complementary data set $\mathscr{Z}[k]$. A tracker using both measurement sets will now generate $\mathscr{G}[k]$-statistics where $\mathscr{G}[k] = \mathscr{Y}[k] \vee \mathscr{Z}[k]$. When we wish to emphasize the availability of the regime data, we will refer to a $\mathscr{G}[k]$-estimate; e.g., the $\mathscr{G}[k]$-mean location of the target. Though not available in every application, we will augment $\mathscr{Z}[k]$ with such idiosyncratic information we might have that bears upon the engagement: path maps, forbidden regions, speed constraints, etc. This latter collection will be called the set of para-measurements

At a fundamental level, all of the tracker/classifiers that follow are $\mathscr{G}[k]$-trackers. They all use $\{\mathscr{Y}[k]\}$. Even if $\{\mathscr{Z}[k]\}$ is trivial, the regime measurement can be included formally with an uninformative $\mathbf{D}$. But we will use the notation in a more evocative fashion. If we talk about a $\mathscr{Y}[k]$-tracker, we mean one that is based exclusively on the $\{\mathscr{Y}[k]\}$ data set. If we talk about a $\mathscr{G}[k]$-tracker, we will be pointing to a non-trivial expansion of $\{\mathscr{Y}[k]\}$. The expansion could be $\{\mathscr{Z}[k]\}$, or the expansion could the para-measurements alone. But the $\mathscr{G}[k]$ modifier points to something over and above $\mathscr{Y}[k]$. Although we could pose problems in which the tracker utilizes only $\{\mathscr{Z}[k]\}$, this is not relevant here.

Conditional expectation with respect to $\mathscr{G}[k]$ (or $\mathscr{Y}[k]$ if appropriate) is denoted with a circumflex: $\hat{x}[k]$ is the $\mathscr{G}[k]$-conditional mean of the kinematic state at time $t = kT$; $\hat{\phi}[k]$ is the vector of $\mathscr{G}[k]$-conditional probabilities of the various modes at time $t = kT$; $\hat{\phi}_i[k]$ is the $\mathscr{G}[k]$-conditional probability of the ith regime at time $t = kT$; and so on. In what follows, the relevant data set will be apparent from context.

To delineate tracker performance in a rational way, we need a quality metric (or metrics) for the estimates generated by the tracker. A hybrid architecture integrates a family of local estimates into a comprehensive estimate of target state. To illustrate, let us ignore the handoff conditions for now. The coefficients in each of the local dynamic matrices and the action matrices are $\mathscr{G}[k]$-adapted random processes that are fundamentally linked to the modal path, $\iota^{\oplus}$. We have taken a reductionist approach in which we keep only the ι prefix of $\iota^{\oplus}$—we maintain only κ–distinct local models. This reduces the complexity of the tracker: the S^k potential models at $t = kT$ are reduced to just S^L models. One such model is given by

$$x_\iota[k + 1] = A_\iota x_\iota{-}[k] + \sqrt{P_\iota^w} w[k + 1]. \tag{1.22}$$

The base-state moves forward from one of the initial conditions, $x_\iota{-}[k]; \iota^- \in \kappa^-$, under the regime proxy ϕ_ι. After extrapolation the modal string expands from ι^- to $\iota = (i, \iota^-)$. Equation (1.22) is a simplification of the actual kinematics, and it ignores the variability of the regime process within the sample interval. Often we will write (1.22) as

$$x_\iota[k + 1] = A_i x_\iota{-}[k] + \sqrt{P_i^w} w[k + 1]. \tag{1.23}$$

In this rendering, the current regime variable is dominant in evaluating the coefficients. For example, if a target moves from CV-motion to CT-motion, the dynamic and action matrices could depend upon the regime state and not the full modal string.

We will assume that the $\mathscr{G}[0]$-distribution of the base-state is Gaussian. The initial regime state is independent of $x[0]$ with distribution $\phi[0] \sim \hat{\phi}[0]$. To translate the regime distribution into a distribution over the modal state, κ^-, set $\mathbb{P}(\iota^- = (jj \ldots j) = \hat{\phi}_j[0])$ for all $j \in \mathbf{S}$ with all other modes having zero probability: we start the encounter by assuming that there have been no regime changes in the recent past. The initial state is assumed to be independent of any exogenous forcing terms in the state and measurement equations. The state histories, $\{x[k]\}$ and $\{\phi[k]\}$, are adapted to the fundamental filtration $\{\mathscr{F}[k]\}$. The exogenous processes that animate the state process, $\{w[k]\}$ and $\{m[k]\}$, are independent $\{\mathscr{F}[k]\}$-martingale difference sequences.

The tracker integrates the local models with the observation sequence to generate the $\mathscr{G}[k]$-conditional distribution function of the target state. The parametric family of distributions used here is based upon the Gaussian distribution. Specifically, a Gaussian random vector, x, with *mean* $\hat{x}$ and *covariance matrix* P_{xx}, is written $x \sim \mathbf{N}(\hat{x}, P_{xx})$ where $\mathbf{N}$ could refer to the distribution or the density as appropriate. The initial kinematic state is assumed to be $x[0] \sim \mathbf{N}(\hat{x}[0], P_{xx}[0])$. Thus, we begin the engagement with an initial state distribution, $\mathbb{P}(x[0], \phi[0]|\mathscr{G}[0])$.

In the applications we will present here, all of the state covariance matrices are assumed to be positive: $P_{xx} > 0$. In this event the covariance has a positive symmetric inverse, the *information matrix*, D_{xx}, with positive symmetric square root, $F_{xx}: P_{xx}^{-1} = D_{xx} = F_{xx}^2 > 0$. The unit Gaussian density with mean m is written: $\mathbf{N}(m, I) = \Phi(m)$.

The kinematic state is a vector in n-space; i.e., $x_t \in \mathbb{R}^n$, and the square of its length is $\|x\|^2 = x'x$. Sometimes it is convenient to look at vectors in a modified coordinate system. If A is a positive symmetric matrix compatible with the x vector,

$x'Ax$ is denoted $\|x\|_A^2$. The matrix A has a positive symmetric square root $\sqrt{A}$. In the modified coordinate system defined by A, the length of the vector $\sqrt{A}x$ is $\|x\|_A$. Indeed, if $x \sim \mathbf{N}(\hat{x}, P_{xx})$, then $F_{xx}(x - \hat{x}) \sim \Phi(0)$: in a coordinate system adapted to P_{xx}, if $x[k]$ is Gaussian, $F_{xx}(x[k] - \hat{x}[k])$ is a unit Gaussian random variable.

The Gaussian distribution has very thin tails and does not well represent the uncertainties encountered in tracking applications. The likelihood of a Gaussian sample deviating from its mean value by more than three standard deviations is so small as to be ignored in most applications. Unfortunately, realistic tracking problems do not lead to compact uncertainty regions. Outliers are far more frequent than predicted by a Gaussian distribution, and the confidence regions generated by the tracker must be flexible to account for outliers.

A class of probability distributions more suitable for tracking applications is the Gaussian sum. To illustrate, let us consider a family of Gaussian primitives of size S^L: the (mean,covariance) pairs are given by $\{(m_\iota, P_\iota); \iota \in \kappa\}$. Let $\{\alpha_\iota; \iota \in \kappa\}$ be a set of non-negative numbers at least one of which is nonzero. A Gaussian sum distribution, $\mathbb{P}$, is written

$$\mathbb{P} = \sum_{\iota \in \kappa} \alpha_\iota \mathbf{N}(m_\iota, P_\iota). \tag{1.24}$$

In forming $\mathbb{P}$ we are adding distributions and not random variables. The resulting distribution is not Gaussian—unless one of the α_ι is one and the rest are zero. The index set κ has been used for illustration. In what follows, we will utilize Gaussian sums over κ^+ and κ^- too.

When the weights are such that $\sum \alpha_i = 1$, (1.24) is said to be normalized. When the $\sum \alpha_i \neq 1$, the distribution is un-normalized. An un-normalized distribution can be normalized with a simple replacement: $\frac{\alpha_i}{\sum \alpha_i} \mapsto \alpha_i$. Many operations within the tracker are more simply displayed with un-normalized distributions, and we will use both normalized and un-normalized distributions without comment.

Sometimes an engagement is such that certain α_ι are intrinsically equal to zero. Because it simplifies the notation and computation, we will work with a fixed dimension Gaussian sum (say of size S^L): we will not reduce the size of the index set by eliminating intrinsically null coefficients. But to avoid singularities, we will require that at least one of the $\alpha_\iota > 0$. The near-tail of the Gaussian sum distribution is broader than that of $\mathbf{N}(\hat{x}, P_{xx})$, though of course, the former reduces to the latter when only one $\alpha_i \neq 0$.

The Gaussian sum is a useful tool for approximating arbitrary distributions. The elemental Gaussian densities are unit vectors in a function space and the usual residual reduction methods can be used to find the best fit of a density given the order of the sum. But here we view the Gaussian sum differently. The size of the index set is that of the cardinality of κ, and each of the elements of the sum has an identity associated with a modal state. So, (1.24) is not the best S^L Gaussian fit to the distribution of uncertain target location. Rather, the Gaussian sum is a distribution composed of densities indexed by the S^L primal motion hypotheses. In what follows, (1.24) will be viewed as the "true" target state distribution instead of an approximation to some underlying, albeit unknown, distribution.

1.5   Assurance Regions

For reasons described above, a Gaussian sum distribution will be our template for representing the uncertainty in target location. To illustrate the implications of this distribution, consider locating a vehicle moving on a rectangular road grid. The path is shown in Fig. 1.6. The vehicle initially moves south on the segment $-550E$. It encounters a junction at $(-550, -400)$. Rather than continuing south, the vehicle turns west at the junction. At the time in question, the vehicle is located at $(-600, -400)$ and is west-bound. There is a frontage road at $-500E$ that parallels the north-south road.

A tracker was designed to locate the target. Using measurements of speed, range-bearing and the road map, a tracker presented later generated a 125-term Gaussian sum distribution for the target location, χ:

$$\mathbb{P}(\chi) = \sum_{\iota} \alpha_{\iota} \mathbf{N}(\chi_{\iota}, P_{\chi,\iota}). \tag{1.25}$$

The location density is shown in Fig. 1.7. There are two noticeable peaks in the density. One is in the neighborhood of the true location. Another is near the junction—the tracker is not sure that the target has indeed made the turn. There are smaller peaks back along the $-550E$ segment—the tracker is not sure of the speed of the vehicle. Another peak is found on the $-500E$ frontage road, north of the

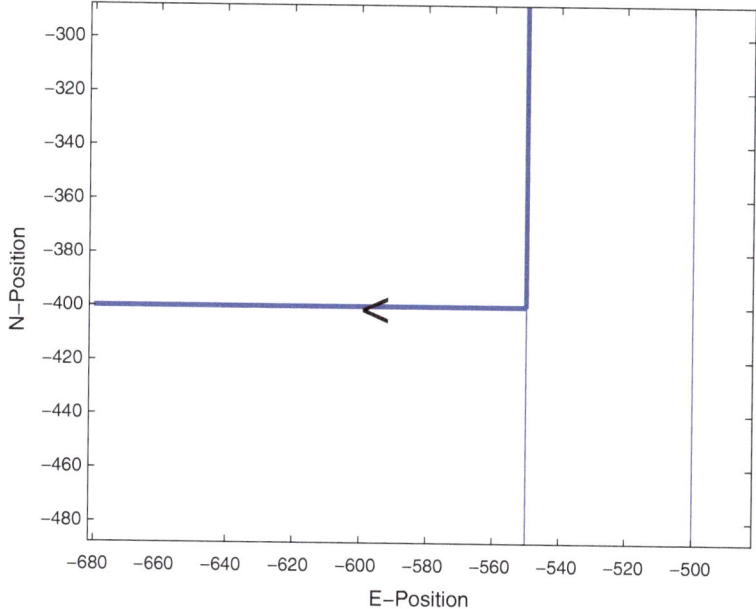

Fig. 1.6 A target is moving on the road path shown. After moving south on the segment $-550E$, the target moves west on the segment $-400N$

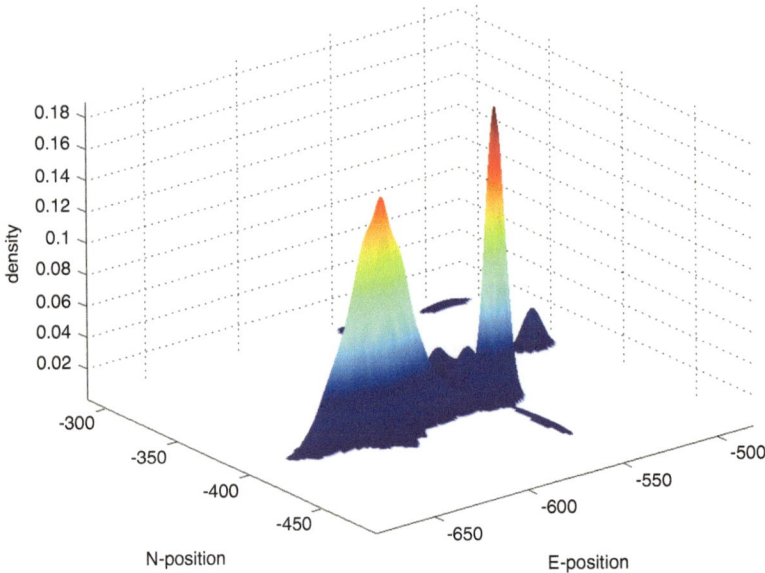

Fig. 1.7 The 125-term Gaussian sum location density shows multiple peaks

junction. There is also some probability assigned to a motion that continues south through the junction and more at various other placements.

The two primary elements of the density are both located on the $-400N$ segment of the road. The largest (weight 0.95) is that centered near the target. The second largest (weight 0.0077) is that centered on the junction. The spread of the latter is more nearly circular since north-south and east-west motion as permitted from the junction.

The events of $\mathbb{P}$ that correspond to north motion or such as that along the frontage road are individually small: there are not 125 individuated peaks in the figure. Many terms in $\mathbb{P}$ are too small to see—or are subsumed within a nearby element of larger weight. But in aggregate, these terms are relevant.

In command architectures, the Gaussian sum may be difficult to utilize directly: it may be multi-modal or its mean may lie in a prohibited region, whereas what is needed is a single credible estimate of target location along with an indication of location uncertainty. To this end, we can compute the first two moments of $\mathbb{P}$, $(\hat{x}, P_{xx})$ and create a Gaussian approximation that matches these moments. The approximation is in a sense a projection of the Gaussian sum onto a single Gaussian distribution function. We refer to it as the projected distribution, or occasionally as an umbrella distribution. In position, this approximation would be $\hat{\mathbb{P}}$:

$$\hat{\mathbb{P}}(\chi) = \mathbf{N}(\hat{\chi}, P_{\chi}). \tag{1.26}$$

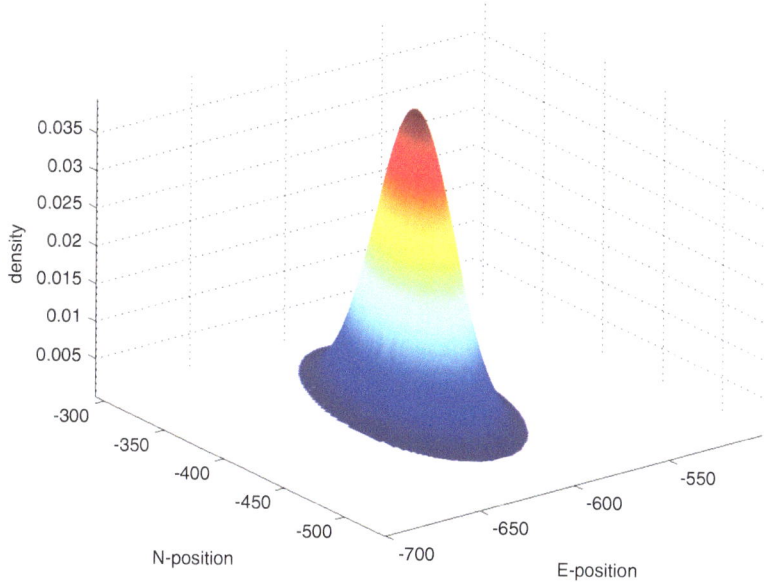

Fig. 1.8 A Gaussian density, $\hat{\mathbb{P}}$, with the same first moments as $\mathbb{P}$

This approximation is displayed in Fig. 1.8. The $\hat{\mathbb{P}}$-distribution has the same (mean,covariance) as does $\mathbb{P}$. The former is unimodal and centered near the true location of the target: the dominant element of $\mathbb{P}$ centers $\hat{\mathbb{P}}$. Interestingly, the major axis of the covariance ellipses of $\hat{\mathbb{P}}$ is not in the direction of longitudinal motion. This is due to the large distance from the mean to the outlying terms in $\mathbb{P}$.

To illustrate the utility of $\mathbb{P}$, suppose we wish to intercept the target. We are not sure where the target is, but do we know the $\mathscr{G}[k]$-distribution of χ. Unfortunately, the support of $\mathbb{P}(\chi)$ is all of $\mathbb{R}^2$ so we can never be sure that any placement of countermeasures will in fact cover the target.

Suppose, therefore, that we are willing to settle for finding a region in the plane that contains the target with probability 75 %. The rationale here is much like that behind CEP, but we do not require the region to be circular. Clearly, the best such region would be that with the smallest area. For a given quantity of the countermeasure resources, the smallest area would yield the highest countermeasure density and the highest probability of target destruction.

The countermeasure allocation problem becomes that of finding the smallest region in the plane, $\mathbf{A}_{.75} \in \mathbb{R}^2$, such that we are assured that $\chi \in \mathbf{A}_{.75}$ with probability $p = 75\,\%$. We will call $\mathbf{A}_{.75}$ the 75 % assurance region for the target. The assurance region can be found by simple search: find the set in position space for which the probability density of position is greater than a threshold h chosen such that the set has aggregate probability 75 %.

The assurance region derived from the density in the example is shown in Fig. 1.9. The true location is displayed in the figure. The assurance region is aligned

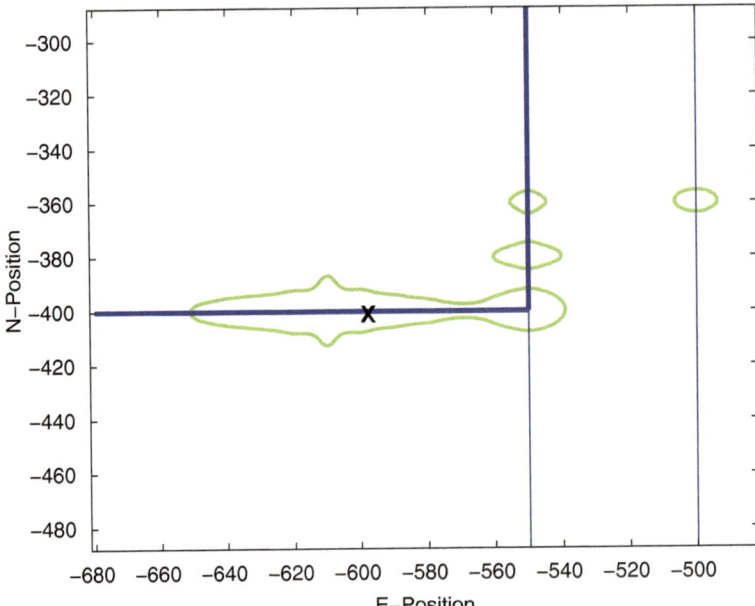

Fig. 1.9 The 75 % confidence region generated by Pr. The true target location, which is on the road, is displayed with a **x**

with the longitudinal motion of the target. There are vestiges of past target locations and even some placements on the $-500E$ road segment.

The 75 % assurance region has an area of $1630\,\mathrm{m}^2$ and is the tightest such region for this example. But its utility in a command architecture is limited by its peculiar shape. It is the union of four disjoint regions, and the primary region is not even convex. If we were using $\mathbf{A}_{.75}$ to assign lethal assets to the target, we would require a very sophisticated apportionment ability. If, for example, the countermeasures were kinetic kill particles, the splitting of the particles in a way that matches $\mathbf{A}_{.75}$ would be beyond the capability of conventional weapon systems.

As an implementable alternative, we could define $\mathbf{A}_{.75}$ in terms of the projected distribution $\hat{\mathbb{P}}$. In this Gaussian approximation, $\mathbf{A}_{.75}$ is the interior of a covariance ellipse centered on the mean, $\hat{\chi}$. The specific 75 % assurance region is shown in Fig. 1.10. The area of this assurance region is $3260\,\mathrm{m}^2$: the area of the assurance region is twice that associated with $\mathbb{P}$.

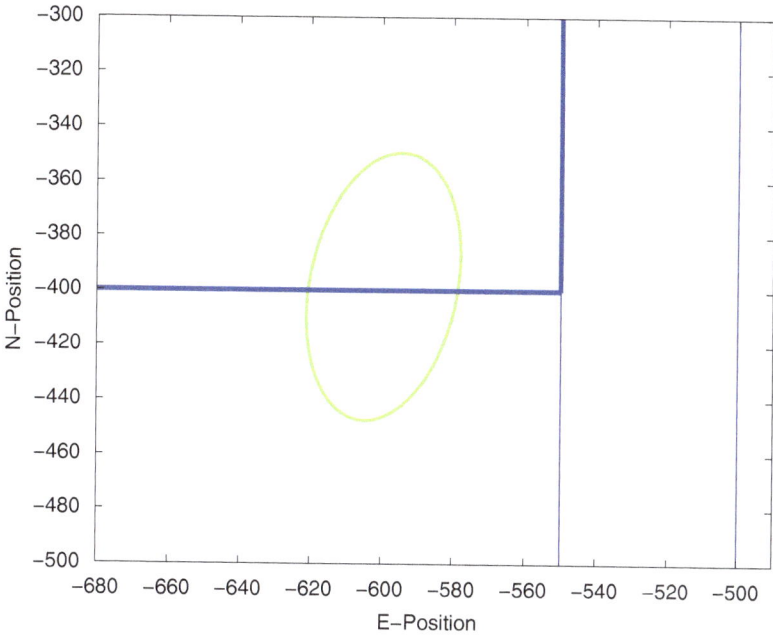

Fig. 1.10 The 75 % confidence region generated by the approximating distribution $\hat{\mathbb{P}}$

The assurance region defined by $\hat{\mathbb{P}}$ has the advantage that $\mathbf{A}_{.75}$ is connected—even convex. But most of the area of the assurance region computed with $\hat{\mathbb{P}}$ is in the direction of lateral motion, and this is not in compliance with the road-width constraints associated with the engagement. If this tracker cued the deployment of kill-particles toward a hostile target, most of the particles would fall harmlessly off the road. Although the notional assurance region of $\hat{\mathbb{P}}$ is 75 %, only 6 % of the Gaussian sum distribution falls within the approximating assurance region generated by $\hat{\mathbb{P}}$.

The contrasting calculations presented in this section provide a warning concerning the use of a simple projected state distribution to determine a notional assurance region. But the example was chosen to highlight anomalous behavior. The target had just made a west turn after a period of south motion. The measurements were not sufficient to clearly distinguish the west motion from: a stop, a U-turn, a transfer to the frontage road. In most of the cases we will study in the sequel, the motion is smooth and the conditional distribution of the kinematic state is unimodal. The assurance regions computed from $\hat{\mathbb{P}}$ and $\mathbb{P}$ will not differ as much as this example would suggest.

1.6   Tracker Architecture

1.6.1   The Conditional Distribution of the Target State

We will study several types of engagements in what follows. A specific hybrid tracker will be developed in each case. But to have a single place that gives an overview of the algorithms, we will present the basic flow here. We will suppose that handoff occurs at the beginning of the sample interval, and there is no latency in the regime observation. Let us focus on the ιth local filter on the interval $[kT, (k+1)T]$. To make the algorithm easier to follow, we will suppose that the dynamic matrix and the action matrix depend only on the regime proxy: $(A_\iota \mapsto A_i, P^w_\iota \mapsto P^w_i)$.

We begin at $t = kT$ with S^L local kinematic estimators, each based upon a distinct modal string from the preceding cycle, $\iota[k] = (j \ldots rlm)$. The local $\mathscr{G}[k]$-kinematic estimate is Gaussian: $x_\iota[k] \sim \mathbf{N}(m_\iota[k], P_\iota[k])$. There is also a $\mathscr{G}[k]$-probability vector that gives a confidence weight to each modal string in κ^+: $\alpha_\iota+[k+1]$. There are three marginal distributions over κ that derive from $\alpha_\iota+[k]$.

$$\alpha_\iota[k] = \sum_p \alpha_{(p,\iota)}[k+1] \tag{1.27}$$

$$\alpha_\iota-[k] = \sum_m \alpha_{(\iota^-,m)}[k] \tag{1.28}$$

$$\alpha_\iota[k+1] = \sum_m \alpha_{(\iota,m)}[k+1]. \tag{1.29}$$

Let us look at the information flow in the ιth local filter in detail. The flow comprises nine steps, performed in sequence, repeated at each time increment T.

1. **Initialize the kinematic state**: The cycle at $t = kT$ begins with the string terminating the preceding cycle, $\iota[k] = (j \ldots rlm)$. First, remove the most remote event from the modal string: $\iota^-[k] = (j \ldots rl) = \iota[k]/m$. We now have a truncated string of length $(L-1)$. Initialize the kinematic-state distribution by averaging over $\mathbf{e}_m \in \mathbf{S}$.

$$m_\iota-[k] = \sum_m m_{(\iota^-,m)}[k]\alpha_{(\iota^-,m)}[k]$$

$$P_\iota-[k] = \sum_m \alpha_{(\iota^-,m)}[k](P_\iota[k] + (m_\iota-[k] - m_\iota[k])(\ldots)')$$

We now have the abridged initial condition on the ιth estimator

$$x_\iota[k] \sim \mathbf{N}(m_\iota-[k], P_\iota-[k])$$

2. **Advance the string**: $\iota[k] = (i, \iota^-[k]) = (ij \ldots l)$ and initialize the kinematic-state.

$$(\alpha_\iota[k+1], m_\iota-[k], P_\iota-[k]) \mapsto (\alpha_\iota[k+1], m_\iota[k], P_\iota[k])$$

3. **Handoff**: As the forward regime moves to $\mathbf{e}_i$, the kinematic-state handoff modifies the initial condition.

$$(\alpha_\iota[k+1], m_\iota[k], P_\iota[k]) \xrightarrow{\mathbf{H}^+} (\alpha_\iota[k+1], m_\iota[k]^+, P_\iota[k]^+)$$

4. **Extrapolate**: The ι-filter extrapolates across the $(k+1)$th interval under regime proxy i:

$$(\alpha_\iota[k+1], m_\iota[k]^+, P_\iota[k]^+) \xrightarrow{\text{Eq. (1.22)}} (\alpha_\iota[k+1], m_\iota[k+1]^-, P_\iota[k+1]^-)$$

5. **Integrate the kinematic measurement**:

$$(\alpha_\iota[k+1], m_\iota[k+1]^-, P_\iota[k+1]^-) \xrightarrow{\text{Eq. (1.21)}} (\alpha_\iota[k+1], m_\iota[k+1], P_\iota[k+1])$$

6. **Integrate the regime measurement**:

$$(\alpha_\iota[k+1], m_\iota[k+1], P_\iota[k+1]) \xrightarrow{\text{Eq. (1.20)}} (\alpha_\iota+[k+1], m_\iota[k+1], P_\iota[k+1])$$

7. **Integrate the para-measurement constraints**:

$$(m_\iota[k+1], P_\iota[k+1]) \xrightarrow{\mathbf{C}^{\mathcal{M}}} (m_\iota[k+1], P_\iota[k+1])$$

8. **Determine the modal probabilities**:

$$\alpha_\iota[k+1] = \sum_p \alpha_{(p,\iota)}[k+1]$$

9. **Finalize the $\mathcal{G}[k+1]$-estimate** :

$$\mathbb{P}[k+1] = \sum_{\iota \in \kappa} \alpha_\iota[k+1]\mathbf{N}(m_\iota[k+1], P_\iota[k+1])$$

We now have the $\mathcal{G}[k+1]$-state distribution: $(\mathbb{P}[k+1], \alpha_\iota+[k+1])$. We can now move to the next cycle.

1.6.2 The Projected Distribution

The $\mathcal{G}[k]$-distribution of the kinematic state is a complicated function of the observation data-set. As pointed out earlier, it is frequently convenient to use an approximation to $\mathbb{P}$ for determining countermeasure placement.

Observe that each of the local filters generates a distinct $\mathcal{G}[k]$-estimate of the kinematic-state: $m_\iota[k]; \iota \in \kappa$. The quality of the local estimate is given by $P_\iota[k]; \iota \in \kappa$. The comprehensive moments of $\mathbb{P}$ can be computed directly. The first two moments are:

$$\hat{x}[k] = \sum_\kappa m_\iota[k]\alpha_\iota[k]$$

$$P_{xx}[k] = \sum_\kappa \alpha_\iota[k](P_\iota[k+1] + (\hat{x}[k] - m_\iota[k])(\hat{x}[k] - m_\iota[k])')$$

The regime estimate, $\hat{\phi}_i = \mathbb{P}(\phi[k] = e_i|\mathcal{G}[k])$, is found by collecting the probability of all modes with current prefix $i \in \mathbf{S}$:

$$\hat{\phi}_i[k] = \sum_{\iota^- \in \kappa(i,\iota^-)} \alpha_{(i,\iota^-)}[k]. \tag{1.30}$$

The $\mathcal{G}[k]$-distribution of the kinematic state is not Gaussian. But there is a natural approximation, $\hat{\mathbb{P}}$, found by matching the first two moments of $\mathbb{P}$. In this abridged representation,

$$x[k] \sim \hat{\mathbb{P}} = \mathbf{N}(\hat{x}[k], P_{xx}[k]). \tag{1.31}$$

The projected distribution $\hat{\mathbb{P}}$ will often be used to give a snapshot of tracker performance. Certainly, $\hat{\mathbb{P}}$ is not as nuanced as $\mathbb{P}$—it has a thinner near tail and is unimodal. But $\hat{\mathbb{P}}$ is simple to compute and suggestive of the broad capabilities of the tracker. In many of the examples that follow, we will present simulations of the performance of the tracker as it follows an uncooperative target. To simplify the display the quality of the estimate, we will often use the statistics of $\hat{\mathbb{P}}$.

The projected estimate given in (1.31) is descriptive and not prescriptive. The projected moments are computed outside the elemental loops that integrate the measurement sequences. Despite its rather arbitrary construction, $\hat{\mathbb{P}}$ is a useful guide to actions taken on the basis of the tracker estimates. For example, if the line-of-sight of an antenna is to be directed toward the target, the antenna pointing control would use $\hat{x}[k]$ to form the command. The EEP calculation would likely be based upon $P_{xx}[k]$.

There are certain caveats that must be observed when $\hat{\mathbb{P}}$ is used as a basis for some action regarding the target. By construction, each of the local estimates satisfies the kinematic constraints. However, the composite $\mathbf{N}(\hat{x}[k], P_{xx}[k])$ need not satisfy these self-same constraints; i.e., a convex combination of estimates each of which avoids a prohibited region may lie within the prohibited region. Thus, constraint compliance at the local level does not necessarily translate into constraint compliance globally. When a violation occurs, the tracker achieves compliance within $\hat{\mathbb{P}}$ in a manner that is appropriate to the application:

$$(\hat{x}[k], P_{xx}[k]) \xrightarrow{C^{\mathscr{H}}} (\hat{x}[k], P_{xx}[k]). \tag{1.32}$$

Again we emphasize that (1.32) takes place outside the local algorithms and does not degrade the fidelity of $\mathbb{P}$.

1.7 Organization of the Book

This book addresses the problem of tracking an agile target from the point of view of a decision maker. In the basic scenario, a target is moving in an irregular manner in the horizontal plane. A defensive system must take an action to protect an object or an area. A real-time algorithm makes decisions regarding: target type, target intent, target location both now and in the future. The target must be identified and countered or ignored.

To accomplish such tasks, the command algorithm must first locate (or track) the target. Sensing devices generally have a limited field of view, and the decision maker must therefore direct them to a window about the target. This window must capture the target at a specified rate and must be smaller than a specified size. Thus, the architecture must extend point placement to determining regions within which the target will be found.

Adding the requirement of faithful assurance regions changes weightings in the common performance metrics. Fortunately, the familiar model-based algorithms actually provide the information needed to determine plausible capture regions. For example, the Kalman filter generates both the best point estimate of the location of a moving target and the error distribution about this estimate.

The Kalman filter is based upon a LGM kinematic model of point motion and measurement. This simple reductionist representation is quite effective in high SNR environments even when the motion and the measurements are far from LGM. But there is a strong tendency for such algorithms to be overly confident in their assessment of the quality of their target placement: the Gaussian error ellipses fail to capture the target as often as would be predicted by the notional conditional distribution.

The failure of the Kalman filter in the engagements to be presented here is not surprising. A Gaussian distribution has thin tails and an abridged model does not display the inherent variability of evasive target motions.

In this book we will explore several distinctive prototypical engagements. Each is chosen to illustrate some feature of the tracking, classifying, countering paradigm.

We will contrast a single-model algorithm with a multi-model (hybrid) algorithm. The former will be based upon a centered EKF. The latter will be based upon a hybrid tracker.

In every case the engagement will be event-driven; e.g., sudden turns, loss of measurements, offensive countermeasures. The kinematic sensors are intentionally chosen to be nonlinear in this coordinate system; e.g., range-only, range-bearing. This is done to force the algorithms to convert the raw data stream into a usable form using a random transformation. The same raw measurements will be viewed differently by different local trackers. This will actually lead to tracking instability in some cases.

Each of the exemplar engagements is investigated using a single target path and a single sample of the raw data. We seek to contrast the personalities of the uni-model algorithms with those of the multi-model algorithms. So, while we will make statements about capture rates and neutralization probabilities, the statements are sample-function dependent. It is insight into the relative performance of the architectures that we seek rather than a validated statistical study.

With theses caveats, consider the following chapters as illustrations of the command architectures needed to operate in complex environments. From kinematic discontinuities to sudden countermeasures, a point defense system must operate well in an unpredictable, event-driven environment. High quality area defense requires robust architectures.

1.8   Conclusion

We have provided an overview of a class of trackers that employs both a kinematic motion model and a regime or modal model for integrating measurement sequences into target placement and for quantifying the quality of the placement. These hybrid trackers are particularly useful for applications involving an agile target in a low SNR environment.

In the hybrid tracker discussed in this book, the fundamental model is replaced with a linked a family of time-discrete LGM-models in a way that approximates the evolution of the underlying nonlinear model. The state space of the hybrid tracker is a combination of the continuous-valued kinematic states (positions, velocities, etc.) and a discrete modal state. In the hybrid formalism, each modal-sequence tracker generates its own placement and quality statistic. These estimates are suitably weighted, and the final distribution of location is determined along with an appropriate assurance region.

In the reductionist model that is used here, the local kinematic sub-models are time sampled. Unfortunately, attaching a single regime tag to the whole of a sample interval ignores important temporal variation. We have chosen to use the regime at the beginning of a sample interval as a proxy for motion over the full interval.

The local LGM-models are convenient because measurement fusion is simply accomplished using well-known algorithms. Further, a large family of such models can be maintained without excessive computational burden. However, an LGM model generates diffuse extrapolations, and this is not compatible with hard constraints. For example, P_t^w may be such as to emphasize motion of a vehicle in the longitudinal direction. The lateral motion of the vehicle is constrained by the road width, but P_t^w does not provide the hard limit that prevents an extrapolation beyond the edge of the roadbed. This must be accomplished in a para-measurement compliance step.

The regime dynamics are nominally Markov. While common practice, this representation is not consistent with some of applications we will to discuss. For example, a car following a road path may only turn north at a congruent junction; e.g., the $\mathscr{F}_t$-event *turn north* is influenced by the event string leading up to the turn. We will adjust the algorithms to account for such modal memory.

In the chapters that follow, we will present engagements that illustrate different tests of tracking and classification. We will contrast a nominal uni-model architecture with a multi-model architecture on a single sample function. This is done to bring out the personality of the individual algorithms. The uni-model and the multi-model algorithms will be provided the same sequence of kinematic measurements. The hybrid algorithms will be provided regime measurements and para-measurements that would be of no value to a uni-model algorithm.

We will not provide a statistical overview derived from a multi-sample function assessment of the various algorithms. This would be useful in corroborating the quality of the classifiers but would mask the distinctive contrast of single model algorithms and their hybrid counterparts.

Chapter 2
Target Location Using the Extended Kalman Filter

Abstract The Kalman filter has earned its fame through its elegance, its compatibility with digital simulation, and its reasonable performance even when the somewhat restrictive assumptions on which it is based are not all met. Extensions of the Kalman filter are varied, and not all estimators dubbed Extended Kalman Filter (EKF) are equivalent. This chapter develops some filter extensions appropriate to target tracking, classification, and command architectures. These include estimation of essentially nonlinear and non-Gaussian motion models for targets, nonlinear conversion of range-bearing measurements into position measurements, and lack of true independence between target state and observation system errors. In target tracking, using models of target motion and a stream of measurements or observations, the Kalman filter provides the conditional distribution of the location of a target following an unpredictable path. From distribution we can not only accurately estimate the location of the target, but we can also place a capture region about the estimate that will contain the target with any pre-specified probability. The Kalman algorithm is based upon a linear-Gauss–Markov (LGM) model of the engagement. Even when there are nonlinearities in the engagement model, an expanded version of the Kalman filter (the EKF) generates plausible position estimates and capture regions. We will contrast realized performance with that predicted by the EKF. The shortcomings of the Kalman filter when used in command architectures are important, but in many cases they can be mitigated. In others alternative approaches like the GWE, explored in later chapters, are much more effective.

2.1 Introduction

To locate an agile target moving on an irregular path requires careful processing of a sequence of noisy measurements. Figure 2.1 shows the path of a ground target moving at speed 28 m/s. Motion begins in a south-east direction from the starting point $(0, 1000)$ m. There are periods of constant velocity interspersed with counterclockwise turns.

There are three range-bearing sensors located near the path. The sensor locations are in the registration table:

$$(X_r, Y_r) = [(0, 600), (100, 600), (1600, 1200)]$$

© Springer International Publishing Switzerland 2016
D.D. Sworder, J.E. Boyd, *Locating, Classifying and Countering Agile Land Vehicles*, DOI 10.1007/978-3-319-19431-8_2

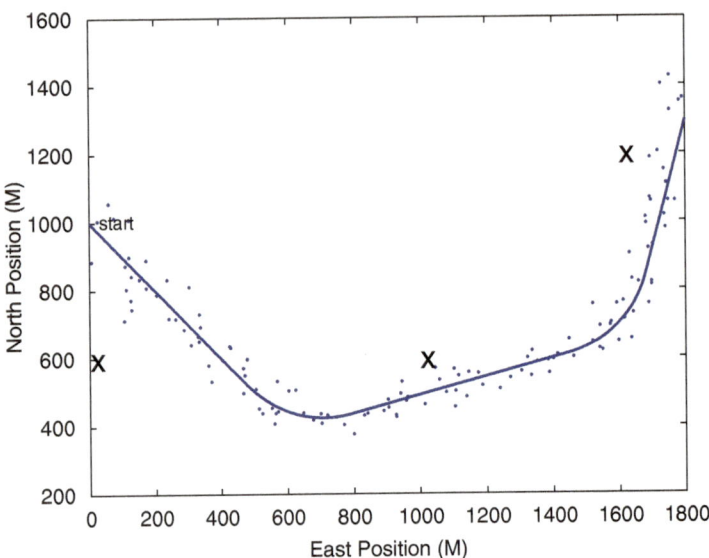

Fig. 2.1 A ground target is moving at 28 m/s. Three range-bearing sensors are shown with a **x** along with the cloud of measurements generated by the sensors

The sensors are displayed in Fig. 2.1 with a *times symbol*. Each sensor provides a noisy range and bearing measurement of the target position every 2 s. The totality of these measurements is shown as the point set in the figure. The measurements are not time-tagged in the figure, but each point used is labeled with a specific sensor and a specific sample time.

Suppose the kinematic state, x_t, of the target is represented by the four-dimensional CT model with position over velocity [see (1.7)], $x_t = [\chi_t; v_t]$. Then

$$\frac{d}{dt}\begin{bmatrix} X\,Y \\ V_X \\ V_Y \end{bmatrix} = \begin{bmatrix} 0\,0\,1\,\,\,0 \\ 0\,0\,0\,\,\,1 \\ 0\,0\,0\,-\omega_t \\ 0\,0\,\omega_t\,\,\,0 \end{bmatrix}\begin{bmatrix} X\,Y \\ V_X \\ V_Y \end{bmatrix} + \begin{bmatrix} 0 & 0 \\ 0 & 0 \\ \sqrt{P^w_{x;t}} & 0 \\ 0 & \sqrt{P^w_{y;t}} \end{bmatrix}\begin{bmatrix} \dot{w}_X \\ \dot{w}_Y \end{bmatrix}, \tag{2.1}$$

where $\{w_t\}$ is a unit Brownian motion and ω_t is a variable turn rate, $\omega_t = \omega_i; i \in \mathbf{S}$ if the turn rate is the ith at time t.

Equation (2.1) can be sampled as was done in (1.10):

$$x[k+1] = \sum_i (A_i x[k] + \sqrt{P^w_i}\, w[k+1])\phi_i[k], \tag{2.2}$$

where $\{w[k]\}$ is a unit Gaussian-white sequence, $w[k] \sim \Phi(0)$. The modal primitive, $\phi_i[k]$, points to the regime state at the beginning of the sample interval. The matrix sequence $\{P^w_i[k]\}$ captures the aggregated uncertainties associated with (2.1).

In this simple LGM model, the path of the target is determined by a sequence of regime events (the turn rates). The state is continuous across event transitions. The state reference is a common origin in state space for all regimes.

Beginning with a Gaussian initial condition, $x[0] \sim \mathbf{N}(\hat{x}[0], P_{xx}[0] > 0)$, Eq. (2.2) generates a Gaussian process, $x[k] \sim \mathbf{N}(\hat{x}[k], P_{xx}[k])$, with moments

$$\hat{x}[k+1] = \sum_i A_i \hat{x}[k] \phi_i[k] \text{ and} \tag{2.3}$$

$$P_{xx}[k+1] = \sum_i (A_i P_{xx}[k] A_i' + P_i^w[k]) \phi_i[k]. \tag{2.4}$$

The mean, $\{\hat{x}[k]\}$, has a drift determined by $\{A_i[k]\}$. At time $t = kT$, the best mean-square estimate of $x[k]$ is $\hat{x}[k]$. The estimation error, $\tilde{x}[k] = x[k] - \hat{x}[k]$, is a Gaussian random vector; $\tilde{x}[k] \sim \mathbf{N}(0, P_{xx}[k])$. The covariance matrix, $P_{xx}[k]$, is a gauge of the (un)certainty associated with the estimate. For example, the estimation error in the X coordinate (or east) has standard deviation, σ_X, equal to $\sqrt{(P_{xx})_{11}}$: $\mathbb{P}(|X[k] - \hat{x}_1[k]| < \sqrt{(P_{xx})_{11}}) = 0.68$. With probability 68 %, we expect that the true value of $X[k]$ is within a centered interval of length $2\sigma_X$: $(\hat{x}_1[k] - \sigma_X, \hat{x}_1[k] + \sigma_X)$ is a 68 % assurance interval for placing the target in the east-direction.

In this application, there are three sensors making simultaneous measurements of target position. Index the sensors with $r \in \mathbf{R}$. We will represent the individual measurements of the kinematic state by $\{y_r[k]\}$:

$$y_r[k] = H_r[k] x[k] + \sqrt{P_r^n[k]} n_r[k], \tag{2.5}$$

where $\{n_r[k]\}$ are all independent unit $\mathscr{F}[k]$-Gaussian white sequences and for all (n, r, k), $P_r^n[k] > 0$.

Equation (2.5) is only an approximation to the actual measurement relation. In the engagement illustrated in the figure, the measurements are presented in a polar coordinate system (range-nearing) while the coordinate system of target motion is presented in a Cartesian coordinate system (east-north). The conversion of the measurement coordinate system to that the motion coordinate system depends upon the sensor-target geometry. But the target location is not known with certainty, and thus the channel noise $P_r^n[k]$ in (2.5) can only approximate the range bearing noise. The coefficient matrices in (2.5) could also depend upon the regime state.

With these caveats, the sensor noise will be assumed to be independent spatially and temporally. Let us stack the individual measurements to form the comprehensive measurement vector $y_\iota[k] = [y_1; y_2; \dots; y_R]$:

$$y_\iota[k] = \sum_i (H_i x[k] + \sqrt{P_i^n} n[k]) \phi_i[k], \tag{2.6}$$

where (H_i, P_i^n) are composed of the matrices in (2.5) in the usual way. Then $\{y[k]\}$ generates the kinematic data-set, $\{\mathscr{Y}[k]\}$. Equations (2.2) and (2.6) yield the elemental LGM-model.

2.2   The Extended Kalman Filter

Equations (2.3)–(2.4) provides a dead reckoning algorithm for estimating the kinematic state of the target. The quality of the state estimate can be improved if we integrate $\{\mathscr{Y}[k]\}$ into $\hat{x}[k]$. Model-free trackers do not have the flexibility required to utilize the motion template underlying the measurements. Rather, they mix $\{\mathscr{Y}[k]\}$ within an elementary point-target extrapolation to infer target location; e.g., an $\alpha - \beta$-tracker[17]. These fixed-form trackers are tuned to specific engagements using a set of adjustable coefficients. Acceptable performance is achieved for smooth motions with high signal-to-noise ratios. Confidence indices like the error covariance are not computed as part of such algorithms

Model-based trackers are able to extract more information from $\mathscr{Y}[k]$ to the degree the model is accurate. The best known of these is the Kalman filter. The elemental Kalman filter ignores the various handoff conditions and biases that occur in applications—or handles them separately.

To delineate the Kalman filter algorithm, suppose we begin with an LGM engagement model and a Gaussian distribution for the kinematic state: $x[0] \sim \mathbf{N}(\hat{x}[0], P_{xx}[0])$. The Kalman filter generates the $\mathscr{Y}[k]$-conditional estimate of the target state along with the $\mathscr{Y}[k]$-covariance of the estimate.

The Kalman filter can be expressed in various coordinate systems. The proper choice depends upon stability and computational advantage [1]; the covariance form is most common in applications. The *covariance filter* computes the $\mathscr{Y}[k]$-(mean, covariance) sequence and presents the $\mathscr{Y}[k]$-distribution of the target state: $x[k] \sim \mathbf{N}(\hat{x}[k], P_{xx}[k])$.

An alternative is the *information filter*. In the engagements discussed here, the error covariance matrix is always positive: $P_{xx}^{-1}[k] = D_{xx}[k] = (F_{xx})^2 > 0$. Define $d[k] = D_{xx}[k]\hat{x}[k]$. The information filter computes the $\mathscr{Y}[k]$-statistics $(d[k], D_{xx}[k])$ [9]. To contrast it with the covariance matrix, which is a gauge of the uncertainty, D_{xx} is called the *information matrix*. The information filter has an advantage when the covariance matrix is ill-conditioned.

As long as we are dealing with positive covariance matrices, $(d[k], D_{xx}[k])$ is equivalent to $(\hat{x}[k], P_{xx}[k])$. The former simply expresses the $\mathscr{Y}[k]$-statistics of the distribution in a different coordinate system:

$$(\hat{x}, P_{xx}) = (D_{xx}^{-1}d, D_{xx}^{-1})$$

While matrix conditioning is not an issue in the applications we will explore, the form of the information filter reveals certain properties of the Kalman algorithm more simply than does the covariance form.

Though not often implemented in this manner, it is possible to express the Kalman filter in a mixed (covariance, information) coordinate system. The algorithm is a two-step process: extrapolate then update.

Extrapolate:

$$\hat{x}[k+1]^- = \sum_i A_i \hat{x}[k] \phi_i[k] \tag{2.7}$$

$$P_{xx}[k+1]^- = \sum_i (A_i P_{xx}[k] A_i' + P_i^w) \phi_i[k]), \tag{2.8}$$

Update:

$$\Delta d[k+1] = \sum_r H_r' D_r^n y_r \tag{2.9}$$

$$\Delta D_{xx}[k+1] = \sum_r H_r' D_r^n H_r, \tag{2.10}$$

where $\Delta d[k+1] = d[k+1] - d[k+1]^-$ and similarly for $\Delta D_{xx}[k+1]$. The final covariance statistics are

$$P_{xx}[k+1] = D_{xx}[k+1]^{-1} \tag{2.11}$$

$$\hat{x}[k+1] = P_{xx}[k+1] d[k+1]. \tag{2.12}$$

The algorithm as presented requires inverting the covariance matrix several times, and this is not proposed for applications. But for expositional purposes, it makes more clear the sequence of transformations that takes place in the filter. First, extrapolation is accomplished without regard to the observation using (2.7)–(2.8). The kinematic estimate at the end of a sample interval is a linear function of the estimate at the beginning of the interval. The dynamic matrix is that associated with regime $\phi[k] = i$. The error covariance is a weighted sum of incremental covariance matrices representing the exogenous accelerations up to the present.

Update to $d[k+1]$ is also a linear function. In this case the increment in $\{d[k]\}$ is found by adding the contribution of the individual sensors in sequence. By isolating the sensors in this way, it is easy to see the consequence of a sensor failure: the associated terms in (2.9)–(2.10) are simply ignored.

The increment in the information matrix is the sum of the information contributed by each sensor. This latter is commensurate with quality sensor—H_r and D_r^n—and the interplay of the directional columns in H_r and the primary directions of the channel noise process.

We emphasize that (2.7)–(2.10) is not the most computationally efficient implementation of a Kalman filter. There are too many matrix inversions required. However, for the examples studied here, the dimension of the kinematic state is small (4D), and the Kalman filter is implemented as displayed. Both P_{xx} and D_{xx} are used internally, and it is convenient to have them available. If a small pseudo-noise augmentation is required to maintain the positiveness of the covariance and information matrices, it will be done without comment. The Kalman filter would be termed "time-variable" since it depends upon the $\{\phi[k]\}$ sequence.

The extended Kalman filter, EKF, broadens the application of (2.7)–(2.10) to engagements in which there are smooth nonlinearities in the motion and measurement models. There are actually many ways nonlinearities can be accommodated; there are actually many EKFs [9]. In what follows, we will use the simplest of these. We will replace the nonlinear engagement model with a local linearization about the estimated state [12]. We have already done this when we took the range-bearing measurement and rotated it and the noise into the (X, Y) coordinate system. The rotation is based upon $\hat{\chi}$ and produces a compatible P_r^n for each of the sensor nodes.

The creation of this linear engagement model, the local model, will be done without comment in the sequel. In contrast to the classical LGM-model, the coefficients of the local model are both time-variable and $\mathscr{Y}[k]$-adapted. It is not possible to predict performance a priori.

We will abbreviate the paired transformations in the EKF as

$$(\hat{x}[k], P[k]) \xrightarrow{\text{EKF}} (\hat{x}[k+1]^-, P[k+1]^-), \qquad (2.13)$$

$$(\hat{x}[k+1]^-, P[k+1]^-) \xrightarrow{\text{EKF}} (\hat{x}[k+1], P[k+1]). \qquad (2.14)$$

In what follows in this chapter, we will study the behavior of the EKF-tracker in some illustrative engagements. The engagements have been selected to reveal the sensitivities of the EKF to off-nominal conditions. We also wish to illustrate the utility of para-measurements used in a formal fashion. This will help us to understand their more integrated role in the hybrid tracker/classifier discussed in later chapters.

2.3   Examples of the EKF-Tracker

2.3.1   The Cooperative Target

Let us look at a specific engagement. We are tracking a ground target moving at a speed of 28 m/s in an SE direction from a starting point at $(0, 1000)$ m. After moving at a constant velocity for 24 s, the target turns left, toward north, at a rate of 5°/s for a period of 12 s. The turn ends for a time before the target turns left again. The sensors sample location every 2 s: $T = 2$. The actual path is shown as the solid curve in Fig. 2.1.

Let us write the kinematic equation as given in (2.1) with: $\phi_t = e_1$ if $\omega = 5°$/s in a counterclockwise direction; $\phi_t = e_2$ if $\omega = 0$; $\phi_t = e_3$ if $\omega = 5°$/s in a clockwise direction. We will refer to these regimes $(S = 3)$ as turn-left, CV, turn-right, respectively. In the engagement shown in the figure, only the first two are exercised.

The time-discrete kinematic model derives from (2.1) with the addition of a white acceleration process with standard deviation $1\,\mathrm{m/s}^2$ in east direction (the preferred direction of motion) and $0.7\,\mathrm{m/s}^2$ in the north direction:

$$\sqrt{P^w}/T = \begin{bmatrix} 0 & 0 \\ 0 & 0 \\ 1 & 0 \\ 0 & 0.7 \end{bmatrix}. \tag{2.15}$$

As the target moves from regime to regime, the kinematic state is continuous. The path is formed by concatenating the CV and the CT segments. The three range-bearing sensors are shown in the figure. Each sensor measures the target location with standard error 56 m in range and 28 m in bearing. The sensor noise is independent temporally and spatially. The measurements with proper time and sensor tags are displayed in Fig. 2.1 and in $\{\mathscr{Y}[k]\}$.

A cooperative target is one that broadcasts its regime to the tracker: $\mathbf{D} = \mathbf{I}$. One can view the cooperative tracker as integrating $\iota[k]$ into $\mathscr{Y}[k]$ to form $\mathscr{G}[k]$. Or one can simply say that the target has known, time-variable dynamics. In either case, we will refer to the engagement as cooperative with $\mathscr{G}[k] = \mathscr{Y}[k]$.

We can implement the EKF-algorithm using (2.7)–(2.10). Suppose we begin with the initial state $x[0] \sim \mathbf{N}(\hat{x}[0], P_{xx}[0])$,

$$\hat{x}[0] = \begin{bmatrix} -5.6 \\ 1040 \\ 20 \\ -21 \end{bmatrix}, \tag{2.16}$$

and $\sqrt{P_{xx}[0]} = \mathrm{diag}([31.6, 31.6, 1, 1])$.

Figure 2.2 shows the response of the cooperative EKF-tracker. The updated estimate of location is shown after each measurement. The estimate is linked to the true location with a solid line, creating what is called a feather plot. The computed error ellipse is then centered on the mean estimate. The interior of the ellipse for a specific multiple of σ gives a tight assurance region for target placement.

The sensor geometry is best in mid-path: one sensor is nearby and others cover the path to the east and to the west. The EKF-tracker is very good. The median tracking error, $\|\chi[k] - \hat{\chi}[k]\|$, is only 11 m. The range error is an order of magnitude bigger.

Figure 2.3 gives an expanded view of three subintervals. The left panel displays the initial response of the tracker. The first error ellipse is large as is the tracking error. This is due to the low quality of the initial placement. But after one measurement cycle, the error has been reduced, and the error ellipses enclose the true position.

The middle panel in the figure shows the response after the first north turn. The LGM-model is fairly accurate, but tracking errors cause the measurement geometry

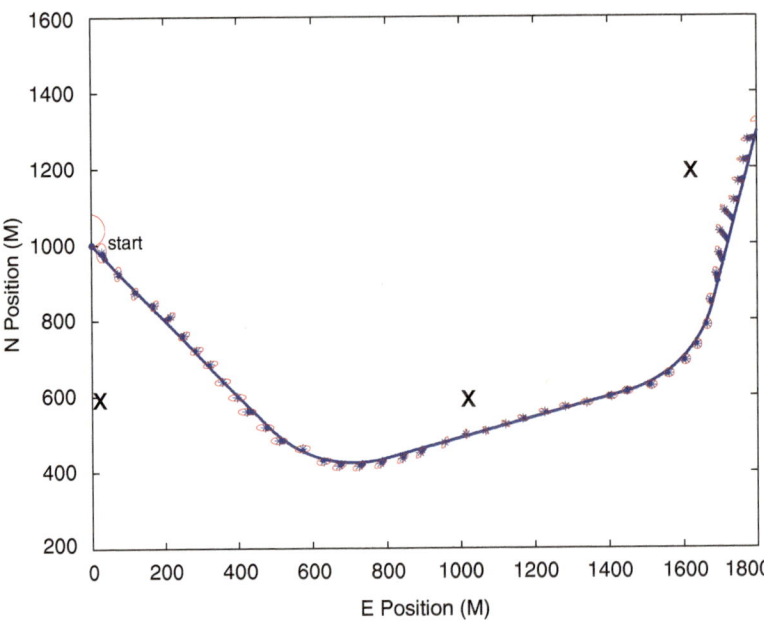

Fig. 2.2 A cooperative target is moving generally east at constant speed. Range-bearing sensors provide location information every 2 s. The EEP error ellipses are shown

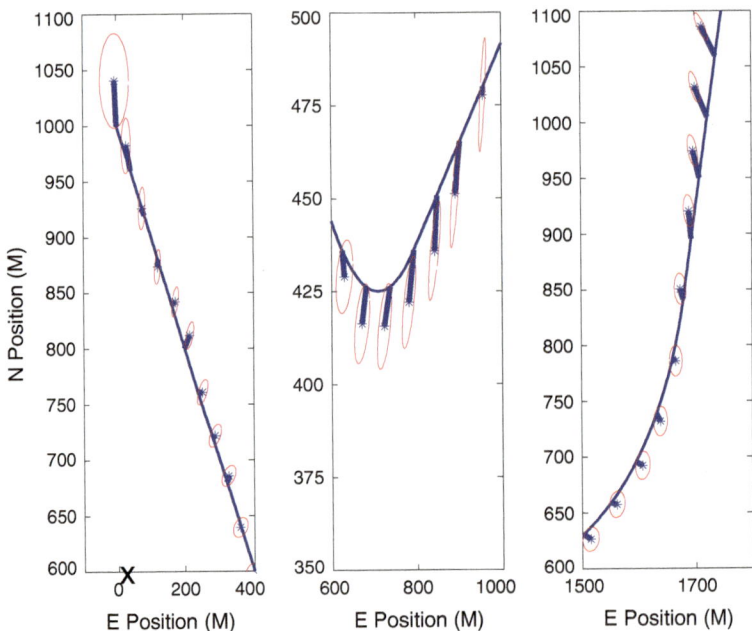

Fig. 2.3 An expanded view of three sections of the path. The plots have unequal axis scales to enhance visibility of the errors. The geometry of the EEP error ellipses is distorted

to deviate from truth. The position and velocity of the target are continuous at the turn. Despite this, the EKF follows the turn with some lag. The error ellipses are small and tend to enclose the target near the north boundary.

The right panel shows the response at the end of the second turn. Here, the maximum errors exceed 50 m because of the lag in recognizing the return to CV-motion. More troubling are the error ellipses. After the turn ends, the error ellipses fail to cover the true position for several time steps.

The mapping from $\mathscr{Y}[k]$ to $\hat{\chi}[k]$ yields a point estimate of target location. However, the EKF does not place the target determinately. Rather, it provides a diffuse likelihood function over the motion space. From this, assurance regions are deduced. In these applications and others like them, the estimator is viewed expansively as providing a likelihood assessment over a spatially distributed domain.

Fortunately, the EKF provides a rational means for evaluating these expanded performance metrics. The position error $\tilde{\chi}[k]$ is approximated with a zero-mean Gaussian random vector. Indeed, $F_{\chi\chi}[k]\tilde{\chi}[k] \sim \mathbf{N}(0, I)$. Hence, in the coordinate system of normalized errors, $\tilde{\chi}_F = F_{\chi\chi}\tilde{\chi}$, the location error has a particularity simple form. The normalized radial error, $\|\tilde{\chi}_F\|$, is Rayleigh distributed with density $p_u = u \exp -\frac{1}{2}u^2$; $u \geq 0$. [10] An engineer can make direct quantitative statements about the size of the tracking error. For example, the probability that $\|\tilde{\chi}_F\| < 1.18$ is 50 %. The 1.18σ-ellipses are the EEP ellipses we seek. The preceding plots are of the 1.18σ-ellipses.

To gauge the performance of the EKF in this example, note that there is no speed change on the path. Thus, the choice of P^w is conservative. On the other hand, the kinematic measurements are not linear in the χ-coordinate system. Consequently, modeling errors are introduced by the repeated linearizations of range-bearing. With all of this, the sample median of $\|\tilde{\chi}_F\| = 1.0$: even with the approximations, the EEP ellipses cover the target half of the time. This is essentially the performance predicted by the Kalman analysis.

The tracker performance shown in Figs. 2.2 and 2.3 is that of a tracker following a cooperative target. In this cooperative engagement, a simple variable-coefficient Kalman algorithm is sufficient to locate the target.

2.3.2 The Uncooperative Target

Let us look now at the same engagement but suppose that the target is uncooperative; i.e., the target does not (or cannot) transmit its turn rate to the data fusion center. In this event, the EKF utilizes a single kinematic model for synthesis. Most of the motion is in the CV-mode. It is natural, therefore, to design a tracker on the basis that $\phi_t \equiv \mathbf{e}_2$. This can be thought of as centering the kinematics.

Figure 2.4 shows the response of the centered EKF-tracker. The raw measurements generating the estimates in the figure are those used in the EKF of Fig. 2.2. But the $\{\mathscr{Y}[k]\}$ data sets will differ because of the changed geometry.

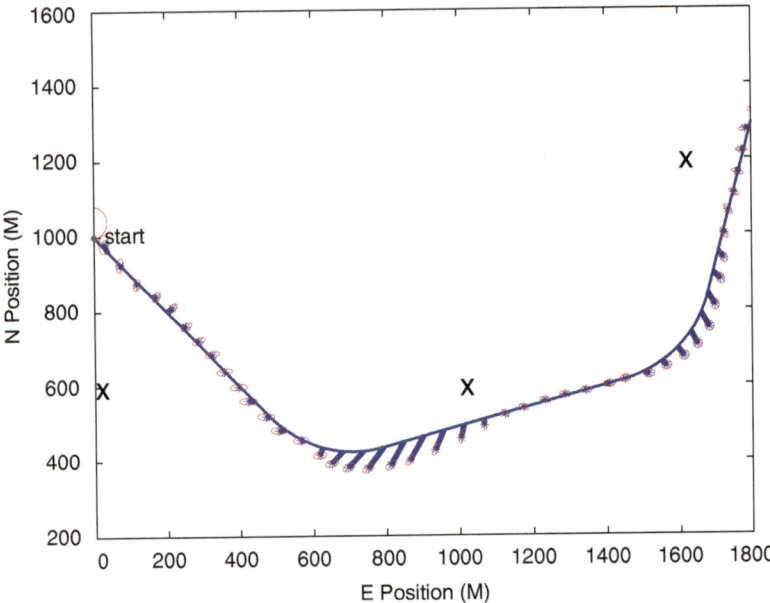

Fig. 2.4 An uncooperative target is moving at constant speed. The performance of EKF-tracker with centered dynamics is degraded

Contrasting the two figures, it is evident that the trackers match initially—for the first 24 s, the two trackers use the correct model. Subsequent to the first turn, this tracker lags, despite the continuity of position and velocity at the regime transition. One reason for the degraded performance is that the white acceleration model does not properly represent the jerk at the turn. Without knowledge of the regime state, this tracker must infer the change in direction from the kinematic measurement. Despite the fact that the tracker has the correct model over the preponderance of the path, the median tracking error is 15.2 m in the uncooperative engagement; a 40 % increase over that observed in the cooperative engagement.

Figure 2.5 shows an expanded view of the response of the centered tracker; contrast with Fig. 2.3. The left panels of the two figures are identical: identical kinematic model; identical $\{\mathscr{Y}[k]\}$. The responses differ significantly in the middle panel: observe that the scales are different in the comparable panels. The tracking error grows to nearly 50 m in the centered tracker without any corresponding growth in the EEP ellipses. The centered tracker sometimes requires a (10σ)-ellipse to capture the target. This is hardly possible under the Gaussian hypothesis that underlies the EKF-tracker. The ratio of $\|\tilde{\chi}\|_{D_{\chi\chi}}$ for the uncooperative engagement to that of the cooperative engagement is over 2.5: the latter is conservative, and the former is hopelessly optimistic.

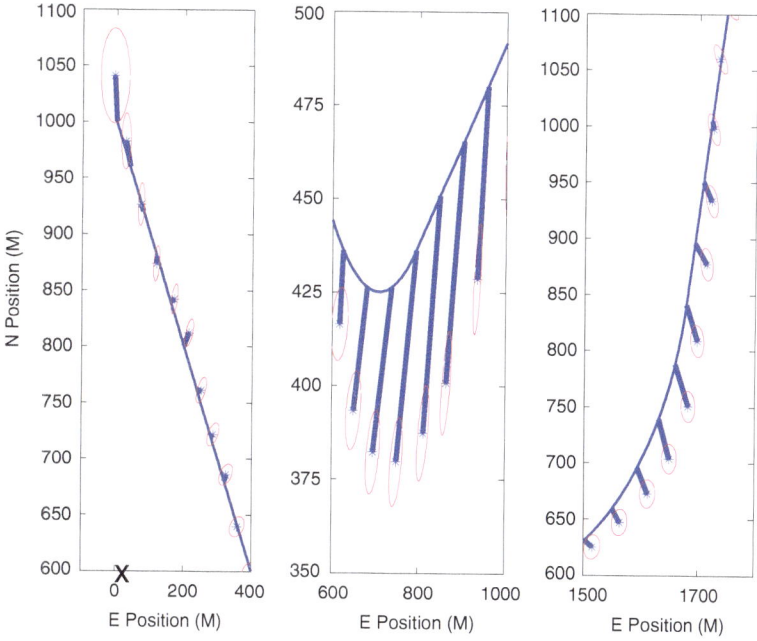

Fig. 2.5 An expanded view of the performance of the EKF-tracker in an uncooperative engagement. EEP ellipses no longer capture the target with consistency. (The geometry of the EEP error ellipses is distorted

2.4   Range-Only Tracking

2.4.1   *The **EKF**-Tracker on a Roadway*

The quality of the location estimates in an EKF-tracker is determined in part by the character of the measurements. In the previous section, a range-bearing sensor produces a putative location of the point target. The current measurement vector is then integrated with those of the past to arrive at an estimate of target position.

Other sensor architectures are more challenging. For example, a range sensor gives only the noisy range-to-target. Lacking bearing, this sensor simply places the target on a circle in the east-north plane. The target must be located using simultaneous ranges from a range array. The geometry of the array modulates the quality of the estimate.

In this section, we will look at range-only tracking. We do this for several reasons. Range tracking is becoming more common in applications because of the simple form of the raw measurements. Range tracking transfers the complexity of the measurements to the subsequent signal processing.

Range tracking is also a major test of the EKF. A single range sensor is of little value in tracking and classification. Only with an appropriate array geometry are the raw measurements useful. The transformation of the range-vector into a Cartesian measurement is fraught with approximations and limitations. As we will see, instability in the EKF-tracker can be produced by poor array geometry and dissimulation by a hostile agent.

To frame the engagement, we will suppose that each sensor transmits a signal toward the target every 0.5 s. The time of the return is processed in a central processor. An individual sensor provides only a circle of possible target locations centered on the sensor itself. To resolve the position ambiguity, more than one such sensor must be used. We will label the sensor nodes by $r \in \mathbf{R}$.

At each sample time, an $\mathbf{R}$-vector of sensor-target directional derivatives generates one column in the LGM gain matrix: $H_r; r \in \mathbf{R}$ [16]. The increment in the information matrix $D_{\chi\chi}[k]$ is determined by the sensor-target geometry, $H_r; r \in \mathbf{R}$, and range noise rotated into the common coordinate system, P_r^n.

As we saw before, the coefficients in the measurement model are $\mathscr{Y}[k]$-random processes. The increment in the information matrix is sensitive to the geometry of the composite H matrix.

To illustrate the influence of transmitter/target geometry on tracker accuracy, consider a simple engagement in which the target is northbound on a north-south roadway with coordinate $-50E$ beginning at $(-50, -60)$ as shown in Fig. 2.6. The vehicle speed is 20 m/s.

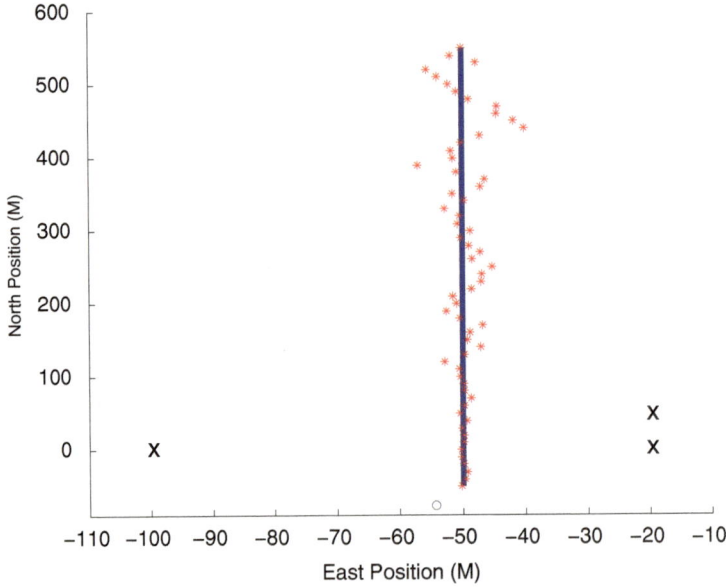

Fig. 2.6 An EKF-tracker for a CV, north-bound vehicle. Only the three sensors in the south array are used. The transmitter locations are shown with **x**, and the location estimates are shown with *asterisk*. The initial position estimate is shown with *open circle*

There are four synchronized transmitters–receivers that generate a noisy range-to-vehicle measurement to the **EKF**-tracker with a standard range error of 1 m. The nodal registration table is:

$$(X_r, Y_r) = [(-100, 0), (-20, 0), (-20, 50), (-100, 500)]$$

The first three nodes are called the south-array with one transmitter west of the road and two to the east. The fourth transmitter is called the north transmitter, and it is used only when necessary. So, if we are tracking using the south-array exclusively, we consider only $r \in (1 : 3)$.

The tracker receives the noisy range measurements, rotates them into an (East, North) coordinate system using the $\hat{x}[k]$-(registration table) geometry. The transformed measurements are stacked as the vector $y[k]$ (in 3D or 4D as appropriate) which in turn generates the data set $\mathscr{Y}[k]$. The channel noise covariance is rotated in conformity with target/sensor geometry to yield the measurement model:

$$y[k] = H[k]x[k] + \sqrt{P^n}[k]n[k].$$

In the stacked model, $H = [H_\chi \quad \mathbf{0}]$ where H_χ consists of the pairs of transmitter-to-target directional derivatives.

Suppose that the **EKF**-tracker is confident that the motion is CV, and that the target is generally north directed. However, the tracker does not know the road coordinates. Nor is the tracker sure of the speed. The engineer expresses the initial uncertainty in the covariance: $\sqrt{P_{xx}(0)} = \text{diag}([10, 10, 0.32, 0.32])$.

Suppose the initial state estimate of the tracker is

$$\hat{x}[0] = \begin{bmatrix} -54.3 \\ -76.7 \\ 0.0 \\ -20.1 \end{bmatrix}. \tag{2.17}$$

The initial positioning is well south of the target and west of the road. Because the target is progressing north, the standard deviation in exogenous acceleration favors the north-south: 6.3 m^2 in north-south and 4.5 m^2 in east-west.

Figure 2.6 shows the sample response of an **EKF**-tracker using just the south-array. The initial target placement is off by 16.6 m. The initial target-sensor geometry is such that the measurement uncertainty favors the north direction. The lateral tracking error is corrected in the first measurement.

South of 100N, the **EKF** does a good job of placing the target. The location estimates follow the road with less than a meter error.

Further north, tracking error becomes more volatile. It grows to 10 m even though the range error is only 1 m. The cause of this change in accuracy is apparent. North of 200N, the directional derivatives of the east sensors become increasingly aligned. As we continue north, the directional derivatives of all of the south-array

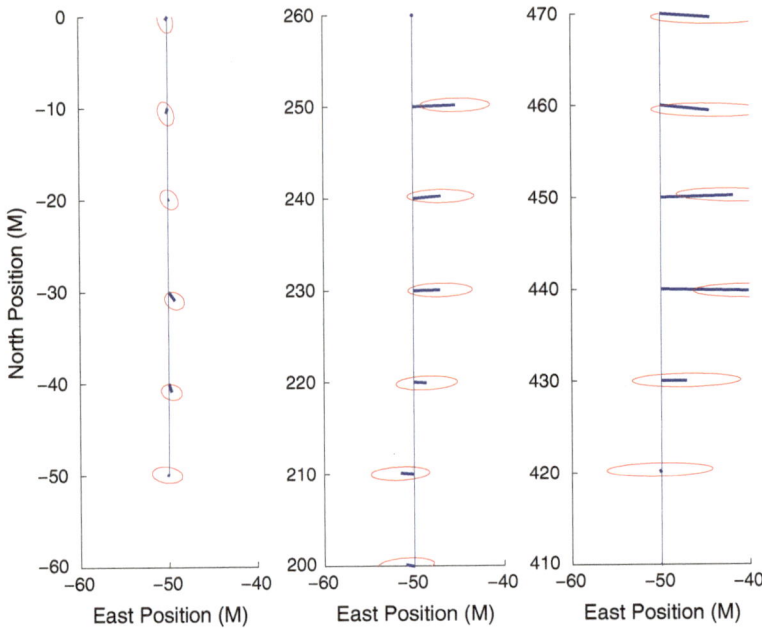

Fig. 2.7 The EEP error ellipses become larger and more eccentric as the vehicle moves north

nodes become nearly parallel. The primary uncertainty is transferred into the lateral direction; i.e., into east-west.

If we wish to classify the target, we must locate it and illuminate it. Suppose we wish to place an illumination ellipse of reasonable size about the target. The EEP ellipses are generated by the EKF, and should suffice for this application. Nominally, the ellipses capture the target half the time, and the capture events are temporally independent. We need only select the interrogation time.

Figure 2.7 shows the illumination ellipses on three segments of the path. Let us begin in the south. After the first measurement, uncertainty is reduced, and the lateral error is almost eliminated. The tiny EEP error ellipses cover the true vehicle location most of the time. We could focus our illumination resources on a small area. Over a small interval, we could be assured of target classification.

As we move north to around 200N (see the middle panel), the EEP ellipses become larger and more eccentric. They still tend to cover the target, within occasional misses, and there are increasing offsets in the estimates. Further, the areas of the ellipses are larger, so when they are used to direct sensor energy, the illumination intensity must be reduced. This in turn reduces the confidence in the classification.

The right panel of Fig. 2.7 shows a sample of the ellipses in the region about 500N. These ellipses are much larger, and although illumination is frequent (i.e., the ellipse frequently contains the true target position), this is aided by the coincidental shift of tracking error to the opposite side of the road. This region would not be a good choice for classification.

In this engagement, the EKF-tracker/classifier would suffice if the illumination region is selected south of 200N. The EKF is simple to implement and interpret. But if system constraints required a longer interrogation interval, another algorithm would have to be chosen.

2.4.2   GDOP and Tracker Accuracy

Figure 2.7 is a sample of the performance of an EKF-tracker using the three-node south array. Not surprisingly, performance degrades when the angles from the target to the nodes narrow. A system architect would have an easier time selecting proper illumination intervals if there were a simple index of path/sensor geometry that was predictive of estimation quality.

One such index has proven useful in error analysis on radiolocation systems, especially GPS. The *geometric dilution of precision*, GDOP, gives a snapshot of the influence of sensor-to-vehicle geometry on location accuracy [16]. GDOP is the factor by which the standard deviation of the range measurement is magnified in the reductive observation geometry: the smaller the GDOP, the smaller the equivalent measurement error. In early GPS architectures, GDOP was used to select the satellites included in the GPS positioning algorithm. For example, with a four satellite architecture, the *best* four transmitters would be selected from the visible satellites on the basis of minimizing GDOP. Newer positioning algorithms do not have the tight limit on the number of satellites—they utilize all available satellites— but a large value of GDOP warns the user of low quality geometries.

GDOP is defined as:

$$\text{GDOP}[k] = (\text{tr}(H_\chi[k]H_\chi[k])^{-1})^{\frac{1}{2}}$$

where $H_\chi[k]$ is the observation gain in position. In this planar application, it would be acceptable to call GDOP, HDOP (*horizontal* dilution of performance), though we will retain the GDOP label.

GDOP is an index of the ambiguity in the raw range measurement vector. We would expect that $\{\text{GDOP}[k]\}$ would increase as the target/sensor geometry weakened. Figure 2.8 shows the plot of the random process $\{\text{GDOP}[k]\}$ as this engagement evolves. Although lacking path memory—except insofar as H_χ depends upon the estimate of vehicle location—the upward slope of $\{\text{GDOP}[k]\}$ presages the growth in tracking error we saw in Fig. 2.6. The anomalous point at the south end is due to the poor initialization of the tracker.

The error covariance is a more important measure of tracking accuracy than is GDOP. The 1σ-error ellipses of $P_{\chi\chi}$ delineate the tightest 68 % assurance region on the sample path. The area of the 68 % assurance regions is $\pi|P_{\chi\chi}|$. The sequence of

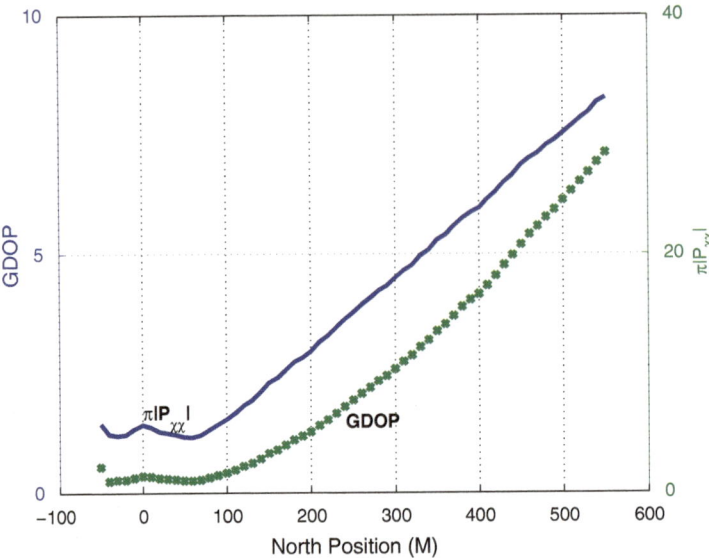

Fig. 2.8 GDOP is predictive of the area of the $P_{\chi\chi}$ error ellipses

inclusion areas is displayed in Fig. 2.8 alongside the graph of GDOP. The areas of
the assurance regions mimic {GDOP[k]}. Beginning with a small value, $\pi|P_{\chi\chi}|$ is
nearly constant for the first 100 m of the path where the sensor geometry is good.
Beyond 100N, the uncertainty area begins to grow almost linearly. As the size of the
1σ-ellipses grows, they become less useful for vehicle location and countermeasure
allocation. Though far simpler in structure, GDOP is seen to be about as good a
predictor of tracking quality as the more complicated $P_{\chi\chi}$. For this reason, in range-
only tracking we will often employ GDOP as a quality metric (and even formally
extend the usage of this term).

2.4.3   The *EKF*-Tracker Including the North Node

In the previous engagement, the degraded tracking performance in the north region
is due to the compressed geometry of the south array. If we wish to intercept the
target in the northern region, the assurance regions should be made more compact
and more faithful to the path. We can expand the region of favorable geometry by
adding the north transmitter to the sensor suite. The ambiguity circle generated by
the north sensor is primarily longitudinal when the target is in the north; contrast the
lateral uncertainty due to the south array.

   If we are to use GDOP as a pointer to regions with compact assurance regions,
we must be confident that the random {GDOP[k]} matches the true process. This
in turn depends upon on the accuracy of $\hat{x}$. That is, the target-node geometry used

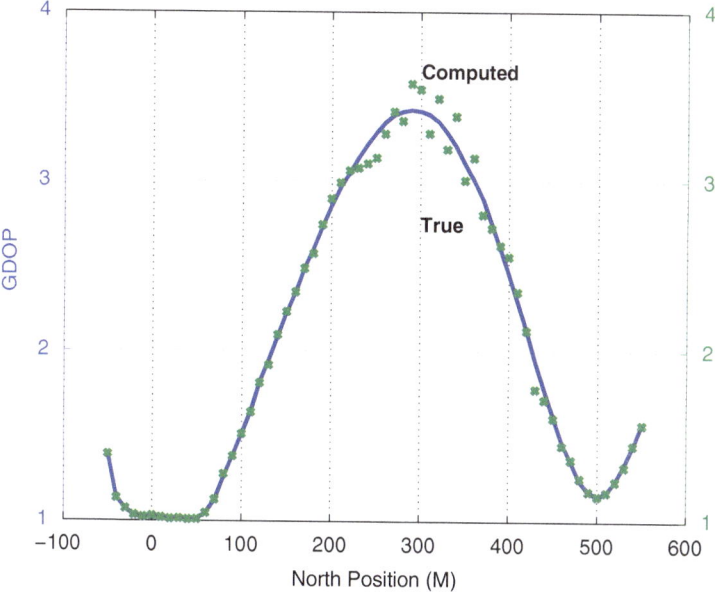

Fig. 2.9 GDOP is computed for at both the estimated target location and the true target location with the full array. The north node reduces GDOP significantly in the north region

in GDOP is that of the sensors and the estimate of target location, and not the true GDOP which is determined by the sensor and the actual target location. To contrast the two, Fig. 2.9 shows the GDOP computed along the true path (the solid curve marked "true"), and GDOP calculated along the sample path (marked "computed"). The two curves match well. The latter is irregular as range noise moves $\{\hat{x}[k]\}$ back and forth across the true path in the middle of the engagement. Both show the small GDOP in the south and small GDOP in the north. Mid-range now has the largest GDOP and would not be the best place to counter the target.

Figure 2.10 shows a sample of the tracking performance when the north transmitter is included. The range measurements from the south-array are identical to those that led to Fig. 2.6. The tracker is much improved north of 100N. The left panel shows the **EKF**-tracker output. It is so good that only one-fifth of the location points are shown.

The right two panels show the EEP ellipses in mid-region and in the north region. The ellipses are compressed significantly: compare Fig. 2.10 with Fig. 2.7. We could classify and intercept the target with good confidence in either the north or the south regions.

Augmenting the south-array with the north transmitter improves the geometry index. The assurance regions are considerably smaller north of 100N. However, the north transmitter involves additional expense. A transmitter has to be purchased and accurately sited. The **EKF** with four transmitters is (slightly) more complex than is

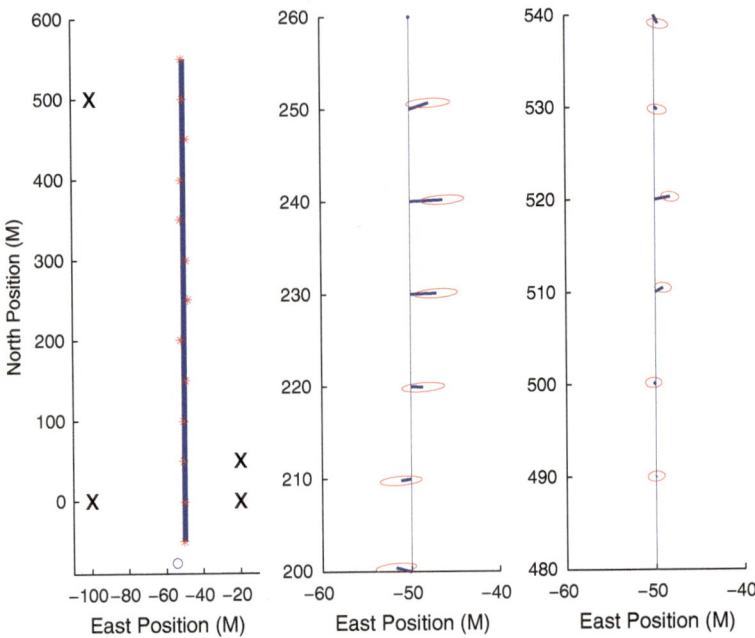

Fig. 2.10 The tracking errors are reduced with the north transmitter. The EEP ellipses are significantly smaller in mid-path and in the north region

the algorithm based exclusively upon the south-array. Still, the error reduction may justify the expense and complexity.

2.4.4　Map-Enhanced Tracking

In this book, we explore the utility of non-temporal data sets. These para-measurements can be used to improve the quality of a basic tracking algorithm. Suppose then that as an alternative to siting another transmitter with its attendant costs and complexity, we provide the tracker with a road map. The map simply provides the east coordinate of the road. The tracker/classifier knows that the target is north-bound at $-50E$. There is some lateral uncertainty due to motion within the roadway and direction irregularities of the road itself. The speed of the vehicle is not known.

The map para-measurement does not fit naturally into the **EKF**. Investigators have proposed alternative ways of integrating such constraints into the Kalman filter structure. [18] For example, we could reduce the kinematic dimension to that of a vertical line with given east coordinate: the kinematic state would be north-position and north velocity. This restriction would however not give a weight to the uncertainty in the lateral motion: motion across lanes, irregularities in road orientation, etc. In another proposed approach, all pseudo-ranges are scaled to have terminus on the road.

In this engagement, let us modify the **EKF** by returning the state estimate to the closest point (with respect to Euclidean distance) on the map and adjusting its velocity:

$$(\hat{x}[k], P[k]) \xrightarrow{\text{EKF}} (\hat{x}[k+1]^-, P[k+1]^-), \tag{2.18}$$

$$(\hat{x}[k+1]^-, P[k+1]^-) \xrightarrow{\text{EKF}} (\hat{x}[k+1]^+, P[k+1]^+),$$

$$(\hat{x}[k+1]^+, P[k+1]^+) \xrightarrow{\mathcal{C}^{\mathcal{M}}} (\hat{x}[k+1], P[k+1]).$$

The handoff detail is

$$\hat{x}[k+1]^+ \xrightarrow{\mathcal{C}^{\mathcal{M}}} \text{diag}(0, 1, 0, 1)\hat{x}[k+1]^+ + (-50; 0; 0; 1), \tag{2.19}$$

$$P[k+1]^+ \xrightarrow{\mathcal{C}^{\mathcal{M}}} \text{diag}(0, 1, 0, 1)P[k+1]^+\text{diag}(0, 1, 0, 1) + 0.1\text{diag}(1, 0, 1, 0).$$

In the map-enhanced **EKF** algorithm, after the measurement update, $\hat{x}[k+1]^+$ is moved to the centerline of the road. To be compatible with the north-bound direction, the expected lateral velocity is set equal to zero. The map doesn't reduce the error in longitudinal position or speed. But the lateral uncertainty is reduced to the lane width: the standard tracking error in east-west is 0.3 m.

Figure 2.11 shows a sample of the tracking performance with a south-array and a road map. The range measurements are those that we have used in the preceding simulations. It is evident that the map obviates any need for the north sensor. The EEP error ellipses in the right two panels of the figure are much smaller than we have computed with north node enhancement. And the tracker infrastructure is far cheaper as well: a stored map replaces a dynamic sensor.

The performance of the map-enhanced south-array tracker given in Fig. 2.11 appears very good because $\hat{\chi}$ is always returned to the road. But a longitudinal error still exists. Figure 2.12 shows a plot of radial position error along the path. In this plot, we contrast three architectures: the south-array only; the map-enhanced south-array; the north-node enhanced south array. In every case, the raw south-array data set is identical.

In the region south of 100N, all three trackers perform well: the radial error is about 1 m with the map enhanced tracker about 3 dB down from that. In the region from 100N to 300N, the trackers lacking a map begin to degrade with errors growing to 3 m or so. The map-enhanced tracker continues to have small errors though they grow to half a meter. Beyond 400N, the tracker based on the south-array sees the errors build to 10 m. The tracker with the full sensor suite keeps the error to a meter or two. The map-enhanced tracker keeps the error well below 1 m.

Adding the map para-measurement to the $\{\mathscr{Y}[k]\}$ data set yields a tracker that is superior to the conventional range-only trackers with either nodal suite. As

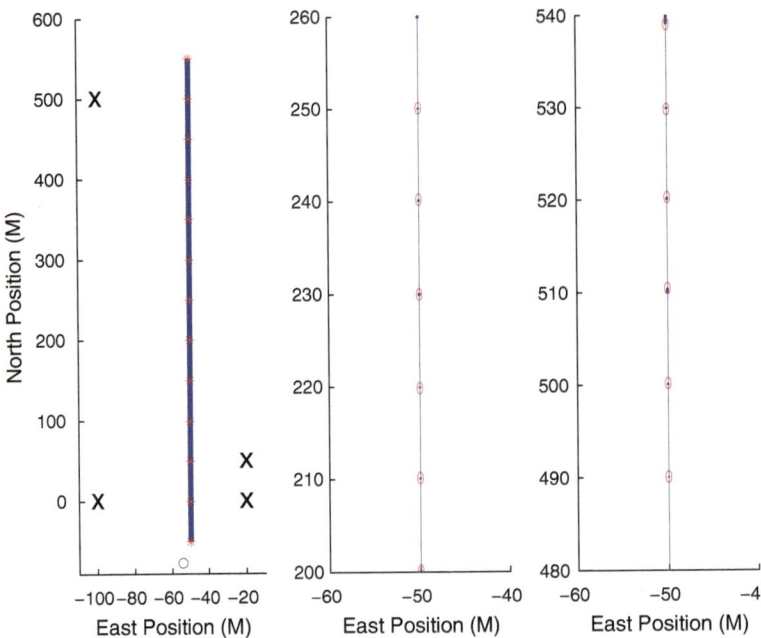

Fig. 2.11 Adding a map to the $\{\mathscr{Y}[k]\}$ data set reduces tracking error even more than does the north sensor

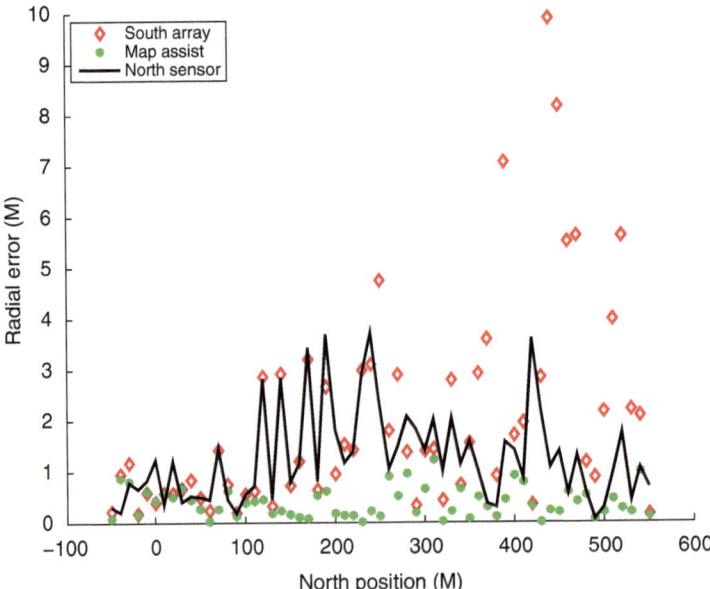

Fig. 2.12 The radial tracking error with the south array is large in the north region. The addition of the north transmitter reduces the error. But map enhancement reduces the error even more

the target moves north, GDOP increases in either case. But with the map para-measurement, the tracking error does not grow as GDOP grows. Map-enhancement avoids the GDOP dilation of the assurance region even though it does not correct the longitudinal error.

2.5 Tracking in the Presence of Spoofing

2.5.1 Registration Errors

The high quality of an EKF-tracker is evident in Fig. 2.11 where a road-constrained, north-bound vehicle is moving at constant speed. A diverse set of four transmitters provides range every 0.5 s. The location error is smallest where GDOP is low—the northern and southern portions of the path. The EEP error ellipses accurately represent the uncertainty in the estimate. Tight assurance regions can be computed on the basis of the mean and variance of the $\mathscr{Y}[k]$-distribution: the error covariance increases for several samples in mid-path just where the target drifts from the road.

A listing of individual transmitter locations is found in a registration table internal to the EKF. Each range measurement is identified with a particular transmitter in the registration table. It may happen, however, that there is an error in the table: this is called a registration error. The error may be a result of a mistake in forming the table, or because the transmitter is not placed where it is supposed to be. In a hostile environment, registration may be wrong because an adversary is intentionally broadcasting false range measurements to confuse the tracker: this is called *spoofing* the tracker. In this last case, the spoofer can broadcast continuously or intermittently.

The small EEP error ellipses are an advantage in finding tight assurance regions, but they magnify the influence of registration error. The $\mathscr{Y}[k]$-distribution of location is notionally Gaussian, and a range outlier is therefore given considerable weight in the EKF. Further, when the covariance is small, the filter gain is small, and any outlier-induced error takes considerable time to correct.

To illustrate the sensitivity of the EKF-tracker to registration errors, consider the following engagement. The target vehicle is moving along the road as before, and the tracker is using the full complement of sensor nodes to place capture regions about the target. An adversary wishes to confound the classifier by transmitting a false range signal that has the north node label. Specifically, the ersatz-north signal is translated in the N-E direction by $(6, 6)$ m: the spoofed node is located at $(-94, 506)$ while north registry is $(-100, 500)$.

Figure 2.13 shows the response of the EKF-tracker during this engagement. Initially, the range from the north transmitter is 5 m too long. But the tracker is insensitive to the registration error because GDOP of the south array is small. The advantageous geometry of the three south sensors compensates for the north bias in $\mathscr{Y}[k]$.

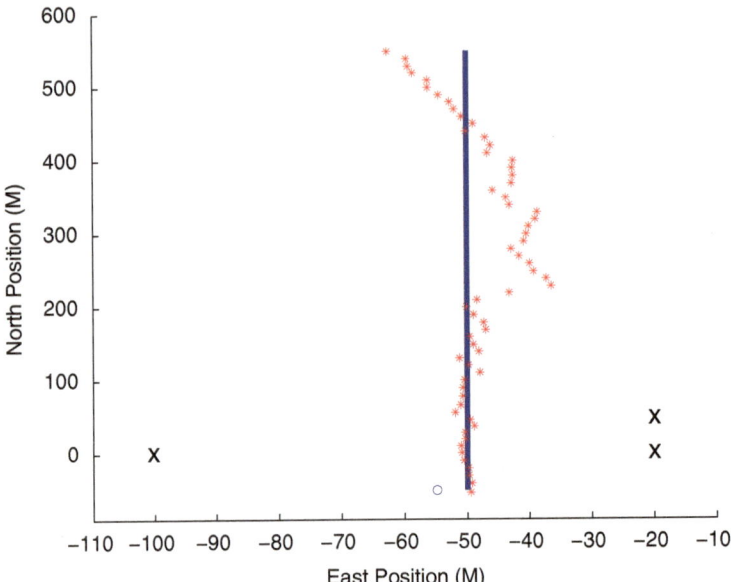

Fig. 2.13 Registration errors cause the estimate of location to degrade. The three valid range nodes are shown. The spoofed node is not shown

However, as the engagement evolves, the south-array GDOP increases. When the target is north of 200N, the error due to spoofing grows significantly. A north range bias of 4.5 m creates an east tracking error of over 10 m. As the vehicle passes 450N, the range offset changes from being too long to being too short. The false registration drags the location estimate west in the north region.

This magnitude of the tracking error is somewhat surprising. The range spoofing offset varies from 5 m to −8.5 m. In the south, with small south-array GDOP, the registration error creates little tracking error—roughly 1 m. However, as the vehicle moves north, the south-array GDOP increases, and the east error grows to over 10 m. Indeed, as the estimate is finally dragged west, the 8.5 m registration bias leads to a tracking error of 20 m.

2.5.2 Flagging Registration Errors Using Assurance Regions

The false north transmitter registration creates the pathology shown in Fig. 2.13. We would avoid the error induced by a false registration table if we eliminated the north transmitter from the registry—or ignored the north measurement. We would return the tracker performance to that shown in Fig. 2.6. Suppose, however, that we do not know *a priori* that the north transmitter is impaired. In fact, we do not know which or when transmitters are corrupted.

We will suppose that the system architect knows the spoofing error is node-specific but is not sure of the temporal policy used by the adversary. With a long observation interval, any impaired transmitters could be identified and excluded from the registration table. Instead of studying the broad identification-isolation problem, we will consider the simpler framework in which the assurance regions computed by the EKF are used to remove offending sensors from the registry at each sample time. We will do this by treating the spoofing errors as if they were temporally and geographically independent.

To tag an offending measurement, let us create a confidence metric. Let X_s be the registration array with 2D columns giving the nodal locations. Consider the transmitter $m \in R$ at time kT at nominal position by $X_s(m)$. Denote by $\varrho(m,k)$ the distance between the mth range measurement, $y(m, k)$, and the notional value of the measurement:

$$\varrho(m, k) = | (y(m, k) - \| (\hat{\chi}[k] - X_s(m)) \|) | . \tag{2.20}$$

Of course, $\varrho(m,k)$ is random because there is a non-zero range measurement error variance, P_m, and there uncertainty in locating the target vehicle, $P_{\chi\chi}$. In the latter matrix, we must acknowledge the $\mathscr{Y}[k]$-dependence of the covariance matrix. A naive approximation to the measurement variance is found by adding the measurement variance to the variance of the range in the direction of the mth transmitter:

$$P_{\hat{\chi}\hat{\chi}}(m, k) = H_\chi(m, k) P_{\chi\chi}(k) H_\chi(m, k)',$$

where $H_\chi(m, k)$ is the direction vector to the transmitter.

A simple gauge of the predicted size of residual is found by adding the variance of the range channel noise to the variance of position in the direction of the transmitter:

$$\sigma_m(k) = \sqrt{P_{\hat{\chi}\hat{\chi}}(m, k) + P_m}.$$

Let us normalize the actual residual by dividing by σ_m:

$$\Delta(m, k) = \frac{\varrho(m, k)}{\sigma_m(k)}. \tag{2.21}$$

When $\Delta(m, k)$ is large, the range is tagged an outlier.

A simple adjustment of the EKF-tracker appears to promise a reduction in sensitivity to registration errors. Let us set a threshold, $\Delta > 0$, for the number of standard deviations that a specific range measurement can deviate from the mean measurement: if $\Delta(m, k) > \Delta$, ignore transmitter m at time kT. Thus, if we observe an 8.4 m deviation in the measurement from the north transmitter that has standard channel noise of 1 m, we would be safe in assuming that the north transmitter was compromised.

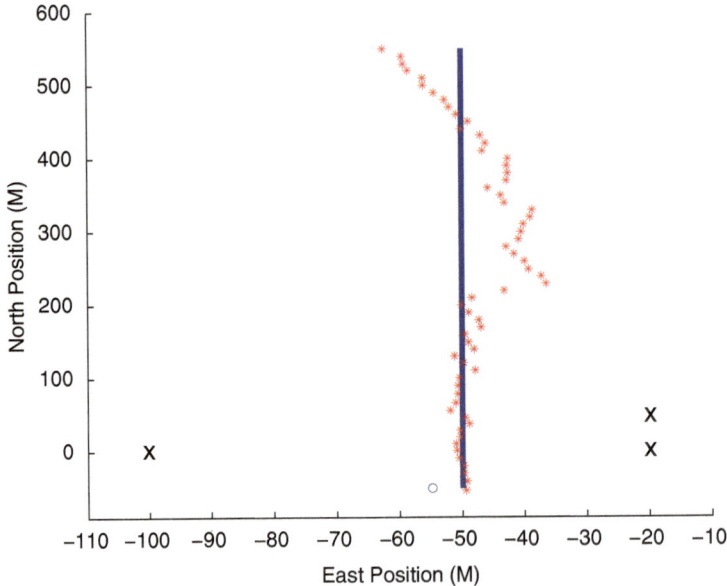

Fig. 2.14 A threshold alert does not improve performance to a significant degree

The threshold-tracker uses a point-wise exclusion logic, and it ignores the fact that errors in the registration table create a bias across time. Further it ignores the fact that specious range measurements compromise the estimate of target location and prejudice $\varrho(m, k)$.

The performance of the **EKF** with threshold exclusion, $\Delta = 2$ is shown in Fig. 2.14. In contrast to the myopic tracker that ignores the registration error, there is some slight reduction in radial error in mid-path. The threshold-tracker is, however, far from replicating the performance of the tracker using the south-array.

The reason for the poor performance of the threshold-tracker is made clearer in Fig. 2.15. The tracker is alerted to an outlier when $\Delta(m, k) > 2$. For clarity, the figure only shows alerts from the north transmitter (impaired) and from the south-west transmitter (not impaired). South of 200N, the impaired transmitter is tagged and ignored several times in the tracker update. The south-west transmitter is tagged only once. In this region the performance of the **EKF** is satisfactory because GDOP from the south array is small—the north transmitter is not needed.

North of 200N, there are few taggings. In fact, the only alert is for a valid measurement from the south-west transmitter. As the south-array GDOP degrades, the error ellipses increase in size, and this increases the size of range deviation required to tag a measurement. Moreover, the **EKF** is drawn into a region where all of the measurements are compatible with the threshold. So, instead of isolating the compromised transmitter, the **EKF** places $\{\hat{x}[k]\}$ so as to accept all of the measurements.

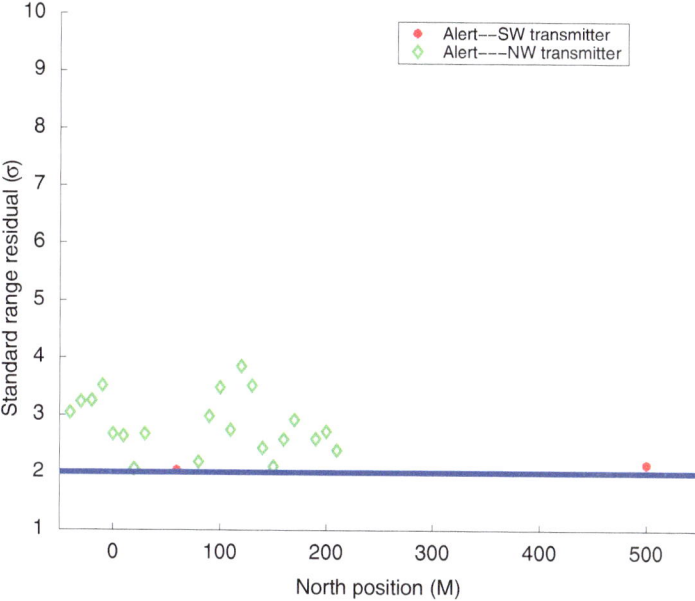

Fig. 2.15 South of 200N, the threshold-tracker correctly flags and rejects many readings of the compromised north sensor. Further north, as GDOP for the south array degrades, the laterally enlarged error ellipses do not support rejection of the compromised north sensor

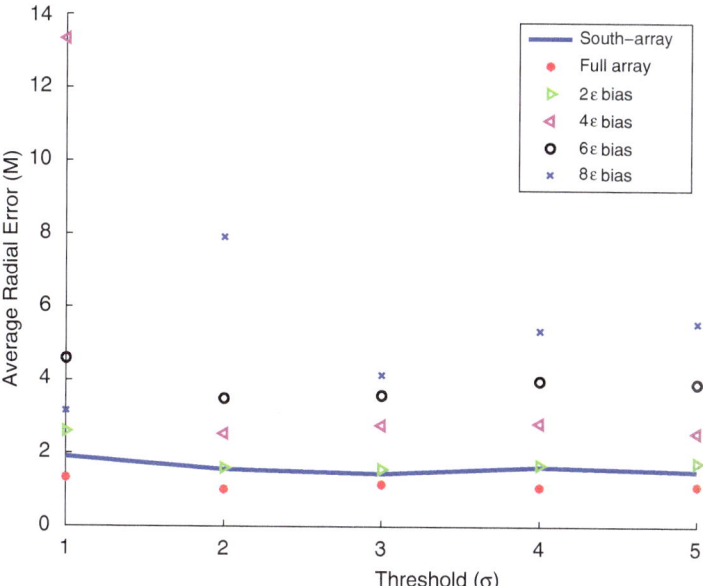

Fig. 2.16 The radial error increases with larger registration biases

The figures shown give the sample-path performance of a threshold tracker with a north-east registration bias in the north transmitter of the size $\varepsilon = 1$ and direction $[1; 1]$. To illustrate the sensitivity for various thresholds, the median radial tracking error was determined along the road path. An average of ten independent simulations was then computed. The use of the median error avoids giving undue weight to the larger error in the north portion of the track. Figure 2.16 shows the average error for trackers with various biases and thresholds.

The figures shown give the sample-path performance of a threshold tracker for a single threshold and single registration bias. To explore the sensitivity of the tracker to registration error, let us consider north-east biases with various multiples of ε. Ten independent simulations were run with specified bias. The median tracking error was then determined—the median avoids undue influence of the larger tracking errors at the north end of the target path. Figure 2.16 shows the mean of the median errors (labeled "average") for the various biases and thresholds.

The curve in Fig. 2.16 labeled with the "red dot" symbol gives the performance of the EKF with the full transmitter suite. As Δ changes from 1 through 5 there is little change in the median tracking error. GDOP is small throughout. The tagging of a single transmitter has little effect on performance. The same can be said when the south array is used—the bold curve. With only three transmitters, performance degrades slightly. But there are few alerts on the path. The slight variation in the two curves is due to sample variations.

The other curves in Fig. 2.16 show the tracking error when there is a registration error in the north transmitter. When there are registration errors, the threshold algorithm suffers from two types of error: First, valid measurements are rejected because they fall too far from $\hat{\chi}$; second, contaminated measurements are accepted because $\hat{\chi}$ has drifted from truth. When $\Delta = 5$, the latter error is dominant. The median error grows with the size of the bias in a congruous way.

As Δ gets smaller, the former threshold error becomes dominant. When $\Delta = 2$, the north transmitter registration bias 8ε causes the median error to grow by 3 dB as compared to the bias 6ε. This is due primarily to the failure to use the range measurements effectively. If $\Delta = 1$, performance is unpredictable with the tracking error for 4ε exceeding 8ε.

Broadly speaking, the performance of a threshold tracker is not as good as expected. The mean estimate drifts into a position compliant with the full set of range measurements. If Δ is small enough to tag the impaired transmitter, it will also tend to reject valid range measurements. Of course, performance would be improved with an adaptive threshold. Had the anomalous north transmitter been permanently tagged in the south segment of the path, the growth in the tracking errors would have been avoided. Unfortunately, an adversary can easily counter this procedure by sequential spoofing of the four transmitters.

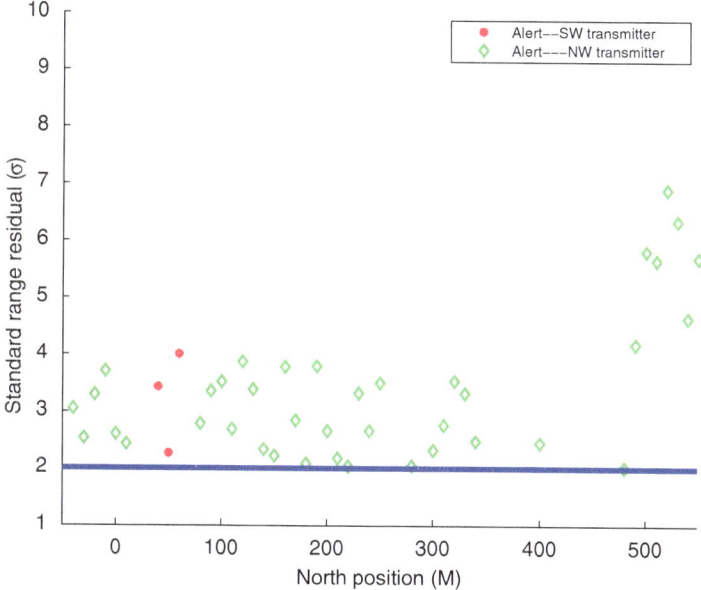

Fig. 2.17 A map-enhanced algorithm tags the impaired transmitter more frequently than does a map-deficient algorithm

2.5.3   *Map-Enhanced Flagging*

A path map has been shown to be useful to an **EKF**-tracker. Let us take a simpler approach in which we use this para-measurement as an adjunct in the threshold tracker. In an earlier section we returned the kinematic state estimate to conformity with the path constraint at each time step. In this spoofing engagement we again use the formula

$$\varrho(m,k) = | \, (y(m,k) - \| \, (\hat{\chi}[k] - X_s(m)) \, \|) \, |, \qquad (2.22)$$

where now $\hat{\chi}[k]$ is the $\mathscr{Y}[k]$-mean *after* the location estimate has been returned to the road.

With this adjustment, consider again the threshold algorithm again with $\Delta = 2$. Figure 2.17 shows the alerts generated along a sample path. The 1σ-ellipses for this tracker are much smaller than those of the **EKF**-tracker lacking map-enhancement; i.e., the region generated by $\Delta = 2$ in this tracker is far smaller than the corresponding region in Fig. 2.15. This is evident in the sequence of alerts assigned the south-west transmitter at about 50 m-north.

The map-enhanced algorithm is much better at tagging the spoofed north transmitter when the target is north of 50 m. Whereas the algorithm sans-map attempted to balance all the transmitters, the map-enhanced algorithm is kept compliant. The impaired transmitter was identified as such throughout the north motion—except

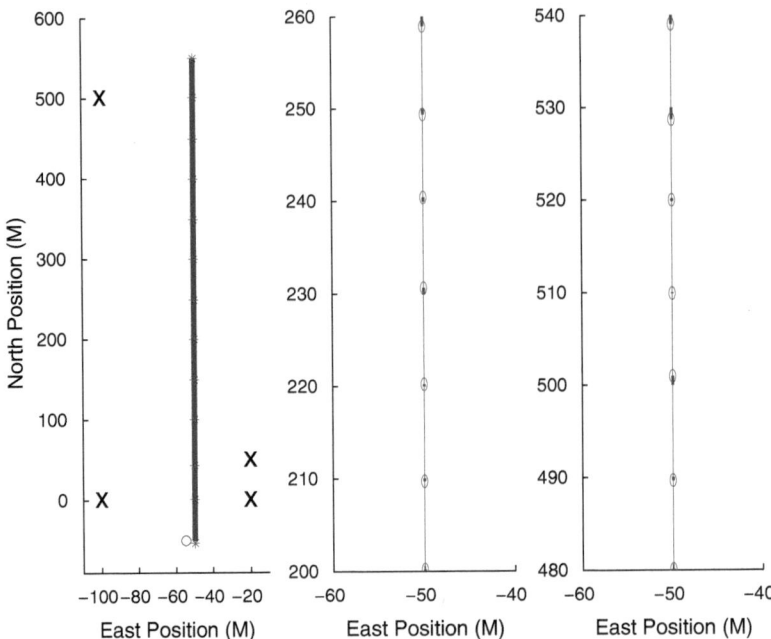

Fig. 2.18 The map-enhanced threshold-tracker has excellent response. The two right panels show the 1σ-error ellipses to be both faithful and small. The north transmitter has an 8 m registration bias

during the crossing at about 450N. Recall that this crossing occurred when the range to the north transmitter became temporarily equal to the true range.

Figure 2.18 shows the response in a spoofing engagement of the threshold tracker with map enhancement. The pseudo-ranges generating this figure are precisely those used to create Fig. 2.14. The horizontal scale of the right panel of the former figure is expanded: the 1σ-ellipses are highly elliptical. At every point, the left panel of Fig. 2.18 shows $\{\hat{\chi}[k]\}$ to be on the road—as it must since the $\mathbf{C}^{\mathcal{M}}$-transformation places it there. A better indication of tracking quality is given by the right panel of the figure. The 1σ-ellipses cover the true position with good frequency.

2.6 Conclusions

The **EKF** is a useful algorithm for locating a target following a smooth path in a high signal-to-noise environment. It uses a paired transformation: an extrapolation using an average kinematic model, and an update using the kinematic observations. Though not discussed in this chapter, prediction can be accomplished using the estimates generated by the **EKF**. The $\mathcal{Y}[k]$-distribution of the target state is $x[k] \sim \mathbf{N}(\hat{x}[k], P_{xx}[k])$; the one-step predictor for a cooperative target is given by (2.7)–(2.8).

Continuing, the two-step $\mathscr{Y}[k]$-predictor to time $(k+2)T$: $x[k+2,k] \sim \mathbf{N}(\hat{x}[k+2,k], P_{xx}[k+2,k])$, where

$$\hat{x}[k+2,k] = \sum_i A_i \hat{x}[k+1,k]\phi_i[k+1], \qquad (2.23)$$

$$P_{xx}[k+2,k] = \sum_i (A_i P_{xx}[k+1,k]A'_i + P_i^w)\phi_i[k+1], \qquad (2.24)$$

and so on.

If the measurement/kinematic model deviates from the LGM-structure, one of the many EKFs can be used in its place. The EKF presented here simply uses a local replacement for the nonlinear range measurement. The deviations from the true measurement relations are incorporated into the pseudo-noise in the kinematics. When the target is uncooperative, the EKF-tracker lags the true path by a considerable amount.

In atypical engagements, e.g., spoofing, the EKF must be modified in a fundamental manner. For example, registration errors expose an inherent weakness of the EKF: its assurance regions are grossly optimistic. Inappropriately small assurance regions cause the influence of outliers to be outsized. We can try to desensitize the EKF by using measurement gating. But the correct gate locations depend upon the uncertainty in the estimate and that is suspect.

In the case of registration errors, we have seen that

1. Sensor pruning on the basis of $\Delta(m,k)$ may be too inflexible. Once $\{\hat{\chi}[k]\}$ leaves a neighborhood of the path, all of the sensors may look bad (or good). A softer approach would be to increase the pseudo noise in the update adaptively to deweight the measurements with large residuals. This can be done pointwise or recursively. An increase in pseudo noise has the disadvantage that it decreases response to measurement changes.
2. When a path map is available, $\hat{\chi}[k]$ can be returned to the map after every update using the Mahalanobis distance (i.e., standard error). This reduces the tracking error and makes isolating the compromised sensor much clearer.

This chapter illustrates some sensitivities of the EKF in tracking applications. The engagements have been chosen to expose weaknesses in the EKF resulting from its optimistic calculations of its assurance regions. A more complex path would expose additional weakness. For example, a path with road junctions and abrupt turns would alter the basic algorithmic structure. We will explore such engagements in more detail later in the book. For now we will simply say that the EKF-tracker is quite effective in a cooperative engagement. As the conditions of the engagement drift from that represented by the LGM paradigm, the performance of the tracker suffers.

For the architect of a command system, EKF limitation to a single target kinematic model can be a significant disadvantage in tracking a target capable of several different motion regimes. We have seen that EKF accuracy is seriously degraded when target motion mode departs from the single kinematic model. Unrealistic estimator covariances and assurance regions can lead to ineffective decisions, resulting in loss of target track owing to lost illumination and underdeployed or wastefully deployed countermeasures. Latency of estimates and ineffective state prediction associated with target regime changes can spoil the accuracy and reduce the effectiveness of countermeasures.

Chapter 3
Tracking an Agile Target

Abstract Multi-model algorithms have proven effective when the target is not cooperative. A hybrid tracker can tailor its kinematic-sensor model to the current conditions of the engagement. Unfortunately, in applications the correct model is seldom known with certainty. Rather, the command architecture must balance a range of possible options based upon the relative likelihoods of the various alternatives. In this chapter, we explore a specific hybrid algorithm, the GWE, and contrast its performance with a single motion model (CV) EKF. The GWE fuses complementary para-measurements with conventional kinematic data to estimate possible engagement regimes, in addition to current target location and velocity.

3.1 Introduction

In the previous chapter, we studied a specific model-based algorithm for following an agile target. The motion template was a uni-regime LGM model. The random accelerations and noisy measurements represented the unstructured uncertainties in the engagement rather well. If the target is cooperative, an EKF-tracker followed the path with small error, and the notional EEP ellipses enclosed the target position most of the time.

Tracking performance deteriorated in an uncooperative engagement. Figure 3.1 shows the response of the EKF-tracker as the target moves along a multi-regime path in the plane (see Fig. 2.6). When the target is uncooperative the tracker uses a *centered* kinematic model: it assumes target motion is nominally straight ahead, rather than modeling target motion as left turn, straight ahead, or right turn, as was done when the target is cooperative. The high capture rate of the initial EKF is lost when the tracker must estimate the kinematic state without knowing the regime state.

There are various ways to improve tracking performance in an uncooperative engagement. Since it is the lack of regime state information that distinguishes a cooperative engagement from an uncooperative one, one enhancement would be to augment the kinematic measurements with a direct measurement of the regime. Proper data fusion should reduce the tracking error.

© Springer International Publishing Switzerland 2016
D.D. Sworder, J.E. Boyd, *Locating, Classifying and Countering Agile Land Vehicles*, DOI 10.1007/978-3-319-19431-8_3

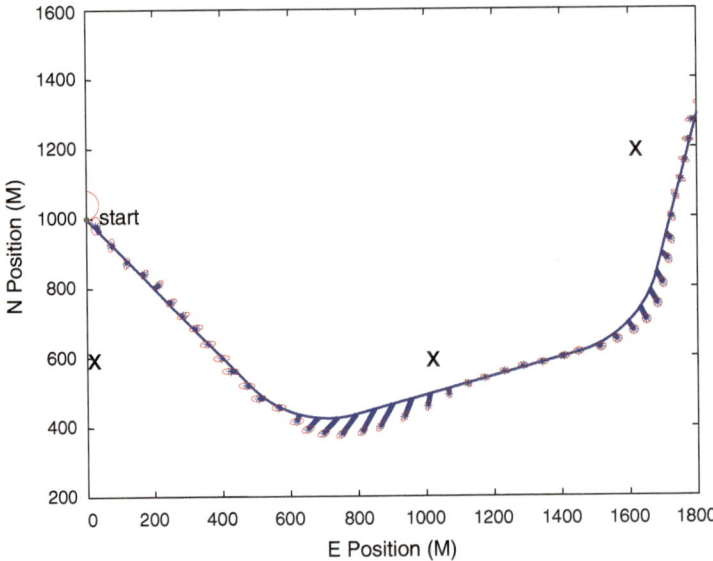

Fig. 3.1 An uncooperative target is moving at constant speed beginning at $(0, 1)$km. A sample function of the response of an **EKF**-tracker with centered dynamics is shown. The EEP error ellipses centered on the location estimates are also shown

In other applications, we have a non-temporal para-measurement. While not related to the current motion of the target, these auxiliary data sets are useful for what they exclude. One engagement in Chap. 2 employed a path map to exclude motion outside a narrow roadbed. The non-temporal information allowed us to isolate a compromised range-to-target sensor.

The utility of auxiliary measurements becomes more apparent when the tracker must respond to more sophisticated queries. In chapters that follow, the tracker may be required to provide target identification, predicted location, allocation regions for countermeasures, etc.

In this chapter and those which follow, we will look at a specific multi-model tracking algorithm, the **GWE**, that fuses a sequence of kinematic measurements, $\{\mathscr{Y}[k]\}$, with direct regime measurements and para-measurements, $\{\mathscr{Z}[k]\}$, into a high quality $\mathscr{G}[k]$-estimate of location. We provide a detailed articulation of the **GWE** algorithm in the Appendix. In the present chapter we revisit the maneuvering-target application presented in Chap. 2 using the **GWE**. The engagement has neither handoff transformations nor path constraints. More convoluted engagements will be studied in subsequent chapters.

3.2 Multi-regime Engagement Model

3.2.1 Kinematic Model

The underlying structure of a multi-regime engagement model is presented in Chap. 1. For the purposes of this chapter, we will return to the first example of Chap. 2. The target kinematics are given by the CT-model of dimension four with position over velocity; $x_t = [\chi_t; v_t]$ [see (1.7)].

$$\frac{d}{dt}\begin{bmatrix} X \\ Y \\ V_X \\ V_Y \end{bmatrix} = \begin{bmatrix} 0 & 0 & 1 & 0 \\ 0 & 0 & 0 & 1 \\ 0 & 0 & 0 & -\omega_i \\ 0 & 0 & \omega_i & 0 \end{bmatrix}\begin{bmatrix} X \\ Y \\ V_X \\ V_Y \end{bmatrix} + \begin{bmatrix} 0 & 0 \\ 0 & 0 \\ \sqrt{P^w_{x;t}} & 0 \\ 0 & \sqrt{P^w_{y;t}} \end{bmatrix}\begin{bmatrix} \dot{w}_X \\ \dot{w}_Y \end{bmatrix}, \tag{3.1}$$

where $\{w_t\}$ is a unit Brownian motion and ω_i is the turn rate.

Equation (3.1) can be abstracted as was done in the (1.10). The time-discrete kinematic equation is

$$x[k+1] = \sum_i (A_i x[k] + \sqrt{P^w_i} w[k+1])\phi_i[k], \tag{3.2}$$

where $\{w[k]\}$ is a unit Gaussian-white sequence.

In this specific engagement, there are three possible turn rates: left, $i = 1$, if $\omega = 5°/s$ in a counterclockwise direction; straight, $i = 2$, if $\omega = 0$ (the CV-model); and right, $i = 3$, if $\omega = 5°/s$ in a clockwise direction. As in Chap. 2, the standard deviation of the exogenous acceleration is $1\,\text{m/s}^2$ east-west, $0.7\,\text{m/s}^2$ north-south. This exogenous acceleration is an aggregation of various things both wideband and structured. The exogenous disturbance was chosen big enough to accommodate the random turns. In a multi-model algorithm that follows, we will handle turns differently and will not need such a large value of P^w_i.

This engagement requires that we follow a target moving in the plane at a speed of 28 m/s in an SE direction from a starting point at $(0, 1000)$ m. After moving at a constant velocity for 24 s, the target turns north at a rate of $5°/s$ for a period of 12 s. It returns to a straight path for a time before turning north again. The path is shown as the solid curve in Fig. 3.1.

3.2.2 Modal Model

The EKF-tracker ignores the dynamics of the regime state. In the cooperative engagement, the regime state is known when needed. In the uncooperative engagement, the motion dynamics are centered on the CV regime. As the engagement evolves, the regime state generates a sequence of target placements. For example, it is apparent from Fig. 3.1 that if we sample the regime state every 2 s, we begin with the string 22 . . . . The initial model state ι is the prefix of length L: if $L = 3$, the mode at $t = 20$ is $\iota = 222$. The set of all three-strings is written $\iota \in \kappa$: κ is the set of all three digit radix three numbers. So $\iota = 122$ represents a left turn at rate of $5°/s$ after a period of CV motion. We see this a transition around $(520, 480)$.

In the GWE algorithm, the regime state is deemed to be a Markov process with ϕ_t a pointer to the current regime:

$$d\phi_t = Q'\phi_t dt + dm_t \tag{3.3}$$

with initial condition ϕ_0. The forcing term, $\{m_t\}$, is a discontinuous martingale. If $\phi_t = \mathbf{e}_i$, the probability of the modal transition $\mathbf{e}_i \mapsto \mathbf{e}_p$ is $Q_{ip} dt$ for $i \neq j$.

The diagonal elements of Q are linked to the mean lifetimes in the individual regimes. From Fig. 3.1, a broad approximation to the regime process might be: the lifetimes are about 25 s in CV; 12 s in each CT regime; a CV regime is equally likely to transition into either CT regime; a CT regime is more likely to transition into a CV regime than it is to transition to a turn of the opposite sense. We will abstract this as the regime rate matrix Q,

$$Q = \begin{bmatrix} -0.08 & 0.05 & 0.03 \\ 0.02 & -0.04 & 0.02 \\ 0.03 & 0.05 & -0.08 \end{bmatrix}. \tag{3.4}$$

If we sample the regime state every 2 s,

$$\phi[k] = \Pi\phi[k-1] + m[k], \tag{3.5}$$

where $\phi[k] = \phi_{kT}$, $\{m[k]\}$ is a time-discrete, martingale difference sequence, and

$$\Pi = \begin{bmatrix} 0.86 & 0.04 & 0.05 \\ 0.09 & 0.93 & 0.09 \\ 0.05 & 0.04 & 0.86 \end{bmatrix}. \tag{3.6}$$

The matrix Π gives the one-step transition probabilities for the regime process:

$$\Pi_{pi} = \mathbb{P}(\phi[k+1] = \mathbf{e}_p \,|\, \phi[k] = \mathbf{e}_i).$$

The columns are probability vectors. Violent jinking; i.e., moving from a left-turn to a right-turn regime—$122\ldots \mapsto 3122\ldots$ occurs only 5 % of the time from a CT regime, and is less likely than moving from left turn to CV—$122\ldots \mapsto 2122\ldots$ occurs 9 % of the time from a CT regime. Of course, the most common event is that of continuing a CV motion: $222\ldots \mapsto 2222\ldots$ occurs 93 % of the time.

In (3.6), Π is constant. There is little change in what follows if Π is slowly varying. The magnitude of Π_{pi} cues the GWE to the likelihood of a regime event.

It may happen that this likelihood is also a function of the kinematic state—which is a function of past regimes. If this is true, $\{\phi[k]\}$ is no longer a Markov process; the probability that $\phi[k+1] = \mathbf{e}_p$ depends upon past values of $\{\phi_t\}$. Although the GWE-tracker is derived on the basis of a Markov regime model, in later chapters, we will adjust the algorithm in situations with a more nuanced structure.

3.2.3   Kinematic Measurement

As in Chap. 2, there are three range-bearing sensors as shown in the figure:

$$(X_r, Y_r) = [(0, 600), (100, 600), (1600, 1200)].$$

Each sensor provides a noisy range-bearing measurement of the target position every 2 s. The same raw data set used in Chap. 2 will be used in this chapter.

The measurement processes conform to the state partition. In this engagement, there is a measurement of range and bearing to the target. The location of the sensors is known, but necessarily, the location of the target is not. Since the coordinate system of the measurements is not that of the local kinematics, we must first convert the polar measurements to an east-north coordinate system using the geometry of the sensors and the estimate of position. Each local estimator has a different geometry, and the coefficients of measurement model must be indexed by the sequence of recent modes, ι, rather than upon i.

The kinematic measurement is written

$$y[k+1] = H_\iota x[k+1] + \sqrt{P_\iota^n} n[k+1] \tag{3.7}$$

with additive, white, unit-Gaussian noise, $\{n[k]\}$. As before, the channel noise matrix, P_ι^n, is positive.

3.2.4   Regime Measurement

In this engagement, the measurement data set consists of two parts: a temporal measurement of some kinematic property of the point target as given in (3.7), and a complementary measurement that indicates in some way the regime of the target.

To illustrate the way in which these regime measurements are constructed, consider the FLIR-image of a tank in profile at fairly close range shown in Fig. 1.4. The silhouette is distinguishable from the background with the rear-mounted engine providing the dominant feature. The contrast of the tank with its background is low over most of the image. Nevertheless, we can infer the orientation of the tank with respect to the sensor. Of course, machine recognition of this orientation requires significant processing. Other features, like target type, are implicit in the image, but may be hard to extract with confidence.

An optical image of a tank is shown in Fig. 1.1. This image is much less ambiguous even though the range is longer. The tank is moving in an open field, and the target-background contrast is greater—though the fine detail in the target is still hard to determine.

In both of these figures, the target image can be processed to yield type and orientation. Target type is important because the mobility properties of the kinematic model differ by type. Orientation is important because orientation cues the tracker to the longitudinal direction; i.e., the direction of extrapolation. Additionally, a sudden change in orientation alerts the tracker to a maneuver long before the acceleration is translated into a recognizable path deviation.

The figures show the kind of unconventional data sets that are included in modern tracking applications. These data sets often display explicit properties of extended targets. To be integrated into the tracking algorithm, the high-dimensional raw measurements must be reduced into a form compatible with the tracker interfaces. For example, this could mean inferring from a $2D$-image a simple statement of type, with or without an orientation classification.

The data compression in a hybrid tracker can take two forms. If the underlying feature is ontologically discrete, the noisy classification has the conventional error structure: target selection from a fixed set of target types has the natural errors of classifying the target as type i when in truth it is of type j.

For example, suppose the tracker is following the target shown in Fig. 1.5. An image processor can deduce the center-of-reflection from the optical image and obtain the bearing angle to a point-target. But to track the target, the tracker must identify the target class to select the correct kinematic model. Shape analysis suggests that this vehicle is more agile than the tank and has significantly different motion constraints. The image-enhanced tracker will utilize a sequence of such images at different ranges and orientations and will classify each image according to a set of stored templates. Because of obscuration and processing errors, differing images might be associated with the same target type. Or a blurred image may change the ostensible target type, an impermissible event.

The mapping of visual image to target classification is fraught with error. Figure 1.3 is an image taken in a region neighboring that shown in Fig. 1.1. It can easily happen that the tracker would *see* targets in the figure. For example, the amorphous but high contrast flora in the figure may appear to be targets. Lacking sophisticated shape analysis, the tracker might mistake a bush for a tank and even give it an orientation. In this application, one of the regime state classifications should be *no-target in image*.

The noisy mapping of a discrete set of regime states to a discrete regime measurement is of the conventional sort. However, when data compression reduces an analog variable to a discrete set of regime states there are inherent errors. For example, target orientation is an analog variable. If we represent orientation digitally (as one of a small number of possible values), we necessarily make an error.

To compensate, we typically must add additional uncertainty into the kinematic equation; i.e., make P^w bigger. But this makes the bandwidth bigger and passes more unstructured noise.

To illustrate this analogue-to-discrete mapping, consider the problem of classifying the longitudinal direction of the tank shown in Fig. 1.1. The image-enhanced tracker proposed in [25] utilizes a comprehensive database that shows targets of the proper type in different orientations; e.g., Fig. 1.2. To classify the rotational state, a tracker would store-and-query a data-file containing several such images, each with a different orientation with respect to the sensor. In an engagement, the tracker matches a silhouette to one of the templates and determines the current target orientation thereby: an image is converted to a discrete measurement of a rotational state. This process places orientation (a rotational kinematic state) into a predetermined set of bins (a discrete regime state): it digitizes the orientation. The image-enhanced tracker is able to accommodate obscuration of the internal features.

With the above caveats, we will assume that there is a sequence of regime classifications coincident with the kinematic measurements. This sequence generates the data-set $\{\mathscr{Z}[k]\}$. Some of the regime states are non-temporal—the target type—and some are temporal—orientation. But all are included in the regime measurement data set. We will assume that the mapping from regime to measurement is fast. Without latency, $z[k + 1]$ measures the forward regime.

In the GWE-tracker, the quality of regime measurements, $\{z[k]\}$, is indicated by the discernibility matrix **D**:

$$z[k + 1] = \mathbf{D}\phi[k + 1] + \eta[k + 1], \tag{3.8}$$

where $\{\eta[k]\}$ is a martingale difference. For engagements in which there is no regime measurement, we will retain (3.8) with an uninformative **D**.

There may be additional elements of $\{\mathscr{Z}[k]\}$ that are non-temporal. For example, the tank shown in Fig. 1.1 can move anywhere within the field of view, but it tends to follow the heavily traveled branching paths shown. In some cases, the topography precludes motions into prohibited regions. This effect was observed in Chap. 2 when the target was confined to a north-south road. These restrictions are collected and added to the para-measurement set.

3.3   The Uni-Model Tracker

3.3.1   The Engagement

Chapter 2 presented an elementary tracking scenario. Repeating the description, a target moves in an east-north coordinate system. The speed on the path is 28 m/s. Motion begins in a south-east direction from the starting point $(0, 1000)$m. After moving at a constant velocity for 24 s, the target turns north at a rate of 5°/s for a period of 12 s. It returns to a constant velocity segment for a time before turning north again at the earlier rate. A sample path of the target and the observations are shown in Fig. 2.1.

The system specifications require both a small location error and a credible assessment of its uncertainty. In Chap. 2 we studied two **EKF** trackers. We will label the first, **EKF** C. This tracker of a cooperative target avails itself of the current turn rate of the target.

The second, the nominal **EKF**, lacks regime information. Instead it uses a centered kinematic model and ignores the regime variation. The nominal **EKF**, or just the **EKF** for short, gives reasonable location estimates for the most part; the tracking errors tend to be smaller than the range errors of the individual sensors. However, the EEP regions are overly optimistic after a change in regime state.

3.3.2   The **EKF**-Tracker

3.3.2.1   **EKF** C

Let us revisit the tracker-classifier algorithms and look at the performance in more detail. The basic time-discrete kinematic model upon which the **EKF** is based is

$$x[k + 1] = A_i x[k] + \sqrt{P_i^w}\, w[k + 1], \tag{3.9}$$

where $\{w[k]\}$ is the usual $\mathcal{Y}[k]$-unit Gaussian-white sequence. In the **EKF** C, $\phi[k]$ is known.

The weighting P_i^w is a composite of a variety of uncertainties. In Chap. 2 we posited a white acceleration process with standard deviation:1 m/s^2 in the east direction and 0.7 m/s^2 in the north direction. Despite the fact that the path of the target is quite regular, the action matrix was made large to stabilize the tracker in the presence of sudden jerks.

In this section, we will contrast the character of several algorithms. Since the endogenous jerk is primarily due to the turns that the **EKF** C recognizes, we will reduce the wide band disturbance and make the tracker less sensitive to unstructured disturbances. For the trackers that follow, the 1σ-exogenous acceleration will be 0.5 m/s^2 in the east direction and 0.25 m/s^2 in the north direction.

The form of the EKF-tracker is dependent upon the information available to it. The raw kinematic observations are transformed to generate a filtration $\{\mathcal{Y}[k]\}$ where $y[k+1]$ is generated by stacking the observations from the three range-bearing sensors:

$$y_r[k + 1] = H_r x[k + 1] + \sqrt{P_r^n}\, n_r[k + 1];\, r = 1, 2, 3. \tag{3.10}$$

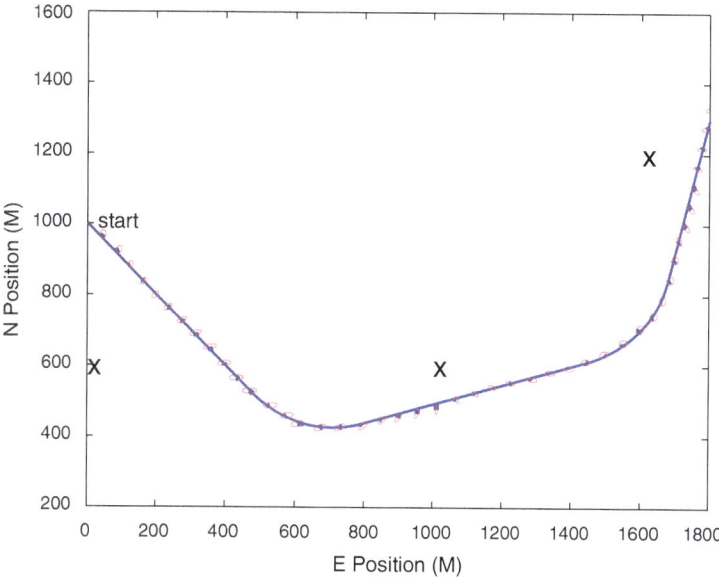

Fig. 3.2 In a cooperative engagement, EKF C gives a high quality estimate of location along with tight estimates of the location of the target. The 1σ-error ellipses are shown centered on mean position

The sensor noise is assumed to be independent temporally and spatially. In each of the trackers that follow, the raw range-bearing measurements are the same. But each has its unique measurement model.

If the target is cooperative, the tracker will avail itself of the regime state as well. In this event, $\mathscr{G}[k] = \mathscr{Y}[k] \vee \sigma(i[k+1])$ where we use $\sigma(i[k+1])$ to represent the data structure generated by the regime sequence. Thus far, the description of the engagement mimics that presented in Chap. 2, with the only difference being the reduction in expected magnitude of exogenous accelerations in view of the target self-reported regime changes.

The response of the EKF-tracker in the cooperative engagement is shown in Fig. 3.2. The performance of the EKF C-tracker is very good. It is able to follow the turns better than the EKF C-tracker in Chap. 2—see Fig. 2.2. The buildup of velocity error before the counterclockwise turn at about (1600, 800) caused the earlier EKF C-tracker to lag behind the target and then over correct. This same effect is not seen in Fig. 3.2 despite the fact that they use the same basic algorithm and the same range-bearing data set.

The EKF C-tracker is the benchmark against which other trackers can be compared. It presupposes small exogenous acceleration and thus smooths the noisy kinematic measurements. It knows the current regime and is responsive to abrupt turns. Figure 3.3 shows a plot of the radial position error. The intervals of CV motion are shown at the bottom of the graph. Despite sensor errors on the order of 60 m,

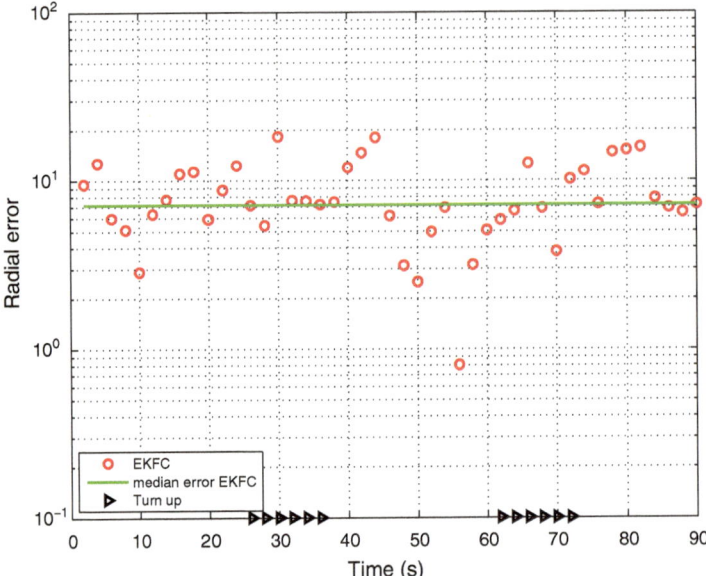

Fig. 3.3 The radial error of the EKF C-tracker varies about 7 m with little sensitivity to the turn events

the median radial position error is only 7.1 m. The tracking error is mildly sensitive to the turn events. The error grows most after a return to constant velocity motion where the velocity errors that build during a turn are resolved.

3.3.2.2   Nominal EKF

An engagement involving an uncooperative target is more difficult. The uncooperative target does not transmit its intent to the tracker. Lacking a regime measurement, the EKF-tracker is unimodal. Since the CV regime predominates, the natural kinematic model would be

$$x[k + 1] = A_2 x[k] + \sqrt{P^w}\, w[k + 1], \tag{3.11}$$

where P^w is as before. This is the *centered* model we have referred to previously.

The measurement sequence is again that shown in Fig. 2.1. The response of the centered EKF-tracker, labeled simply the EKF, is shown in Fig. 3.4. Of course, the performance of the centered tracker is not as good as that of the EKF C. Interestingly, the quality of this tracker is not as good as the centered tracker as shown earlier in Fig. 2.4. That tracker used a centered model and the same raw data set. But the Chap. 2 tracker had greater pseudo-noise excitation and higher

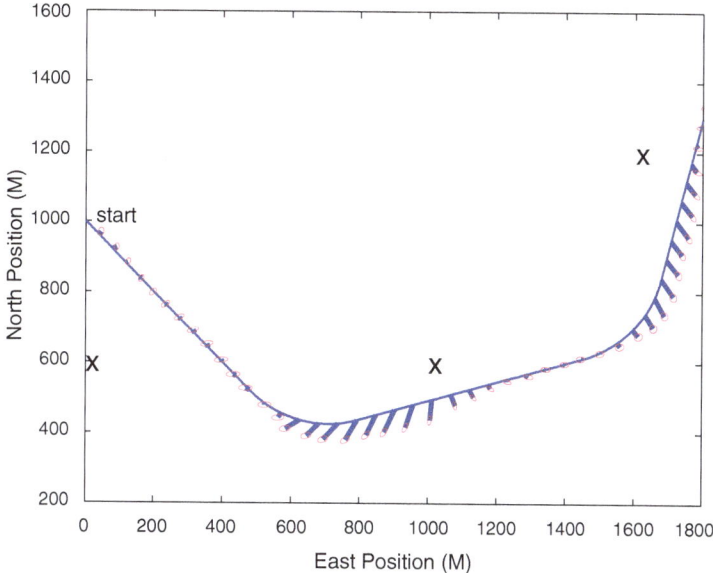

Fig. 3.4 The location estimates of an **EKF**-tracker are shown along with the notional 1σ-error ellipses when the target is not cooperative. The **EKF** smooths the measurements and has small error on the initial CV-portion of the path. After a turn, the tracking error increases without a corresponding increase in the size of the error ellipses

bandwidth. Thus, it was quicker to resolve the velocity errors after a regime change. This example illustrates the general rule that pseudo-noise is an aid when the model is crude and a hinderance when the model is refined.

Figure 3.5 shows plot of the radial position errors for this nominal **EKF**. The tracking error can exceed 71 m and the mean radial error is over 26 m. In contrast with the **EKF** C-tracker, the errors grow significantly after regime transition events. They reach their maximum after a return to CV-motion.

The centered **EKF** is very good on the initial CV segment—the kinematic model in the centered tracker matches that of the **EKF** C-tracker, and they both match truth. The centered **EKF** is less good after the first turn. Even though the tracking error increases after the turn, the notional uncertainty regions fail to adjust. A more detailed view of this uncertainty mismatch is given in Chap. 2. It is interesting to note that in certain regions of the engagement; e.g., (1000, 500), the **EKF** moves away from the path and the nearest sensor. Because of the increase in bearing error from this dominant sensor, we would expect that the area of the 1σ-error ellipses of the centered **EKF** will increase relative to that of the **EKF** C-tracker. This is actually true.

In this chapter, we will refine our study this engagement. We will propose an algorithm that uses the same range-bearing data set and achieves performance close to that of **EKF** C but without the cooperation of the target. We will accomplish this using the hybrid kinematic model and the **GWE**-tracker.

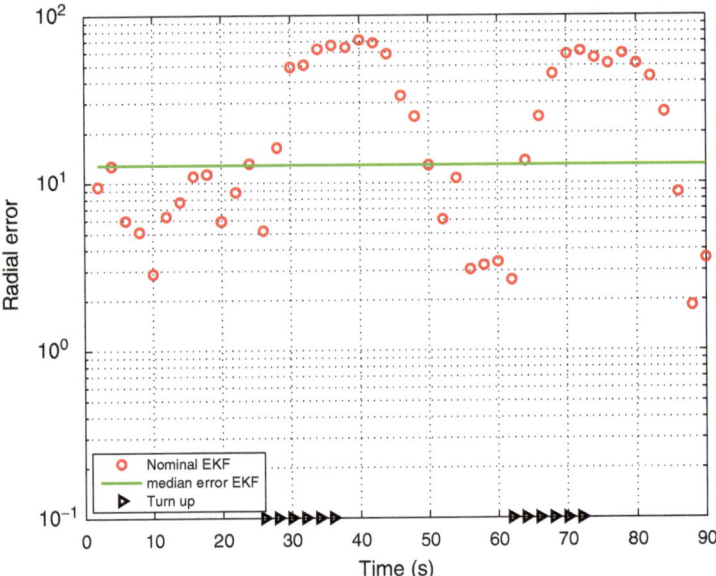

Fig. 3.5 The radial error of the centered EKF-tracker exceeds 71 m with maximum value after a return to CV motion

3.4   The Multi-model Tracker

3.4.1   The Engagement

3.4.1.1   Kinematics

The **GWE** is a hybrid estimator that maintains a set of local estimators indexed by $\iota \in \kappa$. In this engagement, there are the three regime states: left, CV, right. We will chose as the mode a string of length 3: $\iota = ijl$, $S = 3$ and $L = 3$. Thus, the **GWE**-tracker will maintain 27 local filters.

The kinematic model for the **GWE**-tracker derives from (3.1). The form of the ιth kinematic model is

$$x[k+1] = A_\iota x[k] + \sqrt{P_\iota^w}\, w[k+1], \qquad (3.12)$$

where $\{w[k]\}$ is an $\mathscr{F}[k]$-unit Gaussian-white sequence. The **EKF** C-tracker uses a similar form with A indexed by the ι prefix i.

Since the wide band accelerations tend to be more robust longitudinally—drag and speed variations—than laterally, the action matrix in the **EKF** favors the eastbound direction. In contrast with either of the **EKF**s, we will let the action matrix depend upon the mode. Again, we suppose the exogenous acceleration is stronger longitudinally than laterally.

Instead of the constant P^w in the EKF-trackers, in (3.12) we will align the exogenous acceleration along the notional velocity vector of the target: the major axis of P_ι^w is along the local estimate of velocity, $\hat{v}_\iota$, and the minor axis is perpendicular to $\hat{v}_\iota$ in the ratio of ten to one. The intensity of the exogenous acceleration is random both because of its dependence on the regime and its dependence on the local state estimate. To normalize the results which follow, we will maintain the same acceleration intensity for all of the trackers:

$$|P^w| = |P_\iota^w| \text{ for all } \iota \in \kappa.$$

We now have the individuated local models we will use in the GWE. Each is identified by a unique three digit string in mode space: the 222th model is CV for the three sample intervals $k - 2$ through k. There are but three kinematic matrices, but there are a full 27 distinct versions of P_ι^w. The action matrix is random even when the regime process is known because the direction of the major axis depends on the mean kinematic state, $\hat{x}[k]$.

The target operates in one of three regimes indexed by $i \in S$. As the target moves, the regime sequence generates a string. In the GWE, the regime is represented by a Markov process, and $\phi[k]$ is a pointer to the current regime. With a sample interval of 2 s, the discrete transition probability matrix is given in (3.6).

3.4.1.2   Measurements

There are two data sets generated by the sensor suite; $\{\mathscr{Y}[k]\}$ and $\{\mathscr{Z}[k]\}$. The former is the kinematic measurement. The sensor noise is assumed to be independent temporally and spatially. The local model is that given in (3.10) where now P_r^n depends upon ι: the transformation of the range-bearing noise into an east-north coordinate system is based upon $\hat{\chi}_\iota[k]$.

In the GWE there may also be a regime measurement. If latency is short, the regime measurement is linked to the regime proxy in the forward interval; i.e., $\mathbf{e}_p$:

$$z[k + 1] = \mathbf{D}\phi_p + \eta[k + 1], \tag{3.13}$$

where $\mathbf{D}$ is the discernibility matrix.

In what follows, we will explore two alternative sensor configurations. In the first, there is no modal measurement: $\mathbf{D}$ is uninformative. In the second, the regime measurement is of high quality:

$$\mathbf{D} = \begin{bmatrix} 0.8 & 0.1 & 0.1 \\ 0.1 & 0.8 & 0.1 \\ 0.1 & 0.1 & 0.8 \end{bmatrix}. \tag{3.14}$$

In the former case, the **GWE** location estimate is a $\mathscr{G}[k] = \mathscr{Y}[k]$-adapted process. In the latter case, the **GWE** location estimate is a $\mathscr{G}[k] = (\mathscr{Y}[k] \vee \mathscr{Z}[k])$-adapted process. We will label the trackers as $\mathscr{Y}[k]$ or $\mathscr{G}[k]$ as the case might be.

3.4.2   The **GWE** Algorithm

3.4.2.1   Extrapolation and Update

The **GWE** is a predictor–corrector estimator. It fuses 27 local estimates. Let us start with the ιth filter at time $t = kT$ and initial condition: $\alpha_\iota, m_\iota-, P_\iota-$ with $\iota = ijl$ and $\iota^- = jl$. The kinematic extrapolation and update is that of the local **EKF**.
Extrapolate:

$$m_\iota[k + 1]^- = A_i m_\iota-[k], \tag{3.15}$$

$$P_\iota[k + 1]^- = A_i P_\iota-[k]A_i' + P_\iota^w-; \tag{3.16}$$

Update

$$\Delta d_\iota[k + 1] = \sum_r H_{\iota,r}' D_{\iota,r}^n y_r[k + 1], \tag{3.17}$$

$$\Delta D_\iota[k + 1] = \sum_r H_{\iota,r}' D_{\iota,r}^n H_{\iota,r}, \tag{3.18}$$

where $\Delta d_\iota[k + 1] = d_\iota[k + 1] - d_\iota[k + 1]^-$, and similarly for $\Delta D_\iota[k + 1]$.
The update of the modal state proceeds in two steps. First the kinematic measurement, $y[k + 1]$, is used. The covariance of the rth kinematic measurement is given by $P_{\iota,r}^y$,

$$P_{\iota,r}^y = H_{\iota,r}' P_\iota[k + 1]^- H_{\iota,r} + P_{\iota,r}^n. \tag{3.19}$$

Then $P_{\iota,r}^y > 0; \iota, r \in \kappa \vee R$. All of the following matrices are well defined and positive: $(P_{\iota,r}^y)^{-1'} = D_{\iota,r}^y = (F_{\iota,r}^y)^2$.
The first part of the modal update is

$$\alpha_\iota^-[k + 1] = \alpha_\iota[k] \prod_r |F_{\iota,r}^y| \exp(\tfrac{1}{2}\Delta \|m_\iota\|_{D_\iota}^2 - \textstyle\sum_r \|y[k + 1]\|_{D_{\iota,r}^n}^2). \tag{3.20}$$

The form of (3.20) is particularly convenient. If a sensor is obscured, we need only reduce the sum (and product) by the index of the obstructed sensor.
We complete the update with the regime measurement. The increment in the modal probability mass function is

$$\alpha_\iota+[k + 1] = \alpha_\iota^-[k + 1]\Pi_{pi}z[k + 1]'\mathbf{D}_{\cdot p}. \tag{3.21}$$

Equation (3.21) generates a mass function on κ^+. The premise in (3.21) is that the measurement is that of the forward regime proxy. If this is not the case, (3.21) must be adjusted appropriately.

It should be observed that if a regime measurement is lacking, (3.21) has a similar form without the final factors:

$$\alpha_\iota + [k + 1] = \alpha_\iota^-[k + 1]\Pi_{pi}. \tag{3.22}$$

3.4.2.2   Reconciliation and Projected Statistics

At $t = (k + 1)T$, we have an unnormalized probability mass function on κ^+. We can normalize $\alpha_\iota + [k + 1]$ and obtain $\hat{\phi}_\iota + [k + 1]$. From this we can find the modal probabilities,

$$\hat{\phi}_\iota[k + 1] = \sum_p \hat{\phi}_\iota + [k + 1]. \tag{3.23}$$

Write the final estimate as

$$\mathbb{P}[k + 1] = \sum_{\iota \in \kappa} \hat{\phi}_\iota[k + 1]\mathbf{N}(m_\iota[k + 1], P_\iota[k + 1]). \tag{3.24}$$

The projected distribution, $\hat{\mathbb{P}}$, is derived from $\mathbb{P}$:

$$\hat{\mathbb{P}}[k + 1] = \mathbf{N}(\hat{x}[k + 1], P_{xx}[k + 1]), \tag{3.25}$$

with

$$\hat{x}[k + 1] = \sum_\iota \hat{\phi}_\iota[k + 1]m_\iota[k + 1] \cdot$$

$$P_{xx}[k + 1] = \sum_\iota \hat{\phi}_\iota[k + 1](P_\iota[k + 1]$$

$$+(\hat{x}_\iota[k + 1] - \hat{x}[k + 1])(\hat{x}_\iota[k + 1] - \hat{x}[k + 1])').$$

To reconcile the final state estimates in the kth interval with the initial conditions for the $(k + 1)$th, make the transformations

$$\hat{\phi}_l[k+1] = \sum_{ij} \hat{\phi}_\iota[k+1] \quad \text{progenitor regime}, \tag{3.26}$$

$$m_\iota - [k+1] = \sum_l \hat{\phi}_l[k+1]m_\iota[k+1], \text{ and} \tag{3.27}$$

$$P_\iota-[k+1] = \sum_l \hat{\phi}_l[k+1](P_\iota[k+1]$$

$$+(m_\iota-[k+1] - m_\iota[k+1])(m_\iota-[k+1] - m_\iota[k+1])'). \quad (3.28)$$

This completes the GWE-tracker for this example.

3.4.3 The GWE-Tracker

3.4.3.1 The $\mathcal{Y}[k]$-GWE

To illustrate the advantage of the hybrid model, let us continue our study of the earlier engagement. We will first look at the performance of the $L = 3$ tracker when **D** is uninformative, $\mathcal{G}[k] \equiv \mathcal{Y}[k]$, and the kinematic measurements are as before. We will refer to the centered tracker simply as the uni-model tracker to contrast it with the EKF C.

The kinematic data is ambiguous regarding the time of a regime change. Several time increments pass before the path deviation conclusively signals a turn. It might therefore be supposed that the $\mathcal{Y}[k]$-GWE would offer little advantage over the $\mathcal{Y}[k]$-EKF. This is not the case: contrast Fig. 3.4 with Fig. 3.6. The $\mathcal{Y}[k]$-GWE

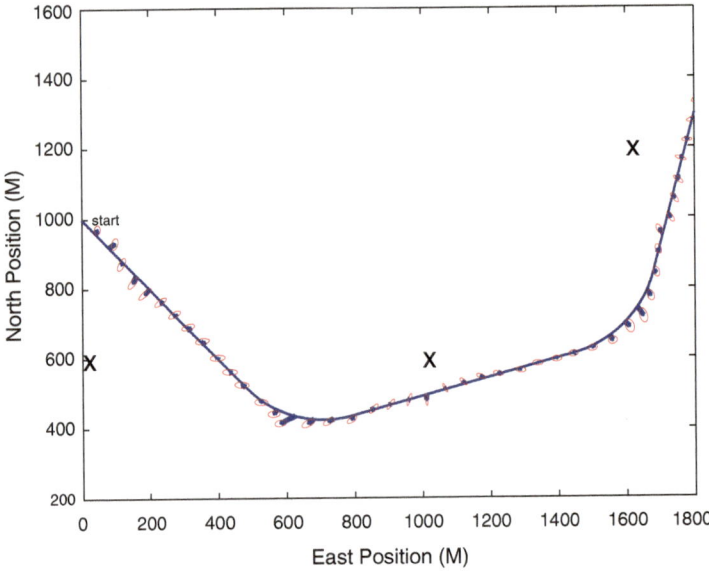

Fig. 3.6 The $\{\mathcal{Y}[k]\}$ data is used exclusively in this GWE-tracker. The projected distribution, $\hat{\mathbb{P}}$, is used to determine the notional tracking statistics. The location estimates of an GWE-tracker are shown along with the 1σ-error ellipses

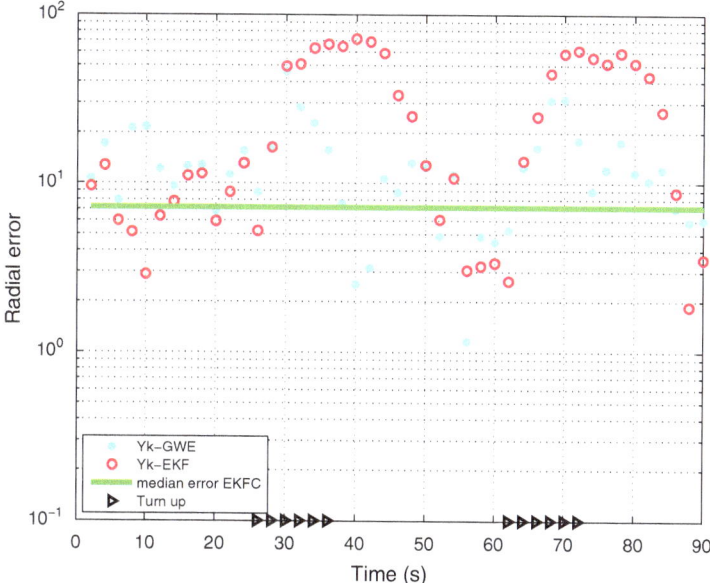

Fig. 3.7 Temporal plots of the radial errors show both trackers are confused by a turn event. The $\mathscr{Y}[k]$-GWE is quicker to reduce the error during quiescent periods than is the $\mathscr{Y}[k]$-EKF-tracker

gives much better estimates of vehicle location than does the $\mathscr{Y}[k]$-EKF-tracker—though not so good as the EKF C-tracker. The projected distribution, $\hat{\mathbb{P}}$, was used to generate the 1σ-error ellipses in Fig. 3.6.

Figure 3.7 shows the radial error of the $\mathscr{Y}[k]$-GWE-tracker along with the error of the $\mathscr{Y}[k]$-EKF-tracker. For reference, a plot of the median error is EKF C is also shown.

The median error for either $\mathscr{Y}[k]$-tracker exceeds that of the EKF C. However, the GWE is more nimble than the EKF. The maximum measurement error for each of the range-bearing sensors is of the order of 65 m and there are three such sensors operating independently. Any tracking error that exceeds 65 m must be thought of as excessive. The tracking error of the EKF exceeds 65 m several times during the engagement. The $\mathscr{Y}[k]$-GWE-tracker has an error of 10–30 m after a turn. There are individual excursions to 30 m and even 50 m. The median tracking error is 10 m, only a 20 % increase over the EKF C.

The $\mathscr{Y}[k]$-EKF-tracker actually does well in the first segment of the engagement. The implicit kinematic model for the EKF is $\iota = 222\dots$ and this tracker assigns probability one to the true modal string—initially, the EKF and the EKF C are identical. Both perform better than the GWE because the GWE weights a full collection of alternative modes.

However, when the motion regime changes, this advantage is lost. After the first turn, both the EKF and the GWE lag—the radial error increases to 50 m. But the

GWE-tracker quickly recovers: the error in the GWE decreases to 20 m while that of the EKF grows toward 80 m. The same behavior is displayed subsequent to the second turn.

Tracker quality is a function of both the error and the notional fidelity of that estimate. A single quality metric is difficult to prescribe. But one plausible element is the area of the computed 1σ-error region. The tighter the target placement region is, the more useful the estimate will be seen to be. Ideally, the $\hat{\chi}$-centered, 1σ-error region will have a very small area.

The $\mathscr{Y}[k]$-distribution of $\chi[k]$ is nominally Gaussian for both the $\mathscr{Y}[k]$-EKF and the EKF C. The conditional distribution generated by the GWE is a multi-term Gaussian sum. To simplify the comparisons of the trackers, we will use the statistics of the projected distribution, $\hat{\mathbb{P}}$, when looking at the $\mathscr{Y}[k]$-GWE-tracker.

The EKF C-tracker bases its estimate on the true regime sequence. It is, therefore, the most accurate of the trackers—the GWE attempts to identify the regime from $\{\mathscr{Y}[k]\}$, and the EKF ignores regime variation. The 1σ-ellipses for the EKF C provide a baseline for comparison of the putative accuracy of the location estimate.

Figure 3.8 shows the ratios of the area of the 1σ-error ellipse of the $\mathscr{Y}[k]$-GWE-tracker and the $\mathscr{Y}[k]$-EKF-tracker normalized by the area of the 1σ-error ellipse of EKF C. Initially, the $\mathscr{Y}[k]$-EKF-ratio is identically one: the same algorithm is processing the same data to arrive at the same conclusion. The GWE-ratio is much more conservative because of the initial uncertainty regarding the regime state. The GWE-ratio increases rapidly to 50 % above the baseline before dropping to near the median value of about 25 % above the baseline. A larger error ellipse translates into a higher gain and in turn an increase in the bandwidth of the tracker.

The trajectory of the assurance regions for the EKF-tracker is surprising. The 1σ-error ellipses for both the EKF and the EKF C are independent of $\{\mathscr{Y}[k]\}$ to first order. Thus, one might expect the EKF-ratio to be identically one. This is not the case. Figure 3.8 shows the increase in size reaches 60 % as the target passes the middle sensor and rises again near the east sensor. The $\mathscr{Y}[k]$-EKF error becomes large as the estimate drifts away from the true path. The $\mathscr{Y}[k]$-EKF reflects the larger range as an increase in the size of the uncertainty region.

Of course, this area ratio is a very coarse measure of estimation quality since $\hat{\mathbb{P}}$ does not provide the fine detail of $\mathbb{P}$. At $t = 64$, the vehicle is in the midst of a north turn. The true location is $(681, 436)$. The EKF C places the vehicle at $(673, 425)$ and the GWE places the vehicle at $(659, 408)$. Both trackers lag the true path. Figure 3.9 shows both $\mathscr{Y}[k]$-densities of $\chi(32)$. The EKF C density is Gaussian and nearly centered on the target. The $\mathscr{Y}[k]$-GWE conditional density is broader but includes the true location within the flange of $\mathbb{P}$. The $\mathscr{Y}[k]$-density generated by the EKF-tracker (not shown) completely misses the target at the level displayed in the figure. Even though the $\mathbb{P}$ generated by the GWE has 27 terms, it is Gaussian-like.

The GWE-tracker has an average tracking error of 12.6 m. This is only 48 % of the median error of the EKF-tracker—though 50 % more than the average tracking error of 8.2 m for the EKF C-tracker.

To illustrate the ability of the $\mathscr{Y}[k]$-GWE to isolate the regime from the range-bearing measurements alone, Fig. 3.10 shows a sample path of $\{\hat{\phi}_1\}$; $\mathbb{P}(\phi[k] = e_1)$.

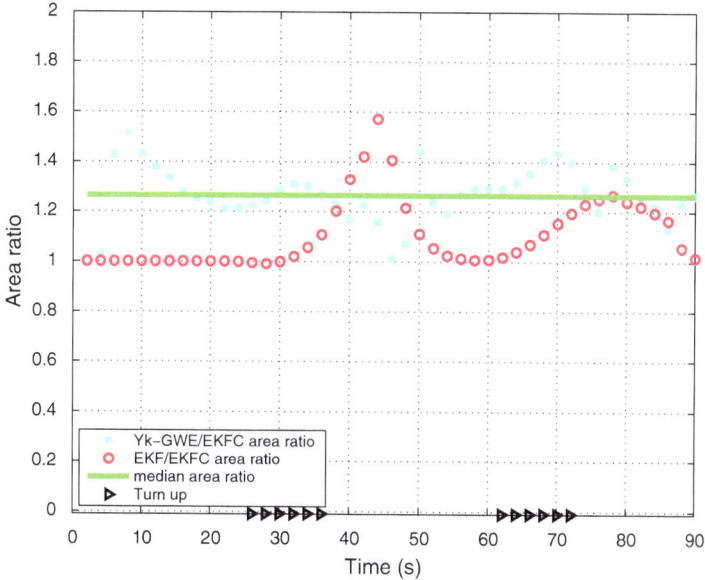

Fig. 3.8 The area of the 1σ-error ellipse of the $\mathscr{Y}[k]$-EKF-tracker is about a quarter less than that of the $\mathscr{Y}[k]$-GWE-tracker

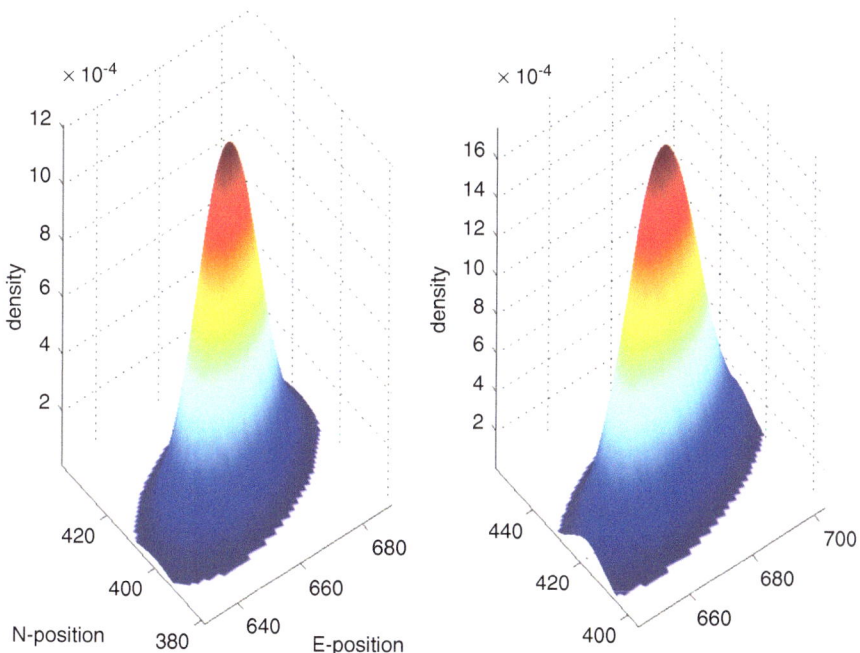

Fig. 3.9 The *left panel* shows the notional $\hat{\mathbb{P}}$ for $\chi(32)$ for the $\mathscr{Y}[k]$-GWE. The *right panel* shows the density for $\chi(32)$ as computed by the EKF C. The $\mathscr{Y}[k]$-GWE has a larger error since it is unsure of the modal sequence. It also has a smaller modal peak

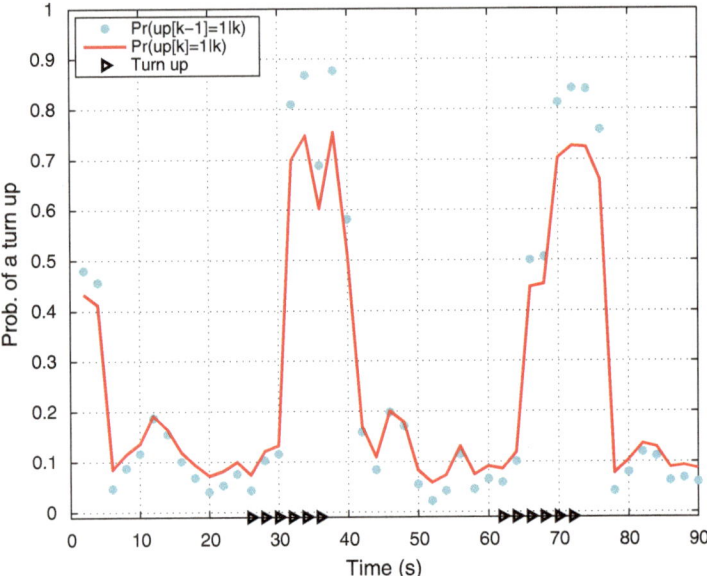

Fig. 3.10 The computed probability of a left (north) turn lags the regime event. Intervals of CT left-turning motion are shown on the time axis

The figure shows both the estimate of the regime proxy, $\mathbb{P}(\phi[k-1] = e_1|\mathscr{Y}[k])$ and the estimate of the forward regime, $\mathbb{P}(\phi[k] = e_1|\mathscr{Y}[k])$. It is evident that turn events are not well differentiated in $\{\mathscr{Y}[k]\}$, and the regime estimates lag the relevant event.

For example, during the first 24 s of the engagement the target moves south-east at a constant velocity. The probability of a counterclockwise turn drifts down from the initial value to about 0.1.

A turn begins at $t = 24$. But the first kinematic measurement indicative of a turn is at $t = 26$, and it is rather ambiguous. The **GWE** does not recognize the regime clearly until $t = 32$. Throughout, $\{\hat{\phi}_i\}$ is seen to be a lagging indicator of the regime state.

Despite the failure of the $\mathscr{Y}[k]$-**GWE** to identify the regime state in an expeditious manner, the tracking quality of the algorithm is quite good. It is much nearer the **EKF** C than it is the centered **EKF**. The advantage that the $\mathscr{Y}[k]$-**GWE**-tracker has resides in its multimodal nature. Of course, this is also a disadvantage insofar as 27 estimators must be managed instead of one. Each local filter has a weight, or notional probability, attached to it. The probability of any one of them being true is small in the evasive segments of the engagement. This uncertainty is partially reflected in the spread of the regime probabilities $\hat{\phi}_i$. This uncertainty spreads the assurance regions and makes them more reflective of the broad uncertainty in the target placement.

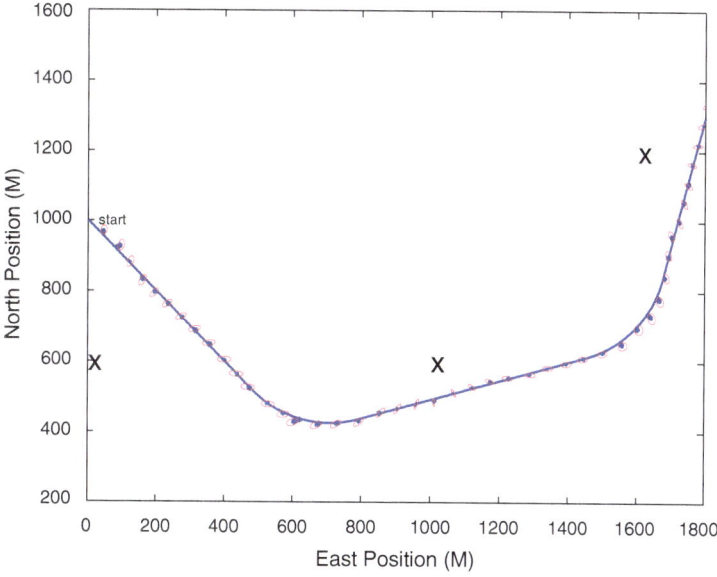

Fig. 3.11 The $\mathscr{G}[k]$-estimates of location are more responsive to a change in regime than are those of the $\mathscr{Y}[k]$-GWE displayed earlier. The projected distribution, $\hat{\mathbb{P}}$, was used to determine the notional 1σ-tracking statistics

3.4.3.2 The $\mathscr{G}[k]$-GWE

As this engagement is described, when the $z[k]$ is available, it is of high quality. The filtration $\mathscr{Z}[k]$ is merged with $\mathscr{Y}[k]$ to yield $\mathscr{G}[k]$. We use the modifier $\mathscr{G}[k]$ to refer to the tracker with the regime sensor, though it is strictly true that $\mathscr{G}[k] \equiv \mathscr{Y}[k]$ when $\mathbf{D}$ is uninformative. But the choice of symbols is evocative.

To illustrate the utility of the regime measurements, look at the performance of the $L = 3$ tracker when $\mathbf{D}$ is that given in (3.14) and $\{\mathscr{Y}[k]\}$ is as before. Figure 3.11 shows the response of the $\mathscr{G}[k]$-tracker with the common range-bearing data set.

As noted, the kinematic measurements are slow to indicate the time of a turn. The regime measurements reduces this latency. The $\mathscr{G}[k]$-GWE algorithm gives much better estimates of vehicle location than does the $\mathscr{Y}[k]$-GWE—though still not so good as the **EKF** C-tracker. The projected distribution, $\hat{\mathbb{P}}$, was used to generate the 1σ-error ellipses.

The radial tracking error of the $\mathscr{G}[k]$-GWE is shown in Fig. 3.12. This figure contrasts the radial error of the $\mathscr{G}[k]$-tracker with that of the $\mathscr{Y}[k]$-tracker. At nearly every point, the $\mathscr{G}[k]$-tracker is superior to the $\mathscr{Y}[k]$-tracker. Sometimes the improvement is significant; e.g., at $t = 30$. The median tracking error for the $\mathscr{G}[k]$-tracker is slightly over 8 m, a 14 % increase over the **EKF** C. With but one exception, the tracking error of the **GWE** is less than 21 m.

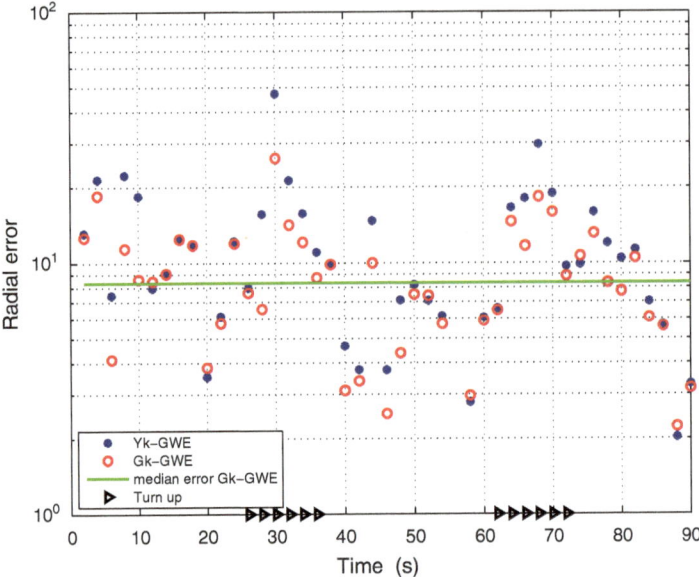

Fig. 3.12 The $\mathscr{G}[k]$-tracker has a smaller radial than does the $\mathscr{Y}[k]$-tracker

The area of the 1σ-error ellipses of the $\mathscr{G}[k]$-tracker are comparable with those of the $\mathscr{Y}[k]$-tracker. But they enclose the true path more often than do those lacking the regime measurement. The performance of the $\mathscr{G}[k]$-tracker is superior to the $\mathscr{Y}[k]$-tracker, and either is far superior to the **EKF**.

Collecting terms with common regime proxy, the $\mathscr{G}[k]$-conditional probability of an up-turn, $\hat{\phi}_1$ is shown in Fig. 3.13. The $\mathscr{G}[k]$-estimates of the regime proxy are more definitive than are the $\mathscr{Y}[k]$-estimates—see Fig. 3.10. The $\mathscr{G}[k]$-adapted version of $\{\hat{\phi}_2\}$ approaches one during turns and is quick to return to near zero during CV intervals. This is in contrast to the range-bound sample of the $\mathscr{Y}[k]$-adapted process. Indeed, the $\mathscr{G}[k]$-version of $\{\hat{\phi}_2\}$ precedes the regime event to some degree (e.g., see $t = 24$), because $z[k + 1]$ is a measurement of the forward regime.

3.5 Conclusion

This chapter presents the response characteristics of several different tacking algorithms. Each uses the same range-bearing data set. This does not translate into the same $\{\mathscr{Y}[k]\}$ measurement model because the mapping from raw data to the linearized algorithm is adapted to the raw data itself. The **GWE** lacking the regime measurement is better than the **EKF**-tracker but less good than the **EKF** C-tracker—as we move from full regime knowledge to partial regime knowledge, and thence to no regime knowledge, we lose confidence in our notional assurance regions.

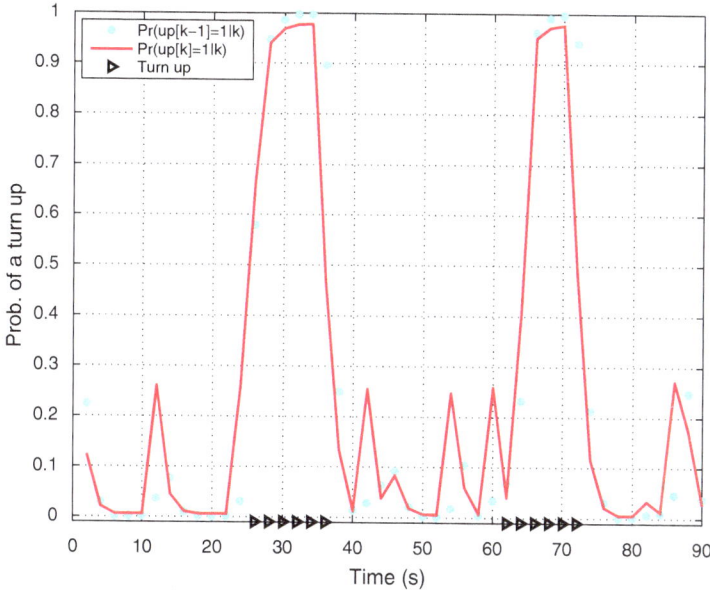

Fig. 3.13 The $\mathscr{G}[k]$-probability of a counterclockwise turn lags the regime event. But the GWE is more confident when it has a regime measurement

The 1σ-error ellipses of the $\mathscr{G}[k]$-GWE are slightly better than the $\mathscr{Y}[k]$-GWE. The former computes $\{\hat{\phi}_i\}$ with more precision. But this is not strongly reflected in the assurance regions. Note that the regime proxy only points to a rotation rate. It gives no information concerning the absolute position of the target in the east-north motion grid. Hence, the improvement in location quality is due much more to $\{\mathscr{Y}[k]\}$ than it is to $\{\mathscr{Z}[k]\}$. A longer string in ι would help to some degree; i.e., we could increase L from three to some larger number. But this would complicate the GWE algorithm and would not correct for absolute errors.

The modal model is deficient in several respects. For example, the regime state is assumed to be a Markov process. But this seems not to be the case in the engagement. Intervals without changes in regime have a minimum time interval: CV motion is uninterrupted for 10–20 samples. The high likelihood of brief CV periods is not seen in the engagement. In Chap. 5, we will look at the modal model in more detail.

While an improved ability to identify the mode is useful, improvement in the location estimate is not commensurate with the improvement in regime identification. There are applications in which the improved fidelity of the modal estimate will translate into significant improvement in performance; e.g., prediction. For many command architectures the ability to predict is crucial. Most target countermeasures have significant latency; e.g., time of flight. The requirement to accurately extrapolate target motion, even for a few seconds, will be found to heavily favor architectures that consider the target motion regime in addition to its current location and velocity. We will study some such applications in the next chapter.

Chapter 4
Intercept and Target Prediction

Abstract In this chapter we further explore the effects tracker design and performance have on the command architecture. Defensive algorithms must both identify a threat and, if necessary, counter the threat. The simplest situation is one in which the defensive resources can be deployed without delay; e.g., when there is a directed energy weapon (a DEW). But a ballistic defense needs to know where the target will be, not where it is. In this chapter we explore the complications that arise when the target must be classified with a minimal expenditure of illumination resources and must be neutralized with a similar minimal expenditure of countermeasure resources. The EKF is contrasted with the GWE. As expected, the ability to accurately predict target motion is greatly aided by the GWE's estimate of the target motion regime. The EKF tracker, lacking even the notion of target motion mode, yields neutralization regions that are inadequate to accomplish the specified tasks.

4.1 Introduction

In the previous chapter, we studied the problem of tracking the motion of an uncooperative target. The target follows a rather benign path with neither handoff transformation nor unstructured accelerations. The longitudinal acceleration is zero throughout. There are unpredictable intervals of lateral acceleration.

In some applications, the command system is designed to do more than simply follow the target. For example, the command algorithm may call for an intercept action, or it may direct an illuminator to interrogate the target and determine intent. A hostile target might try to prevent successful interrogation or intercept by masking its motion.

The regulative algorithm utilizes its limited resources as best it can. For example, if the algorithm must intercept the target, it will maximize the countermeasure density in the blocking region. It does this by limiting the area over which it expends its limited counterforce assets. Similarly, if the intent is to interrogate the target, the illumination energy will be focused on a small area that contains the target. In both cases, the system designer seeks a region that is both small in area and contains the target with high probability.

© Springer International Publishing Switzerland 2016
D.D. Sworder, J.E. Boyd, *Locating, Classifying and Countering Agile Land Vehicles*, DOI 10.1007/978-3-319-19431-8_4

In this book we have focused on the problem of following an evasive target as it moves about in the plane. We have used a model-based philosophy in which a set of stochastic differential equations provide a kinematic template that helps the tracker integrate a sequence of noisy kinematic measurements into a target placement. The nonlinear motion equations are too complicated to use directly, and we replace to them with a *centered linear model* in the EKF-tracker, or a *family of local models* in the GWE-tracker.

The prototypical motion template is the conventional LGM model,

$$dx_t = A_i x_t \, dt + \sqrt{P_i^w} \, dw_t, \qquad (4.1)$$

where $\{w_t\}$ is a unit Brownian motion and $i \in \mathbf{S}$ is the regime state. In the unimodal EKF-tracker, one of the regimes is selected as representative of the engagement, and the index subscript is ignored.

In the engagements we have presented, x_t is a 4-vector with position stacked over velocity. The dimension of the kinematic state could be expanded to include auxiliary variables; e.g., actuator states, acceleration states, jerk, etc., if appropriate. But we will use a simple Newtonian representation for target motion in this chapter.

The tracker/classifier algorithm is based upon a time-discrete approximation to (4.1) synchronized to the sample times of the sensors:

$$x[k+1]^- = A_i x[k]^+ + \sqrt{P_i^w} w[k+1] \qquad (4.2)$$

$$x[k+1]^+ = x[k+1]^- + \Delta x[k+1], \qquad (4.3)$$

where (4.3) delineates the handoff conditions, if any. The regime proxy, $\mathbf{e}_i$, is assumed to be dominant in the kth interval, but we acknowledge that the transition to the forward regime, $\mathbf{e}_p$, usually occurs before the $(k+1)$th measurement. In the engagement that follows, there are no handoff discontinuities.

The sensor suite generates two real-time data streams coincident with the sample times of (4.3). The first is a sequence of range-bearing measurements. The second, $\{\mathscr{Z}[k]\}$, is a sequence of direct regime measurements. The raw location measurements are adjusted to give locations in an east-north coordinate system, $\{\mathscr{Y}[k]\}$. Their combination is $\{\mathscr{G}[k] = \mathscr{Y}[k] \vee \mathscr{Z}[k]\}$.

The model-based tracker generates the $\mathscr{G}[k]$-conditional distribution of the target state. From this, an estimate of $x[k]$ can be made, and some measure of confidence in that estimate given. For example, the EKF-tracker generates a $\mathscr{Y}[k]$-statistic $(\hat{x}[k], P_{xx}[k])$ and offers a Gaussian distribution with the indicated moments as the conditional state distribution.

In important applications, there is a time lag between the time the measurements are taken and the time that the action occurs. For example, suppose we want to intercept a hostile target. If we use a directed energy weapon (DEW), we would focus our counterforce on a region capturing the current target position. But if the counterforce is ballistic with travel time τ we would cover a region capturing the target τ seconds in the future. When there is a delay between measurement and action, we require that the future state of the target be estimated; i.e., we would like to know $\chi_{t+\tau}$, the position of the target at time $t + \tau$.

As another example of where prediction plays an important role, the target might enter a region where the measurements will be obstructed for τ seconds. When we enter this region, we would like to extrapolate forward from x_t to $x_{t+\tau}$ in order that we can reacquire the target at egress and recommence tracking.

Prediction is a form of dead reckoning. Classical dead reckoning assumes we have an estimate of the kinematic state of the target, $\hat{x}_t$. We predict the future state by extrapolating forward by τ seconds using current estimates of the position and velocity. Of course, this extrapolation depends upon the evolution of the regime in the interval $(t, t + \tau)$. Further, classical dead reckoning does not give a clear idea of what happens to the assurance regions over the prediction interval.

There are many forms of the prediction problem [1]. In this chapter we will focus on the fixed-interval problem: given the observation sequence $\{\mathscr{G}[k]\}(or\{\mathscr{Y}[k]\})$, and an integer number of forward time increments, v, estimate the kinematic state at time $t = (k + v)T$. More broadly, we wish to determine the distribution, $\mathbb{P}(x[k + v]|\mathscr{G}[k])$.

4.2 The EKF

4.2.1 EKF-Tracker

To illustrate the prediction problem, let us derive the EKF-predictor using as the example the engagement explored at length in earlier chapters. A target is moving in the plane at a speed of 28 m/s in an SE direction from a starting point at $(0, 1000)$ m. After moving at a constant velocity for 24 s, the target turns north at a rate of $\omega = 5°/s$ for a period of 12 s. It returns to a straight path for a time before turning north again.

There are three range-bearing sensors located in the domain,

$$(X_r, Y_r) = [(0, 600), (100, 600), (1600, 1200)],$$

Each measures the target location every 2 s with standard error 56 m in range and 28 m in bearing. In what follows we will use the same raw measurement sequence we employed earlier.

The basic LGM-model for this engagement is

$$\frac{d}{dt} \begin{bmatrix} X \\ Y \\ V_X \\ V_Y \end{bmatrix} = \begin{bmatrix} 0 & 0 & 1 & 0 \\ 0 & 0 & 0 & 1 \\ 0 & 0 & 0 & -\omega_i \\ 0 & 0 & \omega_i & 0 \end{bmatrix} \begin{bmatrix} X \\ Y \\ V_X \\ V_Y \end{bmatrix} + \begin{bmatrix} 0 & 0 \\ 0 & 0 \\ \sqrt{P^w_{x;\iota}} & 0 \\ 0 & \sqrt{P^w_{y;\iota}} \end{bmatrix} \begin{bmatrix} \dot{w}_X \\ \dot{w}_Y \end{bmatrix}, \tag{4.4}$$

where $\{w_t\}$ is a unit Brownian motion and ω_i is a turn rate with $\phi_t = i \in \mathbf{S}$ if the turn rate is the ith.

There are three regimes: $\phi_t = e_1$ is left turn, $\phi_t = e_2$ is CV, and $\phi_t = e_3$ is right turn. The time-discrete kinematic model derives from (4.4) with the addition of a white acceleration process having standard deviation $0.7\,\text{m/s}^2$ in east direction (the preferred direction of motion) and $0.5\,\text{m/s}^2$ in the north direction. This pseudo-noise is intermediate in the range we have used in other engagements.

The EKF is a unimodal algorithm that presumes that the regime is constant. The path selected for illustration is fairly tame, and the EKF is based upon the CV model.

The EKF is a simple predictor–corrector. The kinematic state estimate at $t = kT$ is notionally Gaussian: $x[k] \sim \mathbf{N}(\hat{x}[k], P_{xx}[k])$. The state estimate is extrapolated forward T seconds using the regime model, e_2:

$$\hat{x}[k+1] = A_2\hat{x}[k] \tag{4.5}$$

$$P_{xx}[k+1] = A_2 P_{xx}[k]A_2' + P^w. \tag{4.6}$$

Upon receipt of the location measurement, $y[k+1]$, the information states of the tracker are updated:

$$\Delta d[k+1] = \sum_r H_r' D_r^n y_r \tag{4.7}$$

$$\Delta D_{xx}[k+1] = \sum_r H_r' D_r^n H_r. \tag{4.8}$$

A sample function of $(\hat{x}[k], P_{xx}[k])$ is shown in Fig. 4.1 (plotted with black +). We observe that the EKF is quite good on the initial CV segment of the path where the kinematic model matches the actual motion.

The left panel of Fig. 4.2 shows the detailed response of the EKF as the target executes the first north turn. The tracking error is small as the target enters the turn, and the realized error is within the 1σ-error ellipses.

As the target executes the turn, the EKF lags by more than 50m. The EKF requires 20 s or so to recover from a change in regime. Further, the error ellipses seem to ignore the north bias in the range-bearing measurements. The ellipses stay far too small to serve as useful assurance regions.

It is not until the target reaches $(1300, 600)$ that the assurance regions are adequate for intercept with a DEW. The EKF is not capable of supporting this tracking/intercept.

4.2.2   EKF-Predictor

The fixed-interval EKF-predictor is an adjunct to the EKF-tracker. The EKF state estimate at time $t = kT$ is $\mathbf{N}(\hat{x}[k], P_{xx}[k])$. The EKF estimate of $x_{(k+v)T}$ is given by $x[k+v] \sim \mathbf{N}(\hat{x}[k+v|k], P_{xx}[k+v|k])$ where $\hat{x}[k+v|k] = E\{x_{(k+v)T}|\mathscr{Y}[k]\}$, and $P_{xx}[k+v|k]$ is the $\mathscr{Y}[k]$-covariance of this estimate. The v-step estimate is

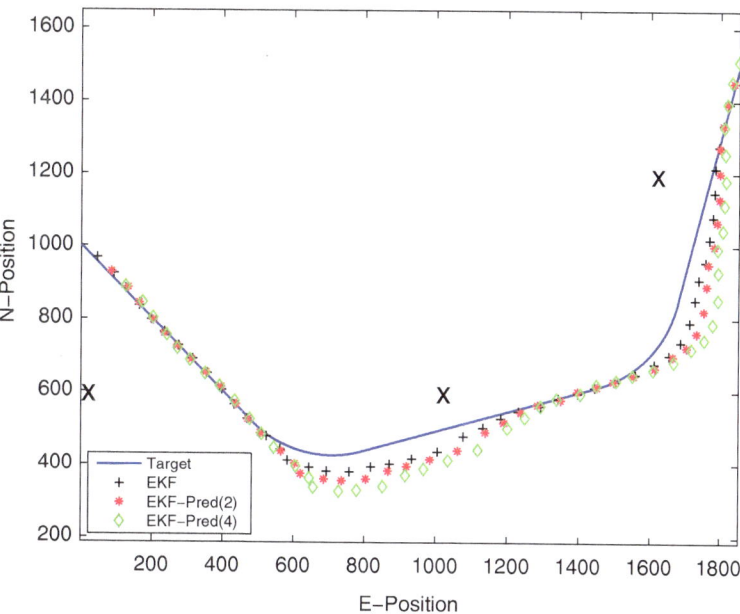

Fig. 4.1 In this engagement, three range-bearing sensors measure location every 2 s. The estimates of the EKF are labeled *black plus*. The response of a two-second predictor is labeled *red open circle*, EKF-Pred(2). The response of a four-second predictor is labeled *green diamond*, EKF-Pred(4)

given explicitly by $\hat{x}[k + v|k] = A_2^v \hat{x}[k]$. The covariance of the predicted location is somewhat more complicated to compute.

In this engagement, we will look at the two-second predictor, EKF-Pred(2), and the four-second predictor, EKF-Pred(4). The specific formulae are:

For EKF-Pred(2):

$$\hat{x}[k + 1|k] = A_2\hat{x}[k] \tag{4.9}$$

$$P_{xx}[k + 1|k] = A_2 P_{xx}[k]A_2' + P^w. \tag{4.10}$$

For EKF-Pred(4):

$$\hat{x}[k + 2|k] = A_2\hat{x}[k + 1|k] \tag{4.11}$$

$$P_{xx}[k + 2|k] = A_2 P_{xx}[k + 1|k]A_2' + P^w. \tag{4.12}$$

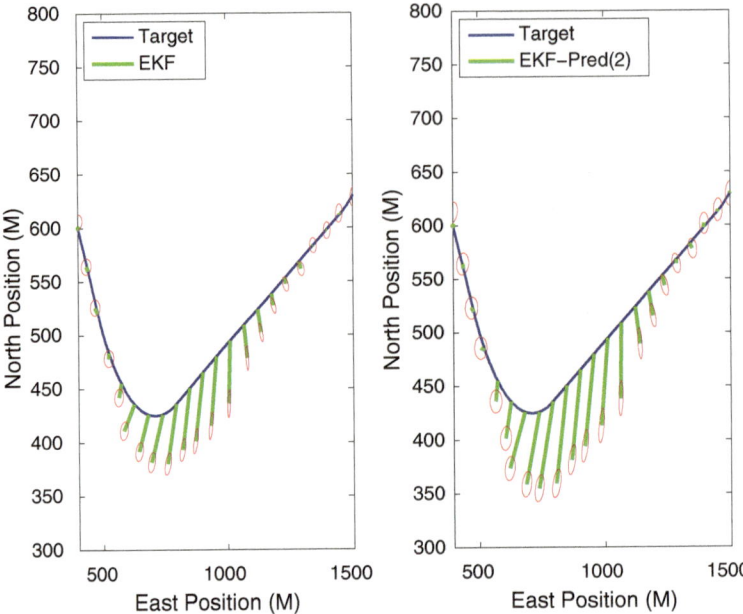

Fig. 4.2 The *left panel* shows the location estimate and 1σ-error ellipses for the EKF as the target enters the first turn—see Fig. 4.1. The *right panel* shows a two-second EKF-predictor using the same measurement data set. The EKF-predictor extrapolates using the CV model. Both the error and the 1σ-ellipses grow in the predictor. But the error covariance does not accommodate the actual uncertainty. Note that axes use unequal scales to improve visibility, at the cost of ellipse distortion

Figure 4.1 shows a sample function of each of the predictors. The two-second predictor is shown with a red mark, and the four-second predictor is shown with a green mark. Not surprisingly, the predictors lag more than does the tracker. This leads to larger radial errors and to a slower error decay rate.

Figure 4.2 contrasts the response of the nominal EKF-tracker (left panel) and the two-second EKF-predictor (right panel) through the first counterclockwise turn. We see that the prediction error increases to twice the tracking error. Initially, the predictor fails to see the turn coming and extrapolates forward using classical dead reckoning. This error is unavoidable within the context of the CV motion model.

More problematic are the 1σ-error ellipses. Before the turn, the error ellipses of the predictor uniformly enclose the true path. But as the turn progresses, the error ellipses drift far from truth. While the error ellipses of the EKF-predictor are larger than those of the EKF-tracker, they do not do justice to the increased uncertainty implicit in $\{\mathscr{Y}[k]\}$.

As we move to a four-second predictor, performance degrades still more. Figure 4.3 shows the response of this predictor in proximity to the first turn. The four-second predictor has larger error ellipses than does EKF-Pred(2). If we are going to illuminate a region that captures the target with a four-second predictor, we

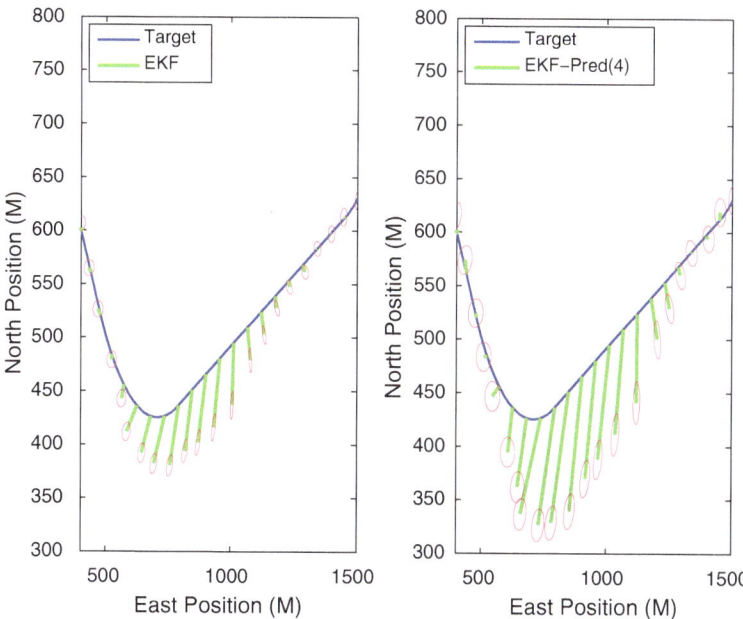

Fig. 4.3 The *left panel* shows the location estimate and 1σ-error ellipses for the EKF-tracker as the target enters the first turn. The *right panel* shows a four-second EKF-predictor. Over the four-second extrapolation, the radial error is an order of magnitude greater than that predicted by P_{xx}. As in the previous figure, unequal axes are used to enhance error visibility

will have to select a focus that distributes the illumination assets over an area that is two orders of magnitude bigger than the 1σ-ellipses that suffice at the beginning of the engagement. The notional 1σ-error ellipses do not have credible in this engagement.

Figure 4.1 shows that the error of the EKF grows as the prediction interval gets longer. The picture of the first turn is just a snapshot of the engagement. Figure 4.4 shows the radial errors for the two EKF-predictors and contrasts them with the tracker error. The maximum error for EKF-Pred(4) is 141 m (not shown on graph) while the maximum error for EKF-Pred(2) is 101 m. The maximum difference between the prediction error of EKF-Pred(2) and EKF-Pred(4) is 50 m. The median error of EKF-Pred(2) is only 59 % of that of the EKF-Pred(4): the median error of the EKF-tracker is only 41 % of that of EKF-Pred(4). Both predictors have errors in excess of the raw sensor error, but unfortunately, this is not reflected in the notional assurance regions.

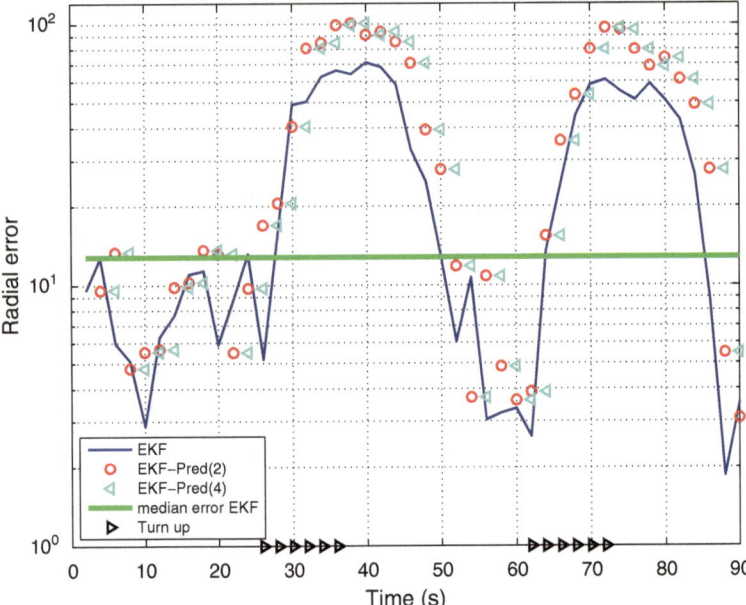

Fig. 4.4 The radial error in an EKF-predictor is greater for a prediction interval of 4 s than it is for a prediction interval of 2 s. The radial error grows rapidly in a turn and decays slowly when the target returns to constant velocity. This decay is quite slow when the prediction interval is 4 s

4.3 The $\mathscr{Y}[k]$-GWE

4.3.1 $\mathscr{Y}[k]$-GWE-Tracker

As was the case of the EKF, the GWE-tracker is based upon a reductionist representation of the kinematics of the target. But instead of the single model used in the EKF, the GWE acknowledges the full complement of three regime states and employs a modal string length of three: $\iota = \{ijl\}$, $S = 3$ and $L = 3$. Specifically, the GWE-tracker uses the time-discrete kinematic model,

$$x[k+1] = A_\iota x[k] + \sqrt{P_\iota^w}\, w[k+1], \qquad (4.13)$$

where $\{w[k]\}$ is an $\mathscr{F}[k]$-unit Gaussian-white sequence.

The dynamic matrix, $A_\iota = A_i$, is determined wholly by the regime proxy. The intensity of the exogenous acceleration is more complex. This wideband acceleration is usually stronger longitudinally than laterally. So, instead of the constant P^w used in the EKF, the intensity of the exogenous acceleration is aligned along the estimated velocity vector of the target in a ten-to-one ratio—see Chap. 3. Even though, $|\sqrt{P_\iota^w}| = |\sqrt{P^w}|$ for all $\iota \in \kappa$: the unstructured acceleration is highly directional since it is aligned with $\hat{v}_\iota$.

Like the EKF, the GWE-tracker is a predictor-corrector. However, it generates 27 individuated estimates. The initial state of the ιth filter at time $t = kT$ is $(\alpha_\iota, m_\iota-, P_\iota-)$. Each local filter uses the Kalman extrapolation and update equations:

Extrapolate

$$m_\iota[k+1]^- = A_i m_\iota-[k] \tag{4.14}$$

$$P_\iota[k+1]^- = A_i P_\iota-[k] A_i' + P_\iota^w \tag{4.15}$$

Update

$$\Delta d_\iota[k+1] = \sum_r H_{\iota,r}' D_{\iota,r}^n y_r[k+1] \tag{4.16}$$

$$\Delta D_\iota[k+1] = \sum_r H_{\iota,r}' D_{\iota,r}^n H_{\iota,r}, \tag{4.17}$$

where $\Delta d_\iota[k+1] = d_\iota[k+1] - d_\iota[k+1]^-$, and similarly for $\Delta D_\iota[k+1]$.

The update of the modal state proceeds in two steps. First the kinematic measurement, $y[k+1]$, is used. The covariance of the rth kinematic measurement is given by $P_{\iota,r}^y$,

$$P_{\iota,r}^y = H_{\iota,r}' P_\iota[k+1]^- H_{\iota,r} + P_{\iota,r}^n, \tag{4.18}$$

with $(P_{\iota,r}^y)^{-1'} = D_{\iota,r}^y = (F_{\iota,r}^y)^2$.

The modal update is given by

$$\alpha_\iota^-[k+1] = \alpha_\iota[k] \prod_r |F_{\iota,r}^y| \exp(\tfrac{1}{2}\Delta\|m_\iota\|_{D_\iota}^2 - \sum_r \|y[k+1]\|_{D_{\iota,r}^n}^2). \tag{4.19}$$

We now extrapolate forward to ι^+:

$$\alpha_\iota + [k+1] = \alpha_\iota^-[k+1]\Pi_{pi}. \tag{4.20}$$

Equation (4.20) generates a probability mass function on κ^+. The terminal state of the ι^+th local filter is $(\alpha_\iota+, m_\iota, P_\iota)$.

Figure 4.5 shows a sample of the path following performance of the GWE-tracker labeled with black +. The placement error is much smaller than that seen in the EKF-tracker (see Fig. 4.1) even though algorithms both use the same raw range-bearing data.

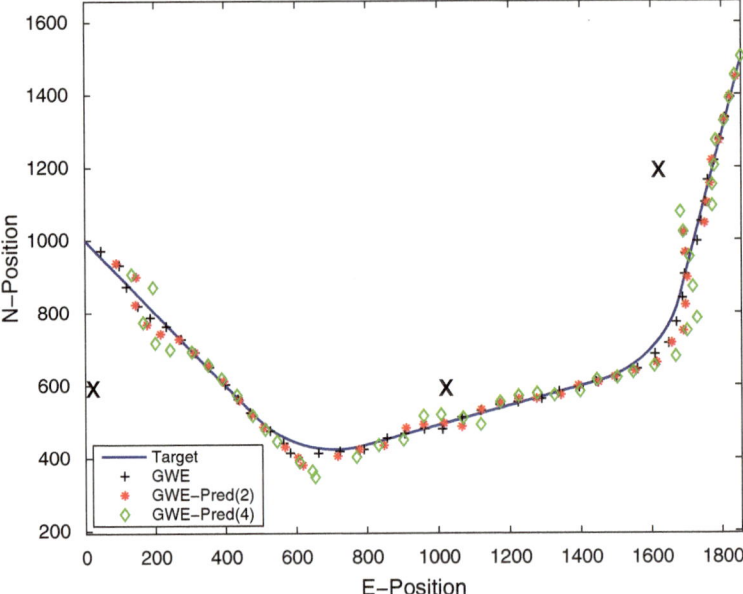

Fig. 4.5 In this engagement, the GWE has no modal measurement. The estimates of the $\mathscr{Y}[k]$-GWE are labeled *black plus*. The response of a two-second predictor is labeled *red open circle*, GWE-Pred(2). The response of a four-second predictor is labeled *green diamond*, GWE-Pred(4)

4.3.2 $\mathscr{Y}[k]$-*GWE-Predictor*

4.3.2.1 $\mathscr{Y}[k]$-**GWE-Pred(2)**

The fixed-interval GWE-predictor is an adjunct to the GWE-tracker, though in this case there are 27 distinct trackers. When we extrapolate forward to $t = [k+1]T$, we use (4.14)–(4.15) to determine the forward estimate of the kinematic state. The modal string now expands by one: $\iota \to \iota^+$. Lacking a position measurement,

$$m_{\iota^+}[k+1|k] = A_i m_\iota[k] \tag{4.21}$$

$$P_{\iota^+}[k+1|k] = A_i P_\iota[k]A_i' + P_\iota^w. \tag{4.22}$$

The modal state space of the two-second predictor is $\iota^{++} = rpijl$, a string of length five,

$$\alpha_{\iota^{++}}[k+1] = \alpha_{\iota^+}^-[k+1]\Pi_{rp}. \tag{4.23}$$

The terminal state of the GWE-predictor at $t = (k+1)T$ is $(\alpha_{\iota^{++}}, m_{\iota^+}, P_{\iota^+})$. There are 243 local estimates now, but all predictors with the suffix ι^+ have the same mean and covariance.

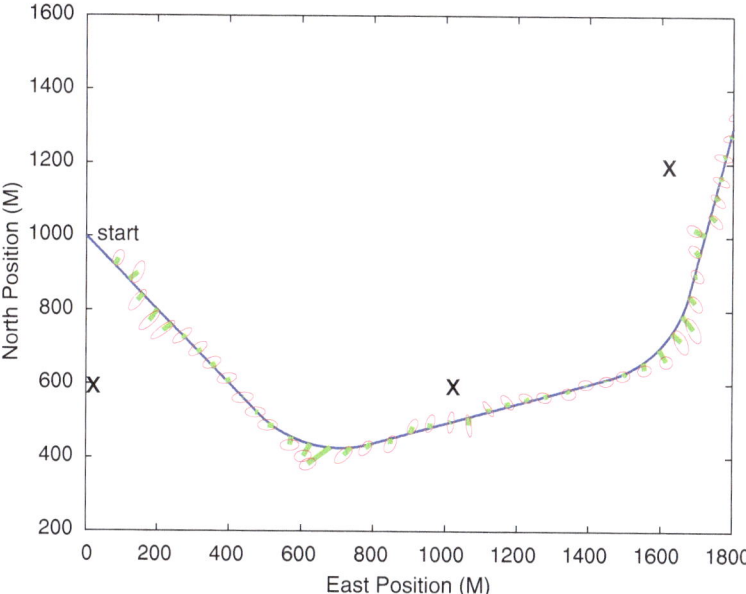

Fig. 4.6 The two-second $\mathscr{Y}[k]$-GWE-predictor has much smaller error than does its EKF counterpart. The projected distribution is used throughout this section

The $\mathscr{Y}[k]$-conditional distribution of the kinematic state of the ι^+th local filter is $N(m_{\iota^+}, P_{\iota^+})$ at time $t = [k + 1]T$. For $\iota^+ \in \kappa^+$, the weight in the GWE assigned to the ι^+th local filter is α_{ι^+}. This completes the two-second GWE-predictor.

Figure 4.5 shows a sample function of the two-second predictor; $\mathscr{Y}[k]$-GWE-Pred(2) shown "red o." The radial error is much less than that of EKF-Pred(2) though more than that of the GWE-tracker. The response of the GWE-Pred(2) is more volatile than is that of the EKF. This is a partial consequence of using the projected distribution for the $\mathscr{Y}[k]$-estimate. Neither the GWE nor the EKF has any way to sense that the first north turn is imminent. But the decay of the radial error in the hybrid tracker is much faster after the turn than it is in the EKF.

Figure 4.6 shows the 1σ-error ellipses for GWE-Pred(2). The error ellipses are bigger than those of the EKF-predictor. They enclose the target most of the time, though many times only at the boundary. If we used this algorithm to direct a illuminator, we could be confident of capture within three samples or so.

4.3.2.2 $\mathscr{Y}[k]$-**GWE-Pred(4)**

If we want a four-second predictor, we extrapolate each of the local predictors forward. The state space of the four-second predictor is $\iota^{+++} = qrpijl$, a string of length six:

$$m_{\iota++}[k+2|k] = A_r m_{\iota++}[|k=1|k] \qquad (4.24)$$

$$P_{\iota++}[k+2|k] = A_r P_{\iota++}[k]A_r' + P_{\iota++}^w \qquad (4.25)$$

along with

$$\alpha_{\iota+++}[k+2] = \alpha_{\iota++}[k+1]\Pi_{qr}. \qquad (4.26)$$

The state of the GWE-Pred(4) at $t = (k+2)T$ is $(\alpha_{\iota+++}, m_{\iota++}, P_{\iota++})$. There are 729 local estimates. All predictors with the suffix ι^{++} have the same mean and covariance. If we do not need to predict beyond 4 s, we can reduce our state to $(\alpha_{\iota++}, m_{\iota++}, P_{\iota++})$. This is of dimension 243.

Figure 4.5 shows a sample function of the four-second predictor; $\mathscr{Y}[k]$-GWE-Pred(4) shown "green *diamond*." The radial error is greater than that of GWE-Pred(2) near the turns. But the decay rate is quite good. The predictor is clearly able to use the $\{\mathscr{Y}[k]\}$ sequence to advantage in modal recognition.

Figure 4.7 shows the 1σ-error ellipses for GWE-Pred(4). The error ellipses are bigger than those of the GWE-Pred(2). They enclose the target at the same rate as GWE-Pred(2). If used to control an illuminator, the 1σ-ellipses would suffice in this application.

Figure 4.7 (and Fig. 4.6) use the projected approximation, $\hat{\mathbb{P}}$, to determine the tightest assurance regions. While this yields a simply connected neighbor-

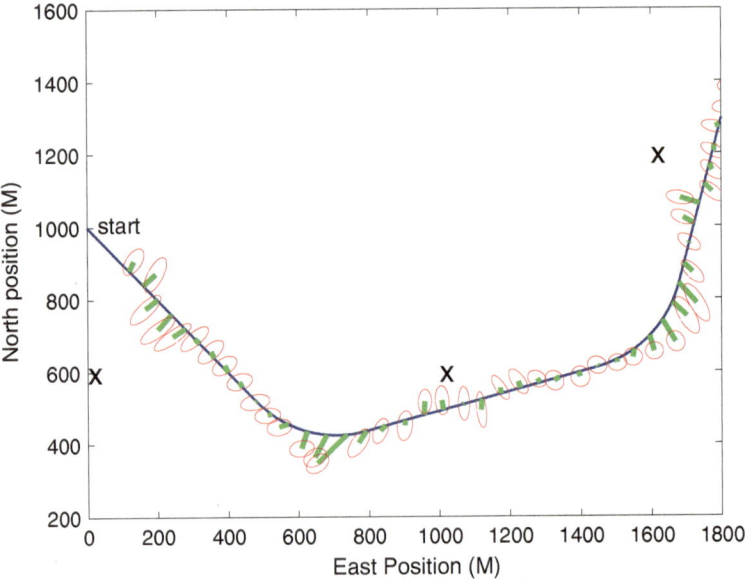

Fig. 4.7 The 1σ-assurance regions of the four-second $\mathscr{Y}[k]$-GWE-predictor are as good as the two-second predictor

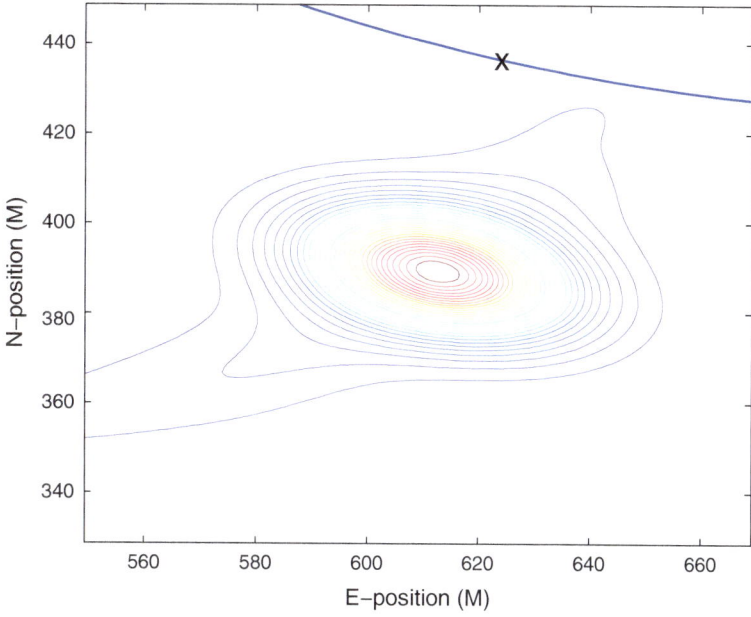

Fig. 4.8 The true target location is shown with a **x**. The location density of the four-second $\mathscr{Y}[k]$-predictor at (625, 436) has significant south bias. An assurance region based upon $\mathbb{P}$ would not likely capture the target unless we set the density threshold very low

hood, it does not utilize the fine structure of the notional $\mathscr{Y}[k]$-distribution. The GWE-Pred(4) produces a 243-term Gaussian sum at each sample time. Actually, a few terms dominate, and $\mathbb{P}$ is often not far from Gaussian.

Figure 4.8 shows a contour plot of the $\mathbb{P}$-density when the target has progressed to (625, 436). The conditional density, though not Gaussian, has a broadly similar form. The predictor points strongly to a CV extrapolation with prediction error on the order 40 m. The density has strong tails perpendicular to the velocity vector because of an increasing weight on modal strings including the e_2 prefix. However, this figure displays a situation in which a tight assurance region would fail to capture the target. On the other hand, the assurance region generated by EKF-Pred(4) is so poor that would not even show on the scale of the graph. Despite its bias, GWE-Pred(4) is far superior to EKF-Pred(4).

Another part of the path that the GWE-Pred(4) finds difficult is subsequent to the second north turn. Let us look at the four-second predictor when the true target location is (1659, 788); $t = 72$ s. The GWE does not have a clear idea of the mode at $t = 68$ s. The result is a bi-modal density with both the turn-north hypothesis and the CV hypothesis clearly represented. In Fig. 4.9 the former is more diffuse than the latter. An assurance region of reasonable size could require two disjoint convex neighborhoods.

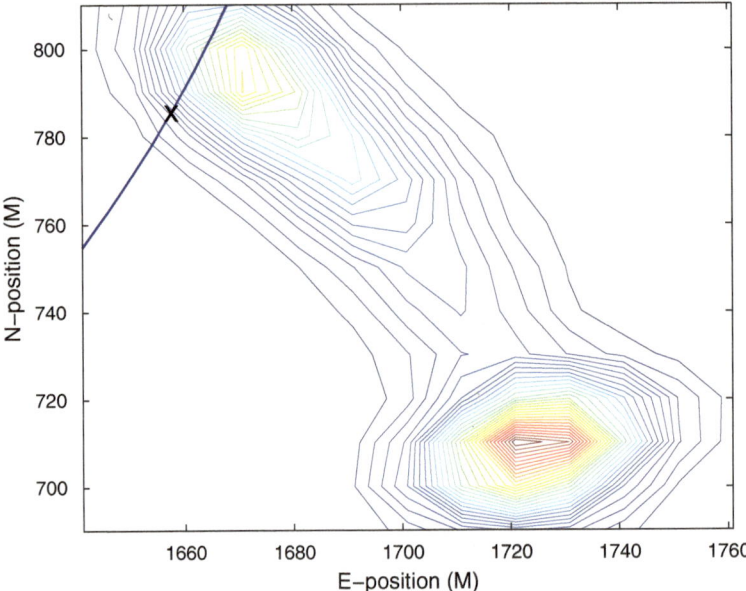

Fig. 4.9 The true target location is shown with a *times symbol*. The four-second $\mathscr{Y}[k]$-predictor at $(1659, 788)$ is bi-modal. A robust assurance region will capture the target in this case though the region is not simply connected

Figure 4.10 shows the radial error for the $\mathscr{Y}[k]$-GWE-tracker, and the two- and the four-second predictors. The error is that generated by the $\hat{\mathbb{P}}$ approximation. The turns generate significant error, and a longer prediction interval magnifies these errors. But the maximum GWE-Pred(4) error is 115 m (as contrasted with 140 m for the EKF-Pred(4)). The median prediction error is 21.3 m for a four-second prediction and 14.6 m for a two-second prediction. The two-second prediction error is roughly equal (both median and maximum) to the error of the EKF-tracker.

4.4   The $\mathscr{G}[k]$-GWE

4.4.1 $\mathscr{G}[k]$-*GWE-Tracker*

In this engagement, we have the same sequence of raw kinematic measurements we have used in the previous examples. But we will assume that we avail ourselves of LIDAR data that provides a noisy measurement of the turn rate. Specifically, we augment $\{\mathscr{Y}[k]\}$ with the coincident sequence $\{\mathscr{Z}[k]\}$ be generated by

$$z[k+1] = \mathbf{D}\phi_p[k+1] + \eta[k+1], \tag{4.27}$$

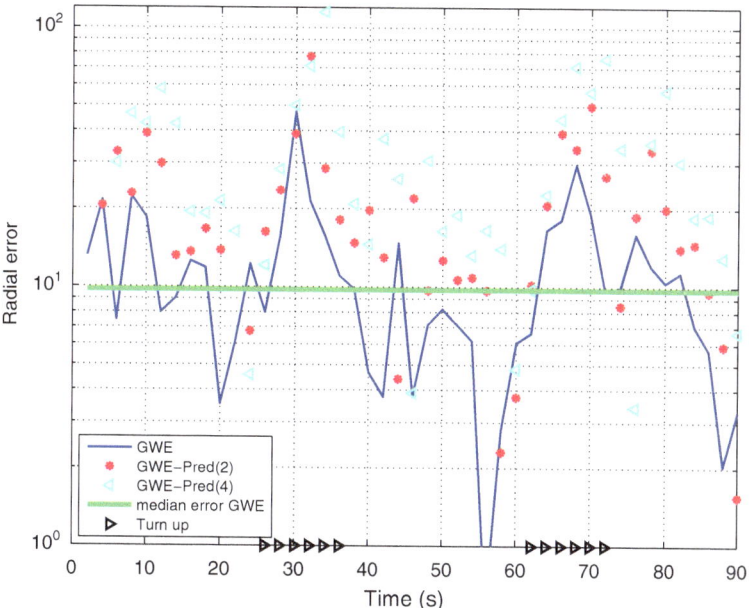

Fig. 4.10 The prediction error in a $\mathscr{Y}[k]$-GWE is far smaller than the comparable EKF

where **D** is the discernibility matrix,

$$\mathbf{D} = \begin{bmatrix} 0.8 & 0.1 & 0.1 \\ 0.1 & 0.8 & 0.1 \\ 0.1 & 0.1 & 0.8 \end{bmatrix}. \tag{4.28}$$

The pointer, $z[k]$, indicates the forward regime proxy with an 80 % confidence. The combined observation sequence is labeled $\{\mathscr{G}[k]\}$: $\mathscr{G}[k] = (\mathscr{Y}[k] \vee \mathscr{Z}[k])$. In this section, we will use the same $\{\mathscr{Z}[k]\}$ sequence we used in the tracking engagement of Chap. 3.

The fusion of the regime measurement sequence leads to improved performance in the GWE-tracker. Because the LGM model in Chap. 3 differs from that used here, the trackers are not directly comparable. Figure 4.11 shows a sample function of the $\mathscr{G}[k]$-GWE-tracker using the common data set.

The quality of the $\mathscr{G}[k]$-regime estimate is superior to its $\mathscr{Y}[k]$-counterpart. Figure 4.12 shows the $\mathscr{G}[k]$-estimate of $\phi_t = e_2$—see Fig. 3.13 of Chap. 3. A high quality modal estimate leads to improved performance of the GWE-tracker: the median radial error for the $\mathscr{G}[k]$-GWE-tracker is 15 % less than that of the $\mathscr{Y}[k]$-GWE-tracker.

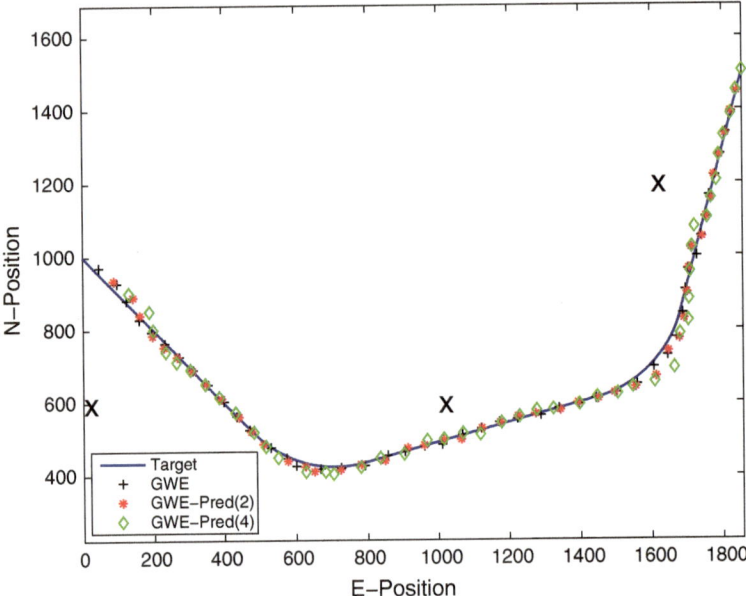

Fig. 4.11 In this engagement, the GWE has a LIDAR sequence that gives a high quality measurement of turn rate. The estimates of the $\mathscr{G}[k]$-GWE are labeled with *black plus*. The response of a two-second predictor is labeled *red open circle*, $\mathscr{G}[k]$-GWE-Pred(2). The response of a four-second predictor is labeled *green diamond*, $\mathscr{G}[k]$-GWE-Pred(4)

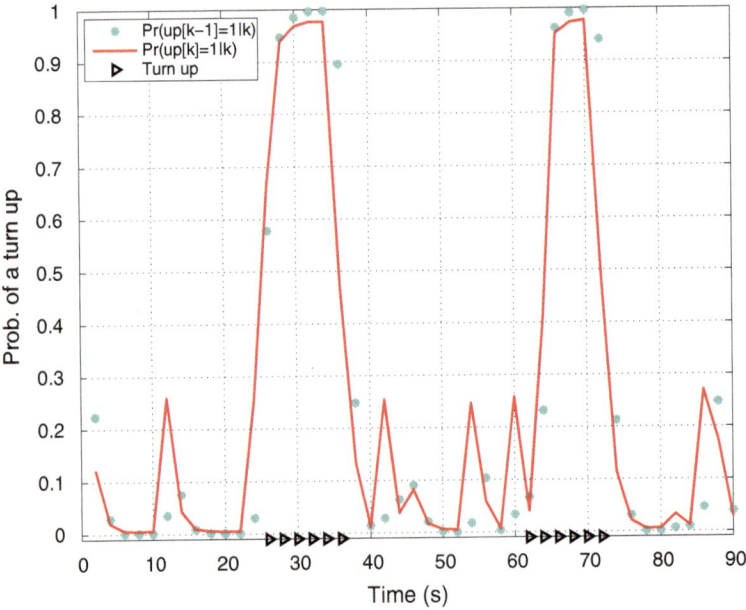

Fig. 4.12 The $\mathscr{G}[k]$ estimate of a counterclockwise turn creates a more responsive tracker

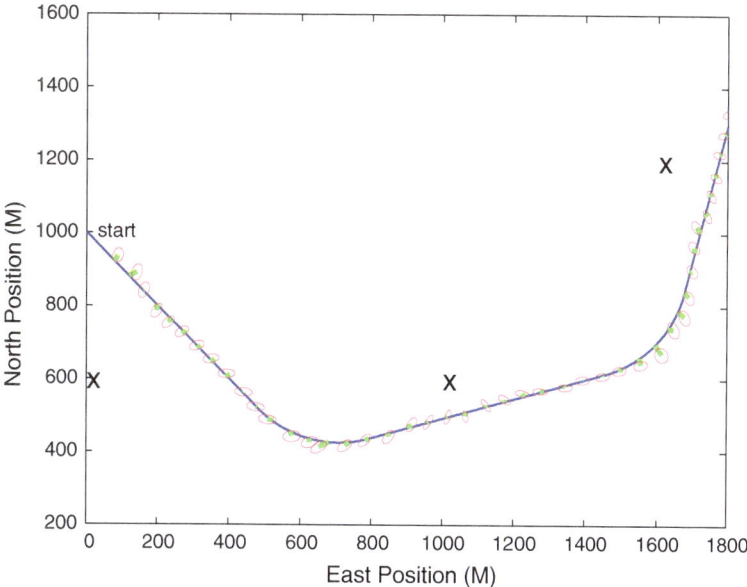

Fig. 4.13 The $\mathscr{G}[k]$-GWE two-second predictor is able to capture the target with small 1σ-error ellipses

4.4.2 $\mathscr{G}[k]$-*GWE-Predictor*

4.4.2.1 $\mathscr{G}[k]$-**GWE-Pred(2)**

As was the case with the $\mathscr{Y}[k]$-GWE-predictor, the $\mathscr{G}[k]$-GWE-predictor extrapolates forward from the terminal state of the $\mathscr{G}[k]$-tracker: $(\hat{\phi}_{t}+, m_t, P_t)$. The $\mathscr{G}[k]$-algorithm is that given in (4.21)–(4.22). The advantage of the $\mathscr{G}[k]$-predictor is strictly due to the superior conditioning from the tracker. The data set $\mathscr{G}[k]$ yields a better estimate of the forward regime $\mathbf{e}_p$ than does $\mathscr{Y}[k]$.

Figure 4.13 shows the response of the two-second $\mathscr{G}[k]$-GWE-predictor. The tracking error is much smaller than that displayed by the $\mathscr{Y}[k]$-GWE-predictor. The first turn is accommodated by the former while the latter drifts beyond the true path. In contrast with the $\mathscr{Y}[k]$-predictor, the 1σ-error ellipses capture the target as it moves through the turns. And the ellipses are smaller. The $\mathscr{G}[k]$-GWE would be a satisfactory algorithm for tracking and intercept.

4.4.2.2 $\mathscr{G}[k]$-**GWE-Pred(4)**

Figure 4.14 shows the response of the four-second $\mathscr{G}[k]$-GWE-predictor. The tracking error is bigger than that shown in Fig. 4.13 because of the longer extrapolation interval. But the 1σ-error ellipses using $\hat{\mathbb{P}}$ are correspondingly larger. There is no run of 6 s during which the target is not captured by the 1σ-ellipses. Contrast that with runs of over 20 s for the EKF-predictor.

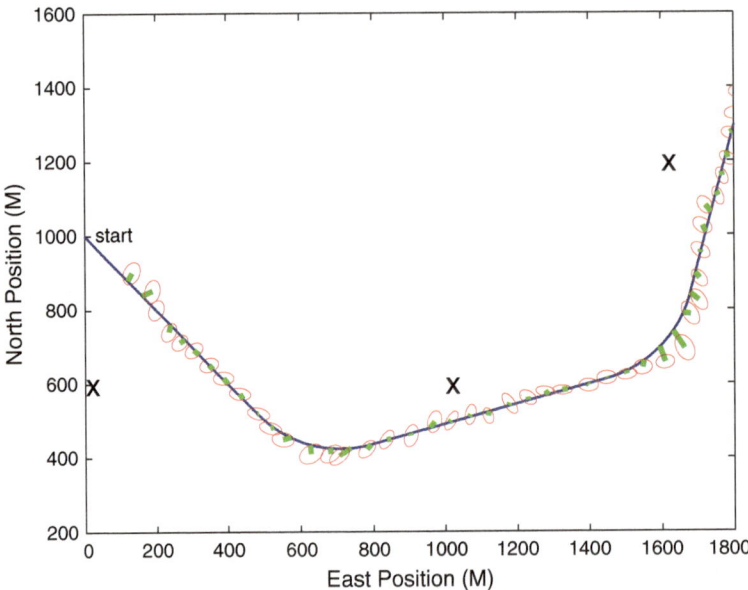

Fig. 4.14 The $\mathscr{G}[k]$-GWE four-second predictor has larger 1σ-error ellipses, but it captures the target essentially every time

Figure 4.15 shows the fine structure of the $\mathscr{G}[k]$-density of the four-second predictor looking beyond the first turn, $(625, 436)$ or $t = 30\,\text{s}$—compare this figure with the $\mathscr{Y}[k]$-predictor shown in Fig. 4.8. The four-second $\mathscr{G}[k]$-predictor has a bi-modal density as contrasted to the uni-modal density of the $\mathscr{Y}[k]$-predictor. The dominant lobe of the former is close to the true target location with significant east error. The dominant lobe of the latter is aligned with a CV-projection. The secondary lobe from the $\mathscr{G}[k]$-predictor is close to the dominant lobe in the $\mathscr{Y}[k]$-predictor. A plausible assurance region for the $\mathscr{G}[k]$-predictor would enclose the target while the same assurance region for the $\mathscr{Y}[k]$-predictor would fail.

When there is regime uncertainty, the position distribution is non-Gaussian but often uni-modal. For example, Fig. 4.16 shows the density of the four-second predictor at $(1707, 952)$, $t = 76\,\text{s}$. The target has completed the second turn and is proceeding in CV motion. The predictor is strongly CV with prediction error a few meters forward on the true path. It is interesting that the primary uncertainty is perpendicular to the path rather than along the path: the uncertainty regarding a regime change is stronger than the uncertainty regarding a speed change.

Figure 4.17 shows the radial error for the $\mathscr{G}[k]$-GWE-tracker and for the two- and four-second predictors. The turns still generate significant tracking error, but not to the degree we observed in the $\mathscr{Y}[k]$-predictor. The maximum GWE-Pred(4) error is only 56 m—as contrasted with 71 m for the EKF-tracker; the $\mathscr{G}[k]$-GWE-predictor has less error that the EKF-tracker.

The median four-second prediction error of the $\mathscr{G}[k]$-GWE-predictor is 71 % of that of the $\mathscr{Y}[k]$-algorithm: the maximum four-second prediction error of the $\mathscr{G}[k]$-

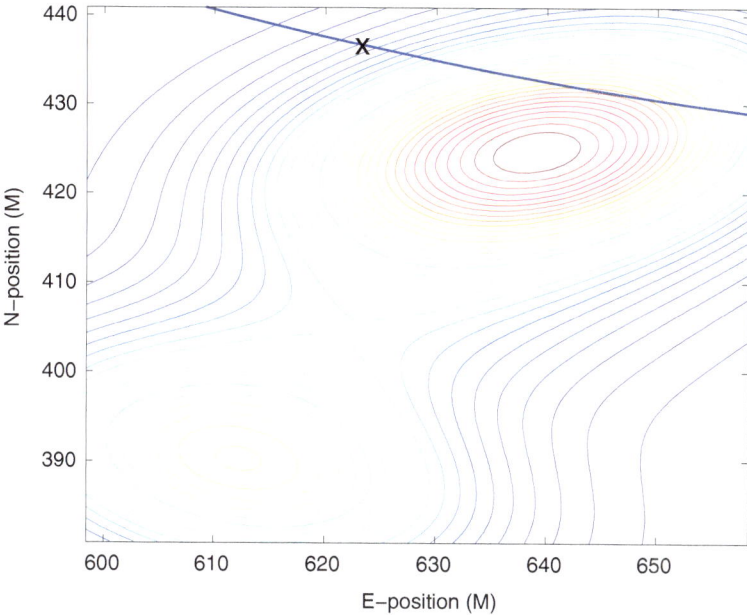

Fig. 4.15 The true target location is shown with a *times symbol*. The four-second $\mathscr{G}[k]$-predictor of position at $(625, 436)$ is significantly more accurate than is the $\mathscr{Y}[k]$-predictor

GWE-predictor is 48 % of that of the $\mathscr{Y}[k]$-algorithm. It is evident that the fusion of the $\{\mathscr{Z}[k]\}$ data sequence is more important in prediction than it is in tracking.

4.5 Conclusion

This chapter presents a command architecture that illustrates the power of the hybrid prediction approach for target intercept. We have used a fixed interval predictor to frame the problem. In other applications; e.g., impact-point prediction, the interval is determined by a terminal condition on the target state. The cooperative EKF-tracker has a great advantage. At time $t = (k+3)T$, the EKF C knows both the modal string and the kinematic measurements over the extrapolation interval. The median radial tracking error of the EKF C tracker is 7.16 m—see Table 4.1. The EKF-tracker has a radial error 2.52 dB bigger than the EKF C. The $\mathscr{Y}[k]$-GWE-tracker has an error 1.2 dB less than that and with LIDAR augmentation the $\mathscr{G}[k]$-GWE has an error only 0.4 dB bigger than EKF C. We see that the multi-model tracker reduces the error attributable to regime events by more than 2 dB even when the same data set is used.

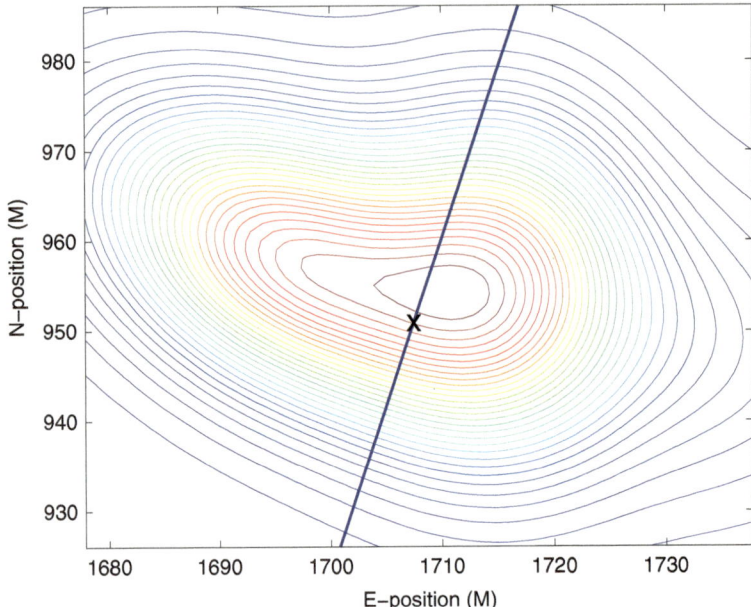

Fig. 4.16 The true target location is shown with a *times symbol*. The four-second $\mathcal{G}[k]$-predictor of position at $(1707, 952)$ is uni-modal and oriented perpendicular to the path

When we use the nominal **EKF** as a predictor, the error ratio grows by over 4 dB for a two-second predictor and 6 dB for a four-second. Table 4.2 gives the median tracking error of the two-second predictor. The error of the $\mathcal{G}[k]$-**GWE** algorithm grows by less than 2 dB at 2 s and a little over 3 dB at 4 s as compared with **EKF** C. The $\mathcal{G}[k]$-**GWE**-Pred(2) has a smaller error than does the nominal **EKF**-tracker.

At 4 s, the nominal **EKF**-error grows by 6.35 dB. This compares with a 4.74 dB growth in the $\mathcal{Y}[k]$-**GWE** (4) and a 3.28 dB growth in the $\mathcal{G}[k]$-**GWE**-Pred(4). The $\mathcal{G}[k]$-**GWE** at 4 s is about as good as the $\mathcal{Y}[k]$-**GWE** at 2 s and the $\mathcal{Y}[k]$-**EKF** at zero seconds. The **GWE**s degrade at 4 s since the **GWE** memory is only 6 s $(L = 3)$. Better prediction would be achieved with a larger L, but that would be at the expense of a more complicated tracker.

The hybrid prediction algorithm is clearly at an advantage in this engagement. The cooperative target not only broadcasts its regime, but the tracker knows it will have the current regime when needed. The predictors must extrapolate from the current data set and know they will not get any more information before intercept. Though dead reckoning is used for extrapolation, the modal measurement cues the $\mathcal{G}[k]$-predictor. This results in a 3 dB improvement over the **EKF**. The implication to the design of a command architecture is clear (Tables 4.1, 4.2, and 4.3).

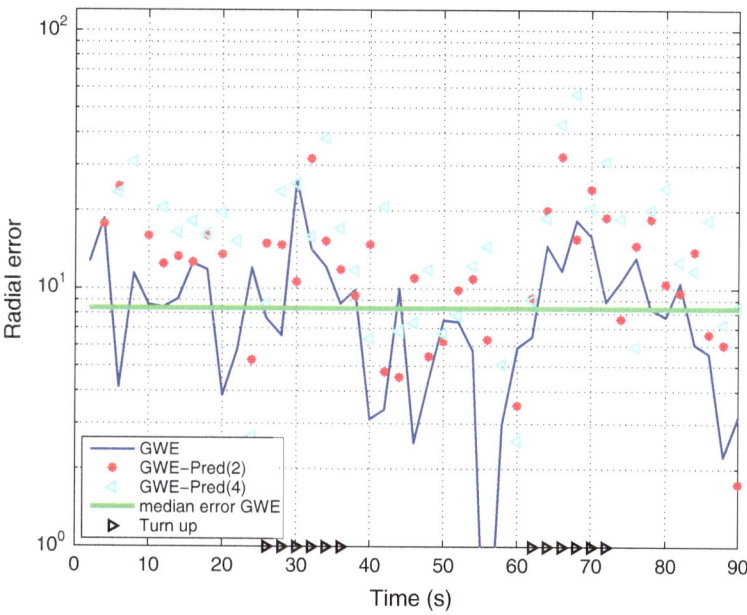

Fig. 4.17 The prediction error in a $\mathcal{G}[k]$-GWE-predictor is smaller than the comparable $\mathcal{Y}[k]$-GWE-predictor

Table 4.1 Median radial tracking error (M)

EKF C	EKF	$\mathcal{Y}[k]$-GWE	$\mathcal{G}[k]$-GWE
7.16	12.79	9.71	8.35

Table 4.2 Median radial 2-second prediction error (M)

EKF	$\mathcal{Y}[k]$-GWE	$\mathcal{G}[k]$-GWE
18.18	14.66	10.91

Table 4.3 Median radial 4-second prediction error (M)

EKF	$\mathcal{Y}[k]$-GWE	$\mathcal{G}[k]$-GWE
30.88	21.34	15.24

Chapter 5
Classification and Target Tempo

Abstract A target can be classified based upon an IFF signal, a visual image, or something similar. But an uncooperative target will not transmit a valid IFF signal and will attempt to obscure its image from the defender. Can the target be identified from its motion pattern; e.g., turn rates, tempo? This might readily be done if the motion parameters were inferred without error. But in actual engagements, the motion tempo can only be approximated. Certainly, classification becomes more sure with a long data stream. But if defense requires quick reaction, sophisticated processing is required. The chapter explores the utility of the GWE for target classification in a noisy measurement environment. We show that in some circumstances the maneuver characteristics of a target are sufficient to give an indication of target type. The result holds promise for the architect of a command system.

5.1 Introduction

The GWE has a hybrid state structure in which the kinematic state, x_t, is a vector in $\mathbb{R}^n$ and the modal state, ϕ_t, is a pointer in $\mathbb{R}^{L*S}$. The tracker generates estimates of both components of the state, $\hat{x}_t$ and $\hat{\phi}_t$, respectively. The former has been used extensively as a high quality estimate of the location. The latter gives an estimate of the current modal state: at time t, $\hat{\phi}_t = \mathbb{P}(\phi_t = \mathbf{e}_t | \mathscr{G}[k])$.

Knowledge of the modal state is an aid in tracking and prediction: contrast the fidelity of EKF C with that of the nominal EKF. We would expect, therefore, that an improvement in the modal estimate would translate into a commensurate reduction in tracking error across the range of trackers we have considered. Improved modal estimates are indeed advantageous, but not always in the ways and to the degree expected. In the rather mild engagement we have used to illustrate the various trackers, the baseline $\mathscr{Y}[k]$-GWE actually yields most of the improvement attributable to the hybrid formalism.

There are problems in which accurate modal designation is important in its own right; e.g., in prediction as discussed in Chap. 4. In this chapter, we will propose ways of improving modal estimation using the GWE to achieve high quality target placement. Further, we will broaden the problem structure to include applications that require real-time target identification.

© Springer International Publishing Switzerland 2016
D.D. Sworder, J.E. Boyd, *Locating, Classifying and Countering Agile Land Vehicles*, DOI 10.1007/978-3-319-19431-8_5

To enhance the quality of modal estimation, it is advantageous to revise the modal model to permit a more nuanced representation of the regime process. The GWE utilizes a Markov model for the regime pointer $\{\phi_t\}$:

$$d\phi_t = Q'\phi_t dt + dm_t \tag{5.1}$$

with initial condition ϕ_0. The forcing term, $\{m_t\}$, is a discontinuous martingale. There are applications in which Q is a function of time, but for the purposes of this discussion, we will assume Q is a constant.

If we look only at transition events; $\mathbf{e}_i \mapsto \mathbf{e}_p; p \neq i$, the regime sequence is also a Markov process. We think of the regime string thus generated as an untimed discrete event process: the times of the transitions are not specified in the string.

Let us look at the regime state i. If $\phi_t = \mathbf{e}_i$, the probability that the next regime is $\mathbf{e}_p$ is $-Q_{ip}/Q_{ii}$. The row sums of Q are equal to zero: the sum over all $p \neq i$ of $-Q_{ip}/Q_{ii}$ must equal one.

The untimed regime sequence can be written as a string; i.e., $ijl\ldots$. In this context, the sequence of prefixes is a Markov process: the probability that $ijl\ldots \mapsto pijl\ldots$ depends only on i. Unfortunately, this language is not defined on a finite space: on any time interval of positive length, the number of possible transitions in $\{\phi_t\}$ is not bounded.

In addition to imposing a probabilistic structure on the untimed modal string, (5.1) dictates a tempo for the regime process. Under (5.1), the mean time interval over which $\{\phi_t\}$ is $\mathbf{e}_i$ is $-1/Q_{ii}$. This is called the mean sojourn time in regime i. The actual sojourn in the ith regime state is exponentially distributed. A plot the sojourn density with mean 1 is shown in Fig. 5.1 labeled with green $\diamond$.

As is evident from the figure, the Markov model predicts a preponderance of sojourns of short duration: the sojourn density decreases monotonically and rapidly from a sojourn interval of length zero. The mean time remaining in a sojourn is actually independent of the elapsed time in the regime; e.g., if the elapsed time of $\{\phi_t\}$ in regime $\mathbf{e}_i$ has been 3 s and the mean sojourn time in $\mathbf{e}_i$ is one second, the GWE expects $\{\phi_t\}$ to remain in regime $\mathbf{e}_i$ for yet another second.

The actual sojourn times in the engagements we have studied are more predictable that the Markov model would suggest. There are no short regime sojourns. Indeed, the value of $-1/Q_{ii}$ derives from a typical sojourn on a typical path. The Markov model blurs the distinction between $\{w_t\}$ and $\{\phi_t\}$. The former represents the unstructured (wideband) target accelerations, and the latter represents the structured (band-limited) target accelerations. The spectral distributions of the two processes tend to overlap in the Markov regime model [19, 26].

A more credible model would reduce the frequency of short-lived sojourns and increase the frequency of lifetimes like those observed on the sample paths of earlier engagements. The regime model should retain the Markov structure of the untimed modal string but introduce a more apt temporal description of the regime transitions. For example, very brief acceleration sojourns result in little change in the target path and would not be efficacious for a vehicle moving toward a designated terminal position.

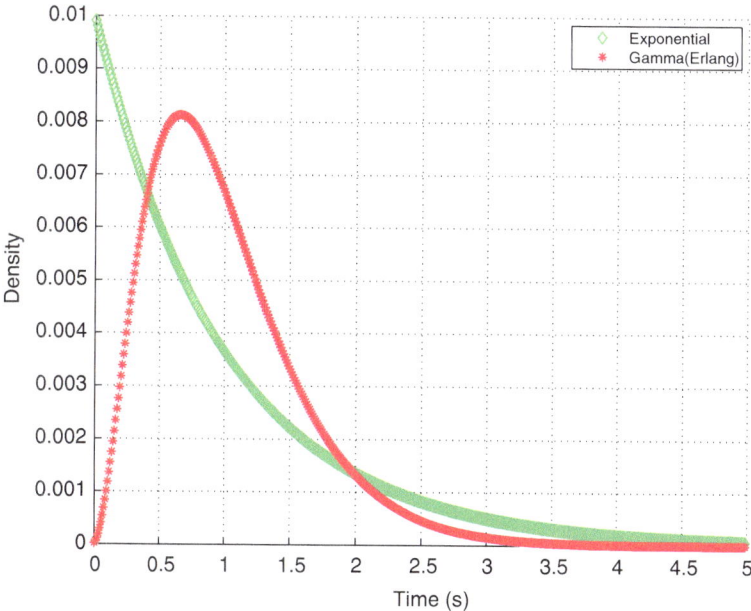

Fig. 5.1 The sojourn density of the regime state is exponential if $\{\phi_t\}$ is a Markov process. The Gamma density is a more plausible alternative in a tracking application

If the motion modes have a quasi-periodic nature, the sojourn density might look more like that shown as the red-* curve in Fig. 5.1. This density has few fleeting sojourns and a thinner tail than that displayed by the exponential density. The likelihood function accentuates intervals with term nearer the mean sojourn time.

The red curve in Fig. 5.1 is a Gamma density with mean 1 s. In contrast with the exponential density, the prevalence of the brief lifetimes is reduced, and the mode of the γ_x-density is shifted toward the mean—though the mode is strictly less than the mean.

Let us create a random process based upon the event structure of the untimed regime process implicit in Q. Instead of the exponential sojourn times of the Markov model, let us use a γ_x-density to delineate the lifetimes. The untimed regime string is still a Markov sequence, but $\{\phi_t\}$ is not a Markov process because the regularity of the sojourn times implies an internal memory. The process thus constructed is a γ_x-renewal process [3, 24].

We will construct the regime model as follows. The untimed regime sequence is generated from Q as before. But the sojourn time in the ith regime is distributed according to the γ_x-density associated with $\mathbf{e}_i$. The sojourn times for a motion path generate a sequence of positive random variables, the ith of which depends only on the ith regime state [19]. The Markov process generated by (5.1) is actually a simple renewal process.

Since the GWE is predicated upon a Markov regime model, we must modify the GWE to accommodate this more nuanced temporal model. The Gamma density, $\gamma(u; r, \lambda)$, is given by:

$$\gamma(u) = \frac{\lambda}{\Gamma(r)}(\lambda u)^{r-1}e^{-\lambda u}; \, t > 0$$

$$\gamma(u) = 0; \, t \leq 0 \tag{5.2}$$

The Gamma distribution is a two-parameter family: $r > 0$ controls the shape of the density and $\lambda > 0$ controls the time scale. The mean of the distribution is $\upsilon = r/\lambda$. The density shown in Fig. 5.1 is for $r = 3; \lambda = 3$. Because it is more evocative, we will parameterize the Gamma distribution with the pair (shape, mean): $\gamma(u; r, \upsilon)$. The functions plotted in Fig. 5.1 are $\gamma(u; 3, 1)$ and $\gamma(u; 1, 1)$. While r can be any positive number, we will restrict attention to γ-densities in which r is a positive integer: such Gamma densities are also called Erlang densities. As r increases, the mode of the density becomes more pronounced, and the mode moves closer to the mean value. The tail of the γ_x-density is narrower than that of the exponential density; the likelihood of long sojourns is lessened.

Suppose $i \in \mathbf{S}$ is an intrinsic regime state with mean sojourn time υ_i. Let r_i be the positive integer that is used to shape the lifetime distribution in ϕ_i. Let us partition regime i into r_i separate sub-bins. Call them $i_m; m = 1 \ldots r_i$. So if $r_i = 3$, we would have three sub-bins (i_1, i_2, i_3). The aggregate, $(i_1 \cup i_2 \cup i_3)$, is identified with the intrinsic regime bin i.

Suppose we enter i_{r_i} at $t = 0$, and that the lifetime in i_{r_i} is exponentially distributed with mean υ_i/r_i. If we necessarily pass from i_m to i_{m-1} for all $m = r_i, \ldots, 2$, we will progress from i_{r_i} to i_1. If the sojourns in each of the bins are independent and exponentially distributed with common mean, υ_i/r_i, then the mean time to traverse the bin aggregate is υ_i, and the lifetime within $(i_{r_i} \cup \ldots \cup i_1)$ is a $\gamma(u; r_i, \upsilon_i)$ random variable.

This construction provides a means for shaping the distributions of the regime sojourns. Suppose that lifetime distribution of each of the intrinsic regime states is $\gamma(u; r_i, \upsilon_i)$. We need only represent each regime state with an aggregate of r_i sub-states, the collection of which are identified with the ith intrinsic regime. The dimension of the regime state space must be expanded to include all of these sub-regimes. For example, if the mean lifetime of the first regime is 1 s, we can achieve the lifetime distribution shown in Fig. 5.1 by selecting $r_1 = 3$ and $\upsilon_1 = 1$. Now $\phi_t = e_1$ is identified within an expanded version ϕ_t in which the first intrinsic state is the union of a set of sub-regimes. If the intrinsic regimes are $\gamma(u; 3, \upsilon_i)$, then: $(i_1 \cup \ldots \cup i_3) \Leftrightarrow$ regime number one; $(i_4 \cup \ldots \cup i_6) \Leftrightarrow$ regime number two; and so on.

To maintain the proper mean sojourn times in each of the intrinsic regimes, the lifetimes of the sub-regimes are shorter by the factor r_i. To accomplish this, the diagonal elements in the extended Q-matrix are increased by the factors $\{r_i\}$. When

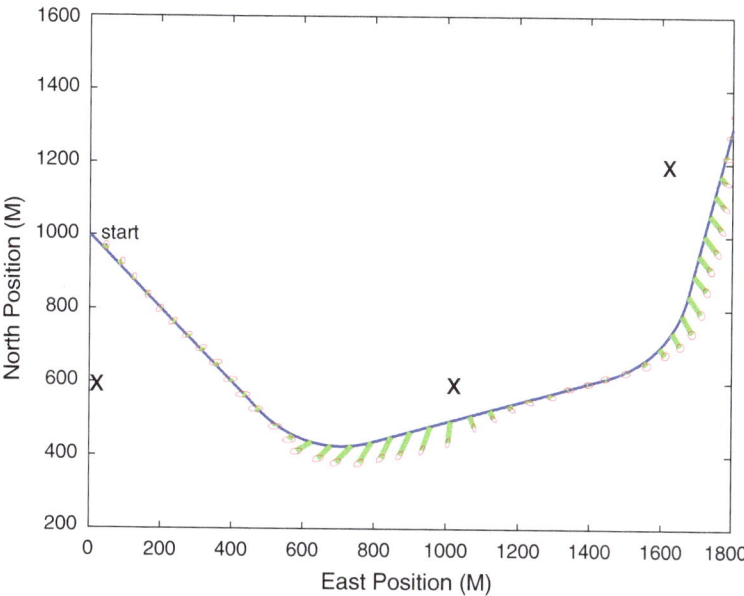

Fig. 5.2 A target is moving at 28 m/s on an irregular path. Three range-bearing sensors are shown with an *times symbol*. The 1σ-error ellipses of the nominal **EKF** are shown

the extended regime state reaches an exit boundary of a regime collective, the $i \mapsto p$ probabilities are preserved by increasing the Q_{nm} elements by the factor r_i when n is the exit from the ith collective and m is the entrance to the pth collective.

In the discussion that follows, we will tend not distinguish between the intrinsic regimes and the sub-regimes; the former is a kinematic primitive and the latter is an analytical expedient. Underlying the discussion is the fact that the ith regime sojourn time has a γ_x-density, and the tracking algorithm utilizes a regime state of dimension $\sum_{i \in \mathbf{S}} r_i$. If we want to describe how the sub-regimes evolve, we will make the regime/sub-regime distinction explicit. In the case where $r_i = 1; i \in \mathbf{S}$, we will just refer to the tracker as Markovian.

The **GWE** is based upon a time-discrete model of the engagement. The expanded dimension of the Q matrix changes nothing of consequence in the algorithm; though the calculations are now more time consuming. The modal state in the time-discrete model will again be labeled $\iota \in \kappa$. Unless otherwise stated, ι is a string of the intrinsic regime proxies derived by aggregating the associated elements of the extended modal state: the internal dimension of ι in the **GWE** is $(\sum_{i \in \mathbf{S}} r_i)^L$ while the external dimension of ι is S^L. In either case, we will label the set of all ι with κ, the size of which will be determined by context. Thus, ϕ_i could point to the ith intrinsic regime or to a sub-regime within an intrinsic regime.

5.2 The Engagement

To illustrate the use of the renewal model, let us return again to the elementary tracking engagement of the previous chapters. The target moves at constant elevation in an east-north coordinate system. The speed on the path is 28 m/s. Motion begins in a south-east direction from the starting point $(0, 1000)$ m. The path is a mixture of CV-motion with intervals of $5°$/s turns. The three regime states are: turn left $(i = 1)$; CV $(i = 2)$; turn right $(i = 3)$.

As before, the three range-bearing sensors are located at

$$(X_r, Y_r) = \{(0, 600), (100, 600), (1600, 1200)\}.$$

The sensors are displayed in Fig. 5.2 with an **x**. Each sensor measures the target location every 2 s with standard error: 56 m in range and 28mr in bearing—a 63 m error at 1 Km. The sensor noise is assumed to be independent temporally and spatially. The target path is shown in Fig. 5.2 along with the response of the nominal EKF as presented in Chap. 4. The raw measurements used in Chap. 4 will be used throughout this chapter as well.

This target path illustrates a shortcoming of the Markov regime model. Rather than exhibiting a series of brief sojourns as predicted by (5.1), the target spends protracted intervals in each motion regime. Look at the initial south-east CV-segment. The interval length is 24 s, and $\phi_i[k] = e_2; k = 1, \ldots, 12$.

The basic regime model we have proposed calibrates the transition matrix based upon empirical studies of typical motions and transitions. For example, we might suppose that the mean sojourn in the CV segments is 24s. Then in our reductionist regime model we might say that the sojourn time in $\phi_i[k] = e_2$ is exponential with mean 24; i.e., the sojourn time is $\gamma(u; 1, 24)$. However, as Fig. 5.1 indicates, this Markov model predicts a preponderance of transition events at time intervals much shorter than 24 s. Moreover, even after a stay of several seconds CV motion, the Markov model predicts that the next regime event be delayed by another 24 s or so.

In Chap. 3 we settled on the regime rate matrix:

$$Q = \begin{bmatrix} -0.08 & 0.050 & 0.030 \\ 0.020 & -0.04 & 0.020 \\ 0.030 & 0.050 & -0.08 \end{bmatrix}. \tag{5.3}$$

The mean sojourn time in the turn regimes is 12 s, and in the CV interval it is twice that.

Let us subdivide each intrinsic regime into three sub-regimes. Specifically, the lifetimes of $\phi_t \equiv e_1$ and $\phi_t \equiv e_3$ are $\gamma(u; 3, 12)$, and the lifetime of $\phi_t \equiv e_2$ is $\gamma(u; 3, 24)$. The dimension of the underlying regime state space is now nine. The rate matrix for the expanded regime process, Q_γ, is 9×9. The calculation of Q_γ is routine. The upper left 4×4 bloc is

$$Q_\gamma = \begin{bmatrix} -0.24 & 0 & 0 & 0 & \dots \\ 0.240 & -0.24 & 0 & 0 & \dots \\ 0 & 0.240 & -0.24 & 0 & \dots \\ 0 & 0 & 0.060 & -0.12 & \dots \\ \dots & \dots & \dots & \dots & \dots \end{bmatrix}. \tag{5.4}$$

Contrasting Q_γ with Q, we see that the diagonal elements of Q_γ are thrice that of Q. The sub-regime $\phi_t = e_3$ in Q_γ is an entry into the turn-left motion. A left turn (a turn up) can be entered from e_4 (with rate 0.06) or from e_9 (not shown). Within the left-turn regime, ϕ_t progresses $e_3 \mapsto e_2 \mapsto e_1$ with exit from e_1.

The time-discrete transition matrix can be computed as before. On the expanded state space,

$$\Pi_{pi} = \wp(\phi[k+1] = \mathbf{e}_p \mid \phi[k] = \mathbf{e}_i),$$

where for the γ_x model

$$\Pi_\gamma = \begin{bmatrix} 0.62 & 0.30 & 0.07 & 0.003 & \dots \\ 0.00 & 0.62 & 0.30 & 0.019 & \dots \\ 0.003 & 0.00 & 0.62 & 0.08 & \dots \\ 0.002 & 0.00 & 0.00 & 0.79 & \dots \\ \dots & \dots & \dots & \dots & \dots \end{bmatrix}. \tag{5.5}$$

In contrast, the transition matrix for the Markov model is

$$\Pi = \begin{bmatrix} 0.86 & 0.04 & 0.05 \\ 0.09 & 0.93 & 0.09 \\ 0.05 & 0.04 & 0.86 \end{bmatrix}. \tag{5.6}$$

The columns of Π_γ are probability vectors. The probability of leaving an exit regime is higher in Π_γ than it is in Π. For example, the probability that the γ_x-GWE target will exit $\mathbf{e}_1$ is 38 %. Contrast this with the same event in the Markov GWE (the M-GWE) which will occur with probability 14 %.

Should $\{\phi_t\}$ exit from CV-motion, it will go to the entrance left-turn sub-regime 40 % of the time, the intermediate left-turn sub-regime 9 % of the time, and the exit left-turn sub-regime 1 % of the time. Note that the time-discrete sub-regime state can bypass intermediate regimes in a way that the time-continuous regime process cannot.

In what follows, we will contrast the performance of tracker/classifiers using the Markov model (5.3) with those using a γ_x-model (5.4). We will prefix the former with M and the latter with a γ_x.

5.3 $\mathscr{G}[k]$-Tracking: Markov or Gamma

The GWE architecture can utilize a direct regime measurement, the $\mathscr{G}[k]$-GWE, or it may be lacking this measurement, the $\mathscr{Y}[k]$-GWE. In this section we will look again at the tracking problem. We will frame the study within the $\mathscr{G}[k]$-architecture since the radial errors for the two architectures are not significantly different.

Let us return to tracking the uncooperative target. As before, the kinematic model utilizes the three intrinsic regimes: left turn, CV, right turn. The time-discrete kinematic local model for the target motion is

$$x[k+1] = A_i x[k] + \sqrt{P_\iota^w}\, w[k+1], \tag{5.7}$$

where $\{w[k]\}$ is an $\mathscr{F}[k]$-adapted, unit, Gaussian-white sequence.

The interpretation of (5.7) is subtle. There are three intrinsic regimes and they index the dynamic matrix. But the action matrix is indexed by the mode label. In earlier chapters, we used a Markov model and a modal string of length three. So there are 27 distinct local models used in the M-GWE of Chap. 3. As long was the prefix of ι is i, we use A_i in (5.7).

In the γ_x-algorithms, we expanded the regime model, and the modal state lives in a new κ, the set of three-digit numbers of radix 9: there are 729 distinct local models represented in (5.7). All of the dynamic matrices in the collective for which ceil(i/r_i) is equal are identical and are written $A_i; i \in (1, 2, 3)$: there are only three values of $\{A_\iota; \iota = ij\ldots\}$ even though there are 27 values of $\iota \in \mathbf{S}^3$. However, the value of P_ι is different for every $\iota \in \kappa$ because each local model has a different history: there are in principle 729 distinct values of P_ι.

There are two sensor groupings that provide motion information. The first provides the location measurements from the three range-bearing sensors shown in Fig. 5.2. These raw measurements are reduced to data sequence $\{\mathscr{Y}[k]\}$ as before. But there are actually 729 such individuated $\{\mathscr{Y}[k]\}$. The exact same raw data set will be used in this section as has been used earlier chapters. But the various measurement models need not be the same.

If we have a sensor suite including what we have called LIDAR, there is also a high quality regime sensor. This sensor measures the maneuver mode of the forward regime with a quality matrix $\mathbf{D}$. In Chap. 3 and following, $\mathbf{S}$ was of dimension three, and $\mathbf{D}$ was a 3×3 matrix: $\mathbf{D}_{ij} = \mathbb{P}(z[k+1] = e_i | \phi[k+1] = e_j)$. The specific $\mathbf{D}$ we used there was

$$\mathbf{D} = \begin{bmatrix} 0.8 & 0.1 & 0.1 \\ 0.1 & 0.8 & 0.1 \\ 0.1 & 0.1 & 0.8 \end{bmatrix}. \tag{5.8}$$

In this engagement, the number of intrinsic regimes is still three, and the actual data sequence $\{\mathscr{Z}[k]\}$ is that we have used before. But in the expanded regime space, $\phi[k+1]$ can take on nine values. So the regime sensor measures the intrinsic regime proxy, which is still of dimension three, while the modal prefix is in the expanded space. For the γ_x-GWE, we use the 3×9 matrix $\mathbf{D}_y$:

$$\mathbf{D}_\gamma = \mathbf{D} \otimes [111].$$

In what follows then, $\{\mathscr{G}[k] = \mathscr{Y}[k] \vee \mathscr{Z}[k]\}$ is unchanged from our earlier work—or more precisely, changed only to the degree that the tracking geometry is changed. We will call this tracker the γ_x-GWE to contrast it with the Markov GWE. The response of the γ_x-GWE is shown in Fig. 5.3 and is little changed from the M-GWE shown in Fig. 3.11.

The γ_x-GWE necessarily has a bigger radial error than does EKF C. This is not surprising since the latter knows the regime and the former can only guess. The renewal model uses $r_i = 3$ for each regime, and the density about the mode is rather broad. Moreover, the γ_x-GWE uses $L = 3$, and this is only 6 s of running time. The regime proxy stays six time samples in the turn mode (12 s), and this may be too long for the memory implicit in the γ_x-GWE to be effective. But the difference is not great. A sample function of the radial error is presented in Fig. 5.4 along with that of the cooperative EKF. The median tracking error of the γ-GWE is 7.69 m. This exceeds the median error of the cooperative EKF by only 0.29 dB: the error increase of the M-GWE is 0.65 dB (not shown).

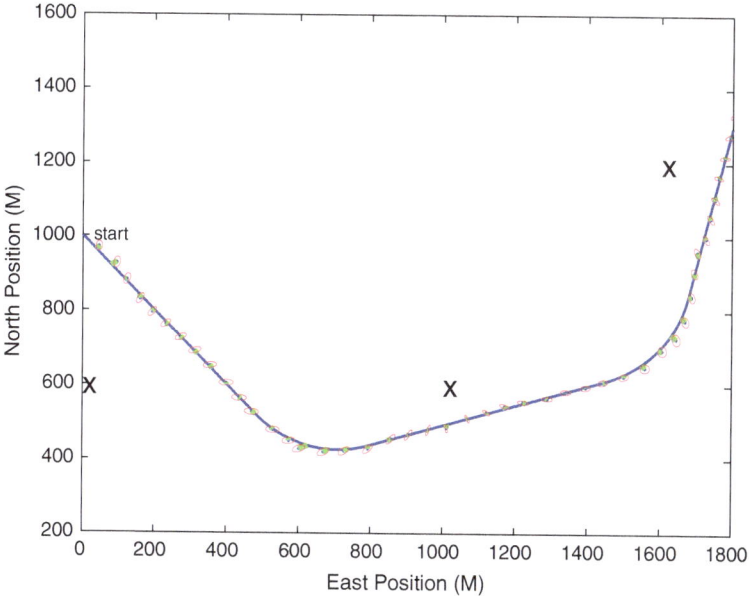

Fig. 5.3 The γ-$\mathscr{G}[k]$-GWE is little improved from the Markov GWE

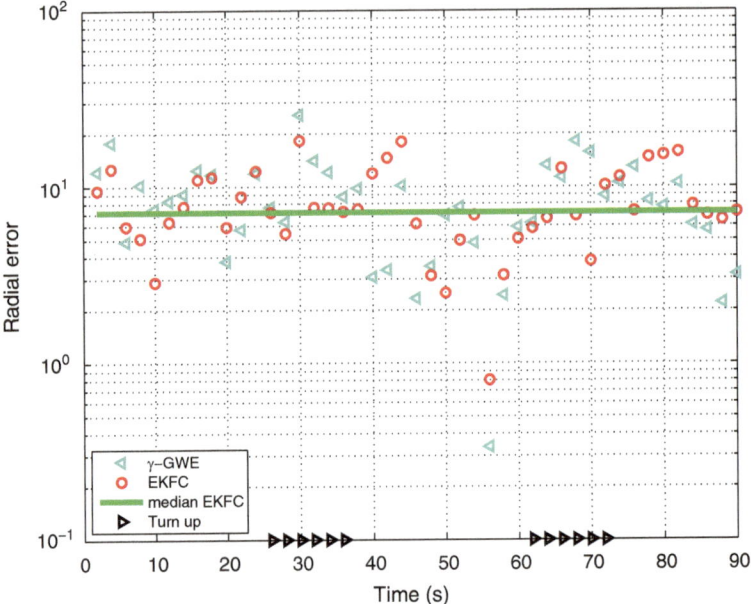

Fig. 5.4 The radial error of the γ_x-GWE is comparable to that of the cooperative EKF

The maximum location error is important in pointing-and-tracking applications because the tracker will loose lock if the placement error is too big. The peak error of EKF C is under 19 m while that of γ-GWE is 25 m, a 1 dB increase. But this large error occurs only once.

As we saw in the preceding chapter, identification of the motion mode is important for generating high quality assurance regions for target intercept [22]. In this regard, the γ-GWE is superior to the M-GWE. Figure 5.5 shows the notional probability of a counterclockwise (left) turn for both the γ-GWE and the M-GWE. The initial portion of the path is CV, and both trackers keep $\{\hat{\phi}_1[k]\}$ below 5 %. The γ_x-GWE has a slight false positive, but this is quickly corrected.

The first turn occurs at $t = 24$: the regime proxy at $t = 24$ s is $\phi_t = e_2$ but the regime is $\phi_t = e_1$ over most of the $[24, 26)$ interval. The γ-GWE recognizes the turn at $t = 24$ s—the GWE has a forward measurement of ϕ_t—and by $t = 26$ s, it assigns a 70 % probability to $\phi[26] = e_2$. The M-GWE is more hesitant. It is not until $t = 32$ s that the M-GWE assigns a probability of more than 20 % to $\phi_t = e_2$.

This indecisiveness in the M-GWE is clear in the turn interval. The γ-GWE assigns a 100 % probability to $\phi_t = e_1$ on the interval $[28, 34)$ while the M-GWE never even gets to a 60 % probability.

The γ-GWE is more volatile than is the M-GWE. The γ-GWE suspects a turn during the initial CV motion more often than the M-GWE, though the level seldom exceeds a few percent.

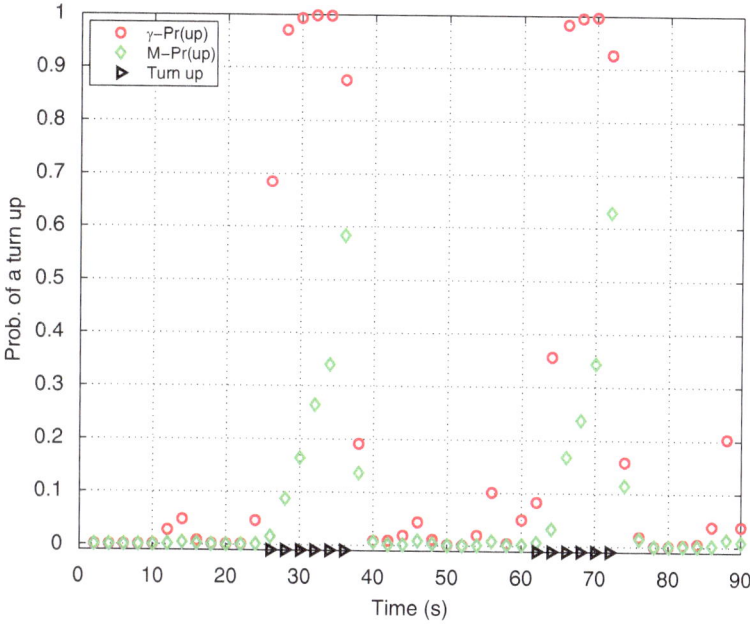

Fig. 5.5 The $\mathscr{G}[k] - \gamma_x$-GWE is quicker to identify a turn than is the $\mathscr{G}[k]$-Markov-GWE

To understand the alacrity of the γ-**GWE**, let us look at the left-turn regime. The intrinsic regime has three sub-regimes which we label: e_1, e_2, e_3. The target enters into a turn from a CV motion. With high probability, the ingress regime is three: $\phi_t = e_3$. Within the left turn, the regime progresses $e_3 \mapsto e_2 \mapsto e_1$—though the time-discrete process can jump across regimes. Again with high probability, the target exits the turn from e_1 and moves to e_6.

Figure 5.6 shows a plot of the $\mathscr{G}[k]$-probability of a left turn (shown with a red $\circ$) along with the probabilities of the three sub-regimes: the former is the aggregate of the latter. The entry regime, $\phi[k] = e_3$ is the quickest to recognize a turn; $t = 24\,\text{s}$. Quickly $\{\hat{\phi}_3\}$ moves to 50 %, and it carries most of the probability of the turn. But $\{\hat{\phi}_3\}$ is also the quickest to signal a false turn; e.g., at $t = 12\,\text{s}$.

As time passes within the turn, the other sub-regimes begin to play a role. Midway though the first turn, $\hat{\phi}_2 = e_2$ becomes dominant, though it never reaches a probability of 50 %.

It is interesting to observe that the probability of $\phi_t = e_3$ declines well before the end of the turn: the mean sojourn time in $\phi_t = e_3$ is only 4 s. However, the lifetime of $\hat{\phi}_2$ is rather broad at a 40 % level. Once the kinematic measurements or the regime measurements suggest a return to CV-motion, the probability that $\phi_t \in e_1 \cup e_2 \cup e_3$ drops precipitously. The γ_x-**GWE** knows the exit from the turn is from $\phi_t = e_1$, and it displays this with a plunge in $\hat{\phi}_1$ at $t = 38\,\text{s}$.

One measure of the notional accuracy of a tracking algorithm is the area of the 1σ-location error ellipses. The cooperative tracker provides a baseline since

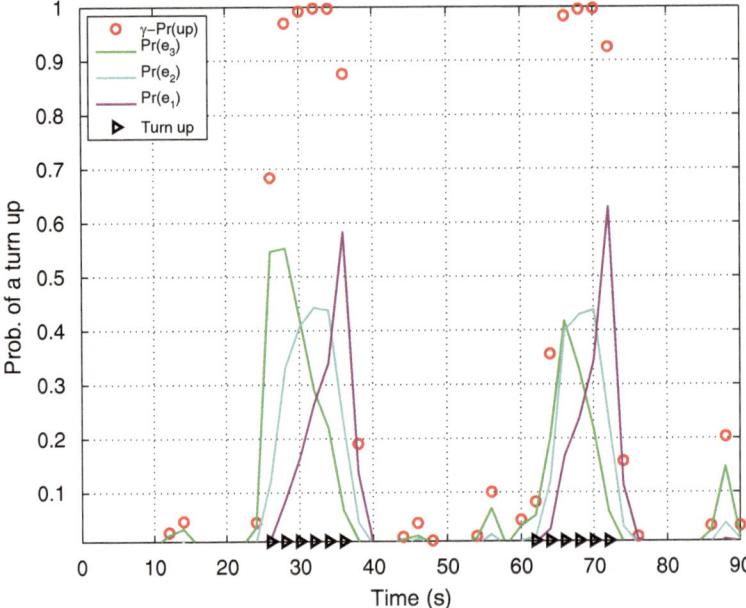

Fig. 5.6 The γ-GWE cycles through a sequence of sub-regimes within an intrinsic regime state. The left-turn motion contains three sub-regimes: (e_1, e_2, e_3)

it knows the regime transitions as they occur. The confusion of the regime proxy and the dominant regime during the interval $t \in [24, 26]$ is avoided: $\phi_{24} = e_2$ and $\phi_{24+} = e_1$. The 1σ-confidence area of the cooperative tracker is thus used as a normalizing factor for sizing the error ellipses in Fig. 5.7. The figure shows the normalized assurance regions for both the γ-GWE and the M-GWE tend to be looser than that of the cooperative tracker. There are certain geometries that favor both GWE trackers. The γ-GWE is more confident than the M-GWE in essentially all cases, though the improvement is only a few percent.

5.4 Target Classification

5.4.1 Friendly or Hostile

In some applications, tracking and target identification are concurrent tasks. For example, a target may be observed to be moving toward a defended position. The target could be friendly, TrF, or it could be hostile, TrH. At detection, there is uncertainty regarding the target type. The defense system must follow the target motion while simultaneously identifying which of the two target classes is the correct one [20, 21].

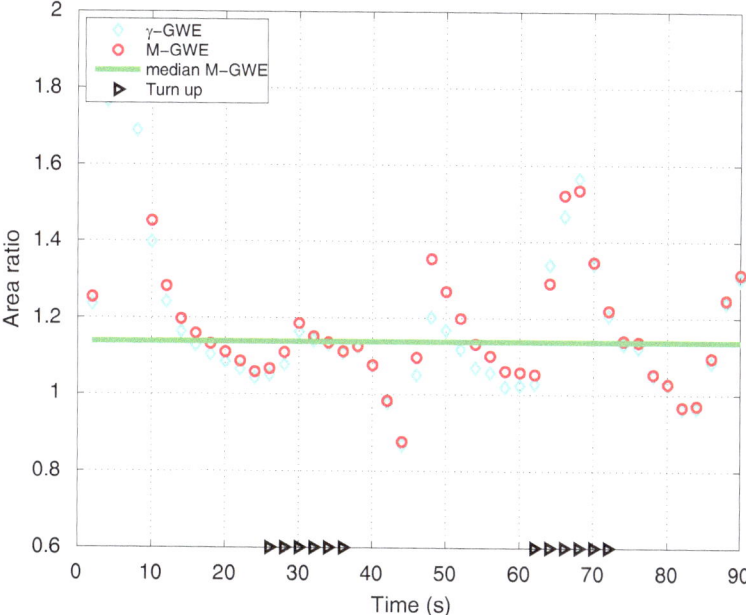

Fig. 5.7 The notional accuracy of the γ-GWE tracker is only slightly better than that of the M-GWE. The cooperative engagement provides the normalizing factor

There are various data sets that, if available, would assist in the tracking/ID problem. There might be an IFF transmitter aboard the vehicle that transmits a designation signal. The IFF signal might be valid, or it might be compromised. There might be a sequence of video images that could be used to separate a friendly vehicle from its hostile counterpart. Either of these could be viewed as a direct measure of target type.

The kinematic data set is also useful for classification. So, while a friendly target might move smoothly and directly toward the final position, a hostile target might try to mask its intent by moving in an intentionally irregular manner. Such jinking makes TrH harder to track with accuracy, but it cues the tracker to purpose. Thus, the variability of the target path points to the kind of the target being followed.

For convenience, we will call the direct measurements of target type "video," and the kinematic measurements of the current regime "LIDAR." A classifier that has a measurement architecture containing both could use the LIDAR to measure jinking events and the video to type the target directly.

The supplemental measurement streams, if they exist at all, will be aggregated into the data set $\{\mathscr{L}[k]\}$. So we might have the point target measurements alone; i.e., $\{\mathscr{Y}[k]\}$. Or the combined data set with LIDAR and/or video; i.e., $\{\mathscr{G}[k]\}$.

The performance metrics in the tracking/ID problem are more nuanced than we have needed in the conventional tracking problem. In order to allocate countermeasures, we must have a high quality estimate of location; e.g., a tight assurance region. But we must also classify the target expeditiously in order to know if countermeasures are required. So, both the tightness of target placement and the quickness with which we can state with confidence that the target is TrH are important—while keeping the probability small that we classify a friendly target as hostile.

In the next sections, we will use the GWE to explore target identification in the context of the TrF vs. TrH dichotomy. In the examples that follow, the target is friendly (though not cooperative), with the congenial motion displayed in Fig. 5.3. We will contrast the influence of the passive motion model for a friendly target with a more energetic motion model for a hostile target. We will be interested in both radial tracking error and the delay in properly identifying the target type.

Initially, the tracker views the target classes to be equally likely, and the motion to be CV: $\hat{\phi}_2 \approx 1$. The accuracy and alacrity of the tracker will depend upon the specific data set. For example, we could use $\{\mathcal{Y}[k]\}$ to track the target and use the $\mathcal{Y}[k]$-mean path to separate a friendly motion from a hostile one on the basis of the perceived volatility. In this architecture, tracking and target ID would be separated. This is not prudent because tracking and classification are synergistic.

In what follows, we pose a tracking/classification problem of the following sort. A friendly target is approaching an objective at 1800E. Without knowing whether the target is friend or foe, the GWE-tracker must follow the target about as well as it did in the single target engagement detailed above. Further, the classifier must properly label the target as friendly with probability 80 % within 20s of detection—this is before the first left turn. The latter specification will be harder to meet.

5.4.2 Common Lifetimes and Regimes

5.4.2.1 Range-Bearing

To illustrate some of the issues in target classification, let us consider a tracking/ID problem in which the target classes have the same kinematic regimes and the same mean sojourns in each motion mode. The target is friendly and follows the path we have delineated above: constant speed with intervals of CV separated by 5°/s turns. We will suppose that the tracker can avail itself of only the range-bearing data sequence we used earlier. This is a hard problem. The raw measurement sequence does not provide a clear picture of the modal state. It is the tempo of the regime string alone that separates the two target classes.

At the intrinsic level, the GWE uses a temporal model for the regime state that is three dimensional. The generator used for the hostile in this example is that we used in the M-GWE—see (5.3). Equation (5.6) derives from the mean sojourn times in each of the motion modes. The M-GWE was shown in Chap. 3 to be an effective estimator. Lacking a direct regime measurement, the M-GWE had a tracking error far less than an EKF using the same measurement data sequence.

The friendly target has the same intrinsic regime structure. Indeed, we will assume that the friendly target has the same mean sojourn times in each of the regimes given in (5.3). We pointed out that the Markov regime model creates an expectation of many brief sojourns along the target path. If the target is friendly, we would expect few such intervals. In this exercise, we will use the more subdued regime model we chose earlier for the γ_x-GWE. The γ_x-tracker is based upon the 9×9 rate matrix, Q_{γ_x}, given in (5.4). The intrinsic modes of the friendly target have γ-lifetimes with the common profile: $r_i = 3; i \in \mathbf{S}$. The γ-GWE views the regime sojourns as more predictable than does the M-GWE. This is reflected in a slightly reduced tracking error and a significantly improved regime estimate.

However, when there are multiple target types, the situation facing the tracker is more complex. Each target type has a distinctive temporal model. The tracker must follow the target without initially knowing which of the modal models is the correct one.

This engagement is particularly hard. Both target classes have the same intrinsic regimes. Both target classes have the same mean lifetimes in each regime. They are distinguishable only by their sojourn density. But the internal memory in the GWE is too short to cover even the most brief sojourn on the sample path.

To address the target ID problem, let us expand ϕ_t into two groupings. The first grouping of components is associated with the friendly target. It is of dimension nine. There are three intrinsic regimes and $r_i = 3$ for each regime. This is the modal model for the γ_x-GWE in the previous section.

The hostile target is expected to have a higher proportion of brief sojourns because of its evasive motion. The state aggregate associated with TrH is three dimensional: three regimes and $r_i = 1$ for each regime. The regimes associated with the hostile target do not communicate with the friendly regimes. Thus, the total number of regimes is twelve: nine are friendly and three are hostile.

We can quantify the comprehensive modal model using the generators we have introduced earlier. In the two target-class problem (2T), the generator, Q_{2T} is

$$Q_{2T} = \begin{bmatrix} Q_\gamma & 0 \\ 0 & Q \end{bmatrix}. \tag{5.9}$$

In contrast with the preceding regime models, the states associated with $\{\phi_t\}$ do not form a single communicating class. Instead, the regime space is partitioned into two disjoint subspaces: TrF is represented by the first nine components of ϕ_t; TrH is represented by the components ten through twelve. Within $\mathbb{R}^{12}$, certain transitions are not permitted; e.g., $\phi_t = \mathbf{e}_i; i \in (1, \ldots, 9) \nrightarrow \phi_t = \mathbf{e}_p; p \in (10, 11, 12)$. The friendly target is differentiated from hostile target only by the paucity of short sojourns in its operational modes (there is actually some difference in the tail events too). This, despite their common mean sojourn time.

The raw dimension of κ is $12^3 = 1728$. As pointed out above, κ is not a single communicating class; e.g., $\iota = (e_1, e_1, e_{10})$ is not permitted. Hence, the number of active local estimators in the GWE is far less than 1728: $9^3 + 3^3 = 756$. While the

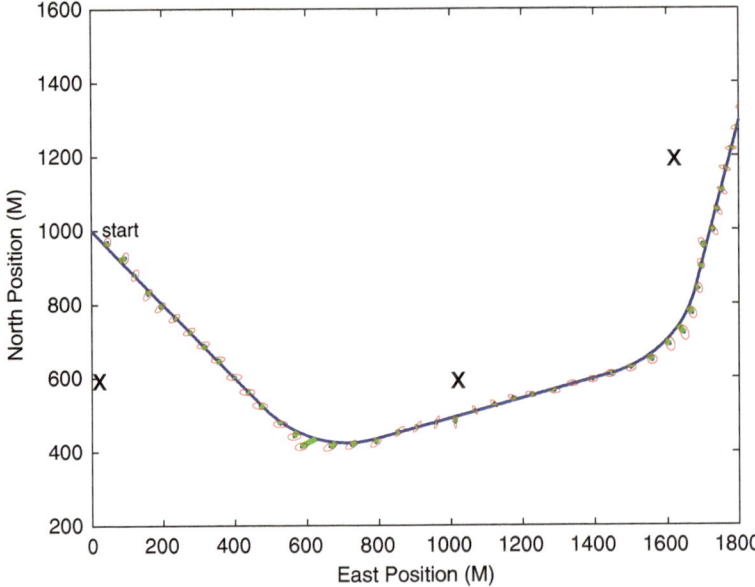

Fig. 5.8 The two target $\mathscr{Y}[k]$-GWE) faces additional uncertainty until it can resolve the target class

number of required local processors in the **GWE** is only 40 % of the maximum size of κ, we will implement the **GWE** using the full 1728 estimators. Of course, over half of them carry a $\hat{\phi}_i \equiv 0$ likelihood. Lacking pruning, the algorithm runs slower than necessary. This would not be a satisfactory operational implementation of the tracker.

In this two-target problem, the first three components of ϕ_i are: TrF-left exit, TrF-left middle, TrF-left entry, and so on through the list of 12 possibilities. The (2T-$\mathscr{Y}[k]$)-GWE (two-target $\mathscr{Y}[k]$-GWE) faces the task of identifying the vehicle type using noisy estimates of path deviations. Figure 5.8 shows the 1σ-assurance regions of $\{\hat{\mathbb{P}}[k]\}$ as the target moves along the path. The median radial error is 9.43 m. This is slightly better than the median error of 9.71 m of the M-GWE using the same data set: the TrF model is better than that used in the M-GWE, but the (2T-$\mathscr{Y}[k]$)-GWE must allow for designation uncertainty. Both are about 1 dB worse than the cooperative **EKF**. Thus the (2T-$\mathscr{Y}[k]$)-GWE meets the system requirement.

In this engagement, the **GWE** must discriminate the friendly target from its hostile counterpart. The basis of this separation is the length of its sojourn intervals: long suggests friendly and short suggests hostile. Unfortunately, the memory in the **GWE** is too short to readily distinguish the sojourn times. Figure 5.9 shows a sample plot of $\mathbb{P}$(target is friendly) $= \mathbb{P}_F[k]$. The engagement begins with the target moving with constant velocity, and $\{\mathbb{P}_F[k]\}$ is approximately 50 %. For the first 30 s,

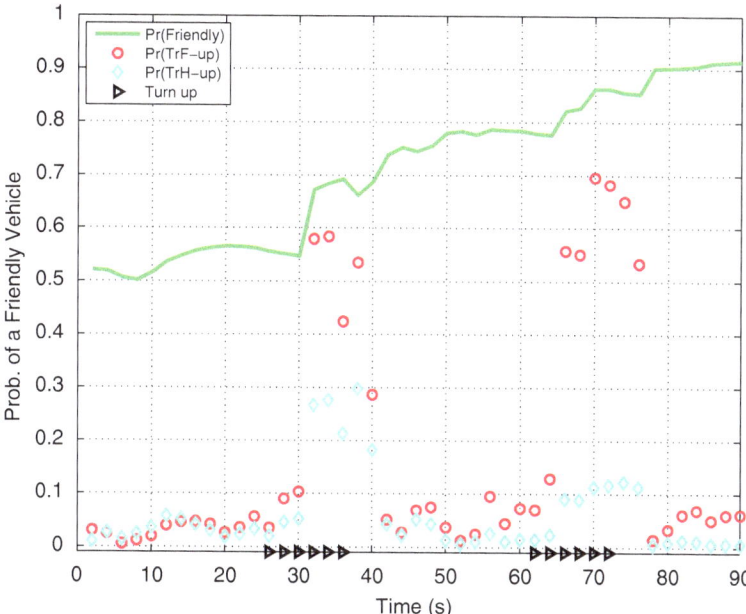

Fig. 5.9 The two target $\mathscr{Y}[k]$-GWE takes over a minute to correctly classify the friendly target at an 80 % confidence level

the **GWE** does not make a clear distinction between alternatives: $\mathbb{P}_F[k] \approx 55\%$. The **GWE** memory is only 6 s and a 24 s sojourn is compatible with both TrF and TrH: the long CV interval is not recognized by the **GWE** as particularly friendly behavior.

It is at the regime transitions that the target types separate themselves. A turn begins at $t = 24$ s but the regime proxy doesn't change until $t = 26$ s. The **GWE** does not begin to favor TrF until $t = 32$ s. Figure 5.9 also shows $\mathbb{P}(\text{left turn}|\text{TrF})$ and $\mathbb{P}(\text{left turn}|\text{TrH})$. The aggregate, $\mathbb{P}(\text{left turn})$ is the sum of the two probabilities. The **GWE** must wait until the path clearly deviates from the quiescent motion 6 s into the turn. At the point of recognition, there is a positive increment in $\{\mathbb{P}_F[k]\}$.

The γ-distributed portion of the tracker is more sensitive to the regime change than is the Markov portion: the transition rate out of ϕ_4 is three times the transition rate out of ϕ_{11}. This differential rate causes the big boost in $\mathbb{P}(\text{left turn})$ when a turn is suspected. This is reflected as a jump in the probabilities at $t = 32$ s and to a lesser extent at $t = 66$ s. TrF is heavily favored over TrH by this latter point. After the turn, $\mathbb{P}_F[k] \approx 75\%$ with a slow drift up as the target returns to a coast path. The influence of the regime transitions events are deweighted in the Markov portion of the **GWE** by the decline in $\mathbb{P}(\text{TrH})$.

Over time, the GWE does an effective job in target ID, but it takes over a minute to identify the target as friendly with an 80 % confidence. If we must counter a threat in an expeditious manner, this delay may well be unacceptable. This architecture does not meet specifications.

5.4.2.2   Range-Bearing and LIDAR

In the previous subsection, we classified the target exclusively on the basis of noisy location measurements. The judgement that the target is friendly is based upon the differential sojourn densities in the two target classifications. This is a subtle distinction since the regime lifetime is never clearly distinguishable in $\{\mathcal{Y}[k]\}$. For example, during the initial section of quiescent flight, $\{\mathbb{P}_F[k]\}$ slowly drifts up, but even at $t = 30$ s, the probability of a friend is only 55 %.

The turn events cause significant increments in $\{\mathbb{P}_F[k]\}$. Unfortunately, the $\mathcal{Y}[k]$-tracker is slow to identify the occurrence of a regime transition and lacks confidence in its own regime estimate. We could expedite the recognition of a turn event if we augment $\{\mathcal{Y}[k]\}$ with a LIDAR measurement: $\mathcal{Y}[k] \vee \mathcal{Z}[k] = \mathcal{G}[k]$. A quicker recognition of a change in motion mode should be reflected in a quicker target ID. We will call this tracker the (2T-$\mathcal{G}[k]$)-tracker to distinguish it from the (2T-$\mathcal{Y}[k]$)-tracker of the previous subsection.

We will use the same range-bearing sequence and the same rotation rate data sequence we use heretofore. We will assume latency is short, and the regime measurement is linked to the regime proxy in the forward interval; i.e., $\mathbf{e}_p$. In this case,

$$z[k + 1] = \mathbf{D}\phi_p[k + 1] + \eta[k + 1], \tag{5.10}$$

where $\mathbf{D}$ is the discernibility matrix.

The data set $\{\mathcal{Z}[k]\}$ derives from the intrinsic regimes of which there are but three: see (5.8). The GWE uses a $\mathbf{D}$ of dimension 3×12 in which the various labels for the intrinsic regime are aggregated.

Let us denote the notional probability of misclassifying the target as hostile as $\{\mathbb{P}_H[k]\}$. Target misclassification is serious as we do not want to use our countermeasures on a friendly target. Figure 5.10 displays the response of the (2T-$\mathcal{G}[k]$)-GWE-classifier along with the earlier (2T-$\mathcal{G}[k]$)-GWE-classifier. When we contrast the (2T-$\mathcal{G}[k]$)-classifier with the (2T-$\mathcal{Y}[k]$)-classifier, we find that the sample functions of $\{\mathbb{P}_H[k]\}$ for both algorithms are quite close initially. The regime data stream simply confirms that the target is not turning. While useful in tracking, this does little to distinguish the target class since both TrF and TrH move in the CV mode.

The turn events cause significant increments in $\{\mathbb{P}_H[k]\}$. Because the (2T-$\mathcal{G}[k]$)-tracker is quicker to identify regime changes, it has an advantage over the (2T-$\mathcal{Y}[k]$)-tracker. This is true for both the CV$\mapsto$turn and the turn$\mapsto$CV events.

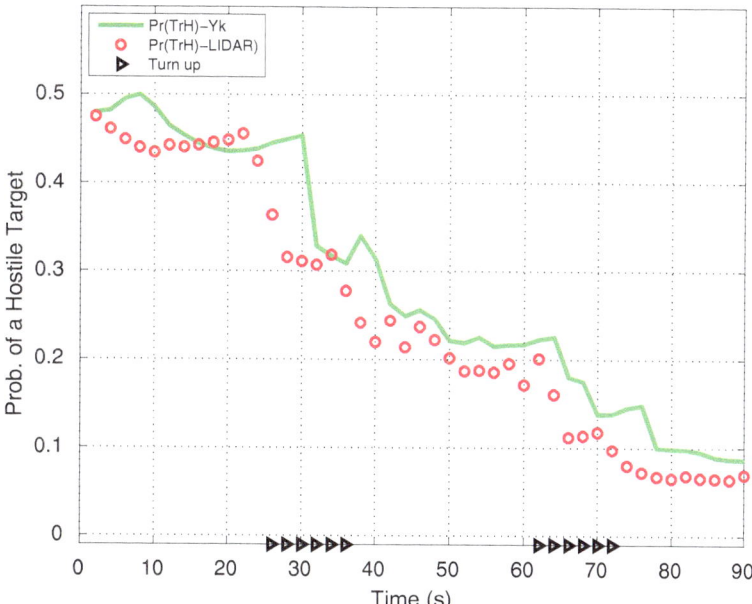

Fig. 5.10 The two-target $\mathcal{G}[k]$-tracker is much faster to reduce the likelihood of classifying the target as hostile at the 30 % level than is the two-target $\mathcal{Y}[k]$-tracker

Because the $\mathcal{G}[k]$-tracker is able to identify and respond to the turn circa 26 s, it is able to reduce the time to classification at the 70 % level from about 40s to 25s. During long sojourns the difference is slight since the (2T-$\mathcal{Y}[k]$)-tracker has time to correct itself. But the $\mathcal{G}[k]$-classifier still fails in this application since it takes 50s to identify the target at 80 % confidence. Though faster than the $\mathcal{Y}[k]$-classifier, the system specifications are not met.

After a minute and a half, the two classifiers agree with 90 % confidence that the target is a friend. Eventually, both algorithms would reach surety, but the figure suggests that convergence is slow.

5.4.2.3   Range-Bearing, LIDAR and Video

Given enough time, the **GWE** is able to distinguish target type from either the location data set, $\{\mathcal{Y}[k]\}$, or from the augmented data set, $\{\mathcal{Y}[k] \vee \mathcal{Z}[k]\}$. Increments in $\{\mathbb{P}_H[k]\}$ are concentrated near regime events. To the degree that the (2T-$\mathcal{G}[k]$)-**GWE** is faster to recognize these events, target ID is also faster. But both trackers have significant delay. If 80 % confidence is required before the target is accepted as friendly and countermeasures are deactivated, both classifiers have excessive delay.

The target classes are hard to distinguish because the motions have little to distinguish themselves. Although the $(2T\text{-}\mathscr{G}[k])$-GWE can identify the sojourn times with more fidelity than can the $(2T\text{-}\mathscr{Y}[k])$-GWE, this is a rather subtle distinction. One way to reduce the identification delay is to have an auxiliary measurement of target class. For example, suppose the hostile target is a jeep or other vehicle of small extent. If this is in contrast to a much larger friendly armored personnel carrier, the range-adjusted target silhouette points toward the target class independently of the motion. An unreliable IFF signal would have this same effect. We call such signals direct category measurement, and refer to them collectively as video. As described, video is a two-dimensional pointer vector: e_1 is associated with TrF and e_2 is associated with TrH.

To frame this identification problem within the context of the GWE, we will retain the same regime generator, Q_{2T}. Recall that the regime process is based upon a friendly target with $\gamma = 3$ sojourn times and a hostile target with Markov sojourn times. The dimension of ϕ_t is 12, with components one through nine associated with TrF and ten through twelve associated with TrH.

Suppose that the regime observation is of two parts: $z_1[k+1]$ points toward the forward motion regime and its dimension is three; $z_2[k+1]$ points toward a target type and its dimension is two. The regime measurement, $z[k+1]$, of the previous sections corresponds to $z_1[k+1]$, while $z_2[k+1]$ augments the data set.

Let us suppose that $z_2[k+1]$ is correct with probability $\mathbb{P}_\diamond$: if $\mathbb{P}_\diamond = 0.67$, $z_2[k+1]$ properly classifies the target 67 % of the time and misclassifies 33 % of the time. We will assume that these probabilities are the same for both target types, though this assumption can easily be weakened. The $\{z_2[k+1]\}$ will be assumed to be temporally independent when conditioned on the true (regime, type) and will be independent of all other measurements with the same conditioning.

The dimension of $z[k]$ is now six: component 1 signifies "friendly left," component 2 signifies "friendly CV," ... component 6 signifies "hostile right." The discernibility matrix for the two-target problem, $\mathbf{D}_{2T}$ is 6×12:

$$\mathbf{D}_{2T} = \begin{bmatrix} \mathbb{P}_\diamond * \mathrm{kron}(\mathbf{D}, [111]) & (1 - \mathbb{P}_\diamond) * \mathbf{D} \\ (1 - \mathbb{P}_\diamond) * \mathrm{kron}(\mathbf{D}, [111]) & \mathbb{P}_\diamond * \mathbf{D} \end{bmatrix}. \tag{5.11}$$

For example, if $\mathbb{P}_\diamond = 0.65$,

$$\mathbf{D}_{2T} = \begin{bmatrix} 0.52 & 0.52 & 0.52 & 0.07 & \cdots \\ 0.07 & 0.07 & 0.07 & 0.52 & \cdots \\ 0.07 & 0.07 & 0.07 & 0.07 & \cdots \\ 0.28 & 0.28 & 0.28 & 0.04 & \cdots \\ 0.04 & 0.04 & 0.04 & 0.28 & \cdots \\ 0.04 & 0.04 & 0.04 & 0.04 & \cdots \end{bmatrix}. \tag{5.12}$$

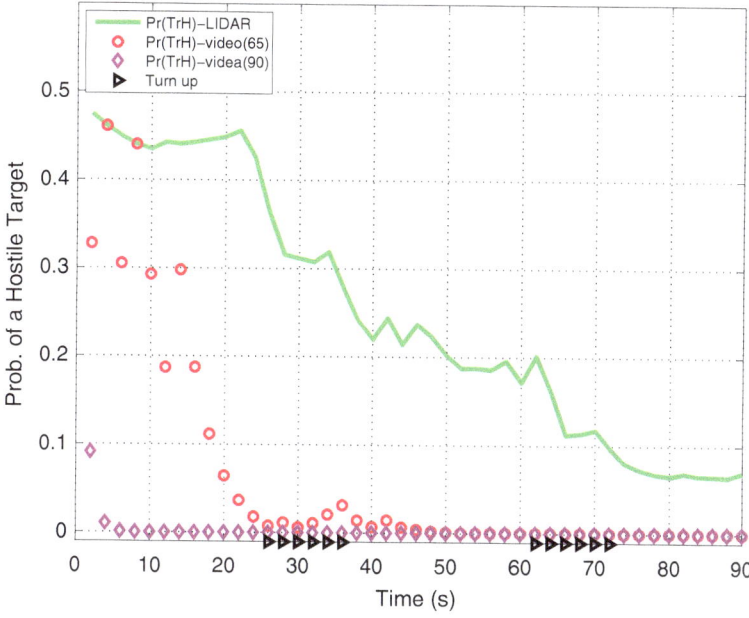

Fig. 5.11 A classificational measurement reduces the time required to identify the target as friendly, and it increases the confidence in the choice

If the target is friendly and turning left (up), it will be so reported 52 % of the time. The classification errors total 48 % with the most common at 28 % being the statement: "hostile-turning-up."

We will call this augmented data set $\{\mathscr{G}[k]^*\}$: $\mathscr{G}[k]^* = \mathscr{Y}[k] \vee \mathscr{Z}_1[k] \vee \mathscr{Z}_2[k]$. Figure 5.11 shows the response of the $(2T\text{-}\mathscr{G}[k]^*)\text{-}\mathbf{GWE}$. All three plots use exactly the same kinematic data set we have used earlier; i.e., $\{\mathscr{Y}[k] \vee \mathscr{Z}_1[k]\}$ is identical throughout. The trackers differ only in the quality of the sensor generating the classificational data.

When $\mathbb{P}_\diamond = 0.65$, the delay in identifying a friendly target at the 80 % confidence level is reduced from 50 to 12 s. Indeed, when $\mathbb{P}_\diamond = 0.90$ the **GWE** achieves 80 % confidence after one observation; the target is identified from the classificational measurement alone. If $\mathbb{P}_\diamond = 0.50$, there is no classificational measurement, and $\{\mathbb{P}_H[k]\}$ is that shown in the previous subsection. Clearly, a tracker based upon $\{\mathscr{G}[k]^*\}$ is far better than one based upon $\{\mathscr{G}[k]\}$. At levels above $\mathbb{P}_\diamond = 0.65$, target ID can be treated without the kinematic measurements with little loss; tracking and target ID are separated.

5.4.3 Distinctive Kinematics, Range-Bearing Sensors

5.4.3.1 Distinctive Sojourns

In this chapter, we have studied engagements in which there are two target types. A friendly target with relatively subdued motions, and a hostile target with more volatile motions. This characterization is evocative, but in the cases studied, the intrinsic motion modes and the mean sojourn times in each motion mode were identical for both target classes. The distinguishing feature of a target class was not the length of time spent in a motion mode, but rather the paucity of short lifetimes predicted for the friendly target. Using the range-bearing data set, the GWE can distinguish friend from foe but only after an extended period. Of course, when direct measurements of target type are available, target identification is much faster.

Let us now consider an engagement in which the kinematic distinctives are more pronounced. Again we have a friendly target approaching its objective at $(1800, 1400)$. We will use the $\gamma(3)$-renewal model to represent the motion of TrF. We will suppose a hostile target would be evasive and would use jinking to avoid countermeasures. We will model this energetic hostile target with a Markov model having lifetimes but a quarter of those of the friendly target while retaining the properties of the untimed modal string. The regime generator of the two target problem is now

$$Q_{2T} = \begin{bmatrix} Q_\gamma & 0 \\ 0 & 4*Q \end{bmatrix}. \tag{5.13}$$

Even with the modified Q_{2T}, the probabilistic structure of the untimed regime sequences for TrH and TrF is identical. Hence, we cannot distinguish one target from the other by its regime string. Rather the timing of the regime transitions is the only distinguishing feature.

Figure 5.12 shows a plot of $\{\mathbb{P}_H[k]\}$ based exclusively upon $\{\mathscr{Y}[k]\}$: there is no regime measurement. This $(2T\text{-}\mathscr{Y}[k])$-GWE achieves target ID solely on the basis of the mismatch of the sojourns of the two target types. But in contrast to the previous section, the sojourns have different lifetimes. The figure contrasts the classificational response when the hostile target has the same mean sojourn times; indicated by "common sojourns," with the classificational response when the hostile target has the shorter mean sojourn times: indicated by "short sojourns." The shorter sojourns of the hostile target arise both because of the exponential distribution of the sojourns times and because the mean hostile sojourn times are a quarter of those of TrF.

The figure shows that $\{\mathbb{P}_H[k]\}$ initially moves above 50 %. But $\{\mathbb{P}_H[k]\}$ quickly falls. In contrast to the earlier example in which the length of the initial sojourn didn't serve to separate friendly from hostile, the longer the target remains in the CV-regime, the higher is the probability the target is a friend. The $\{\mathbb{P}_F[k]\}$ process rises above 80 % before the first turn: the baseline probability is only about 55 % at this time. The $(2T\text{-}\mathscr{Y}[k])$-GWE achieves an 80 % confidence in only 25 s and 90 % confidence in 34 s. Although it does not meet the system specification, the range-bearing GWE comes close.

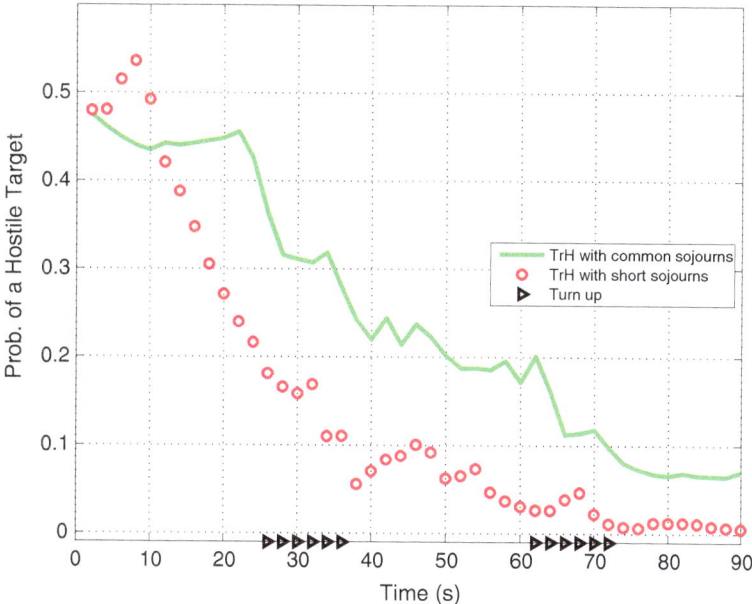

Fig. 5.12   An energetic hostile target is more easily distinguished from a friendly target than is its more sedate kin

5.4.3.2   Distinctive Sojourns and Regimes

Up to this point, the tracking/ID problem has been posed as one in which the motion modes are the same for both target types; e.g., when there is a transition out of quiescent motion, both TrF and TrH turn at a 5°/s . This makes it difficult to use $\{\mathscr{Y}[k]\}$ to disassociate a friendly target from a hostile one since both separate from the CV extrapolation at the same rate. Suppose, however, that the hostile target is both quicker to change motion modes:

$$Q_{2T} = \begin{bmatrix} Q_\gamma & 0 \\ 0 & 4*Q \end{bmatrix} \tag{5.14}$$

and more agile, as evidenced by a higher turn rate: the turn rate of a hostile target is 20°/s as contrasted to 5°/s for the friendly target.

Introduction of a type-specific regime set complicates the tracking portion of the problem. When unsure of the target class, the local models must extrapolate on different paths even when there is agreement on the kind of motion: left turn means a 20°/s turn rate instead of a 5°/s turn rate. Consequently, the extrapolated uncertainty region becomes bigger and more uneven. As the target class is identified, the uncertainty region becomes more regular.

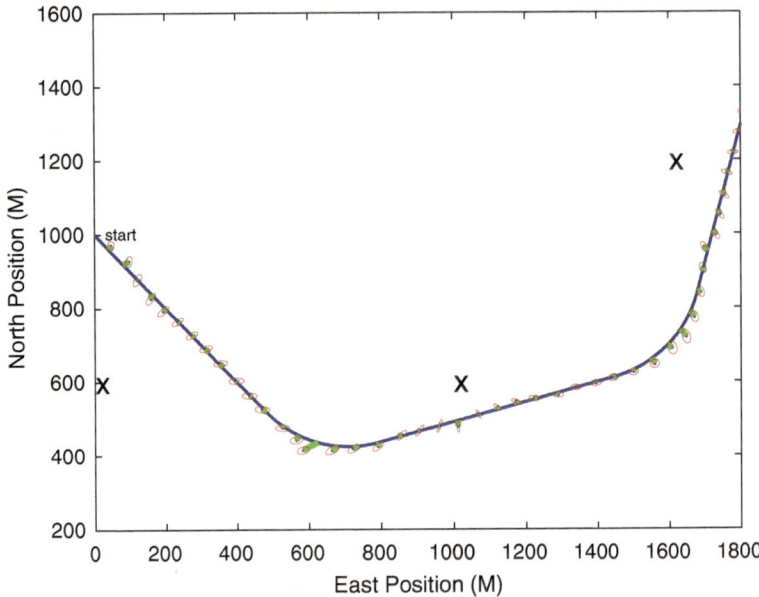

Fig. 5.13 The radial tracking error of the (2T-$\mathscr{Y}[k]$)-GWE differs little from that of the γ_x-GWE

Figure 5.13 shows the response of the (2T-$\mathscr{Y}[k]$)-GWE using the common raw data set. The radial tracking error at the beginning of the engagement is slightly increased over that of the one-target γ_x-GWE. But the change is slight.

While distinctive motion modes make the target harder to track, they are a significant aid to ID. Figure 5.14 shows $\{\mathbb{P}_H[k]\}$ using the standard $\{\mathscr{Y}[k]\}$ data set. The friendly target is contrasted with the baseline hostile target that has the same mean sojourn times. When TrH is agile, the target is identified as TrF at the 90 % confidence level in 18 s. This is in contrast to the baseline hostile target that takes nearly 80 s to achieve the same separation. If the hostile target is merely energetic, the time to identify TrF is still 32 s.

The (2T-$\mathscr{Y}[k]$)-GWE algorithm meets all of the system specifications. A friendly target can be identified quickly from either direct measurements or the elusive strategy of its motion.

5.5 Conclusion

This chapter presents several illustrative examples of use of the GWE where the architect's intent is to track a target, classify it, or both. All of the examples use the same measurement set: either $\{\mathscr{Y}[k]\}$ or $\{\mathscr{G}[k]\}$ as the case may be. The most favorable algorithm for tracking in this application is the EKF C. Target

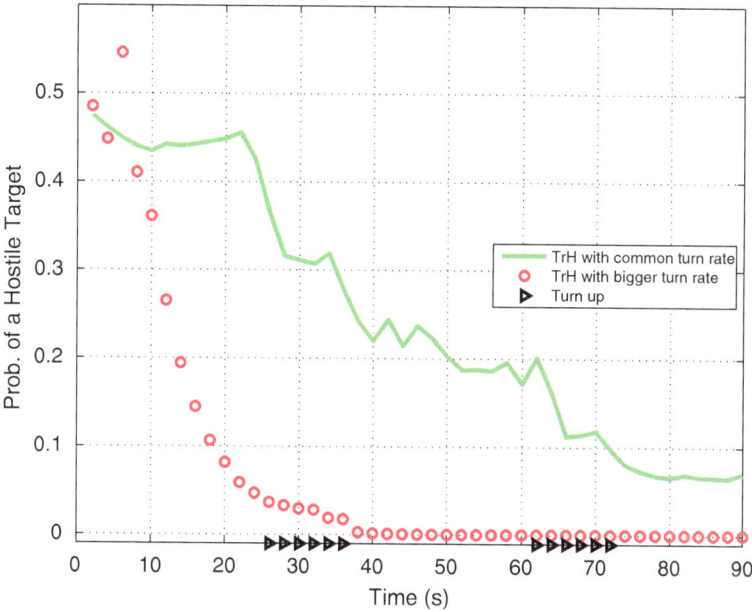

Fig. 5.14 When the hostile target has a high turn rate, it is quickly distinguished from a friend. The more sedate hostile takes much longer to identify

classification is not an issue: the cooperative EKF knows the proper target class, and likewise with target motion mode (it also knows the current motion regime). The tracking error is roughly that of the minimum variance estimator though the nonlinear measurements and the pseudo-noise adjustments argue against taking that too literally.

The least favorable algorithm in this application is the nominal EKF—either with or without pseudo-noise augmentation. The basic EKF-tracker knows neither target motion regime nor the existence of alternative target classes with distinguishing motion characteristics. Instead, the EKF uses a single, *centered* kinematic model within a prespecified target class. In the cases studied here, the specific engagement model is that of conventional CV motion with pseudo-noise introduced to account for the modal uncertainty. The EKF is slow to respond to regime transitions because it has no reason to expect them. The response delays in the nominal EKF can be reduced by increasing the pseudo-noise, but the notional 1σ-error ellipses become correspondingly, and in most cases incorrectly, larger. The response of the nominal EKF gives a credible lower bound on the performance of a tracker in this application.

We would expect the performance of the GWE to lie between normative EKF and the EKF C. The GWE does not ignore the regime changes as does the EKF, nor does it know the regime as does the EKF C. Rather, it acknowledges the possibility of regime events and tries to classify them as they occur. The quality of this classification depends upon the data set at the tracker and on the analytical model that the GWE uses to cue its response to patterns in the observations.

The basic one-target, Markov tracker, (M-$\mathscr{Y}[k]$)-GWE with a string depth of three employs 27 separate local models. We would expect its performance to be superior to the nominal EKF, but less good than the cooperative EKF. The latter can be viewed as a multiple model tracker that always uses the correct local model, reported by the cooperative target, giving it a continuous regime string with length k at time $t = kT$—far longer than the truncated regime histories of length three (6 s) available to the GWE.

There are two useful metrics that contrast the performance of the tracking algorithms. The first is the median radial tracking error. In this application, the median error is superior to the mean error since the latter is overly sensitive to the small number of big errors subsequent to a regime change. The maximum tracking error is also important. In most cases, the tracking window has fixed size. Any error that exceeds the bounds of the window will result in the target being *lost*. A separate search algorithm is required to reacquire a lost target. Search algorithms are time consuming and may fail to find the target in a reasonable time.

Figure 5.15 shows both the median error of EKF C, 7.2 m, and the maximum error, 18.3 m. The nominal EKF has a maximum error that is 6 dB greater than EKF C and a median error that is 2.5 dB greater than the EKF C. Using these two trackers as plausible bounds, the GWE has considerable room for improvement over the nominal EKF.

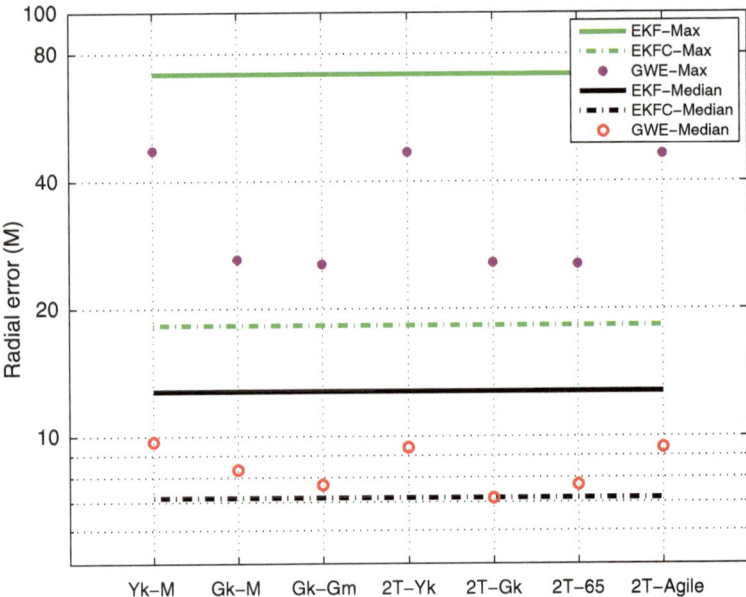

Fig. 5.15 The cooperative EKF and the normative EKF bound the performance of the various GWE architectures

The simplest GWE uses the Markov regime model and the common $\{\mathscr{Y}[k]\}$ data set; it also assumes that the target is friendly. This single-target tracker is labeled $(\mathscr{Y}[k] - M)$-GWE. The $(\mathscr{Y}[k] - M)$-GWE has a maximum error that is 4.1 dB greater than EKF C and a median error that is 1.3 dB greater than the EKF C. This is a significant improvement over the nominal EKF, but not as much as we would like.

If we had a precise regime measurement, we would match EKF C by setting $\hat{\phi}_i = 0$ for those strings that didn't match the regime proxy. The fusion error in the GWE would be avoided because all disparate prefixes would be assigned probability zero. Unfortunately, even with a high quality **D**, the $\mathscr{G}[k]$-GWE is not able to resolve the regime state in real time. The $\mathscr{G}[k] - \text{GWE}$ s are shown in Fig. 5.15 ($\mathscr{G}[k]$-Gm and $\mathscr{G}[k]$-M, with and without the γ_x-model, respectively). Both show a maximum radial error that is about 1.5 dB greater than EKF C and a median error that is 0.7 dB greater than the EKF C for the $(\mathscr{G}[k] - M)$-GWE and 0.3 dB greater than the EKF C for the $(\mathscr{G}[k] - Gm)$-GWE. This is a significant improvement over the $(\mathscr{Y}[k] - M)$-GWE.

In the two-target problem, there are two kinematic hypotheses concerning the target motion. The tracker must identify the vehicle while following it. In the first instance, the friendly target and the hostile have the same regimes *and* same mean sojourn times. The friendly has γ_x-sojourns while the hostile has Markov sojourns. The target classes are difficult to resolve because their kinematic signatures are so close. The two-target, Gamma $\mathscr{Y}[k]$-GWE has a maximum error little different from the one-target, Markov $\mathscr{Y}[k]$-GWE: the advantage of knowing the target class is about equal to the disadvantage of using a Markov regime model. The median error for the two trackers is about the same as well: median error is 1.6 dB greater than the EKF C.

To improve performance, a more complete data set is useful. As we move to a tracker utilizing $\mathscr{G}[k] = \mathscr{Y}[k] \vee \mathscr{Z}[k]$, the performance of the two-target, $\mathscr{G}[k]$-GWE is essentially that of the one-target tracker processing the same data set. The slight improvement over the one-target tracker is simply an artifact of the sample function. Even adding the auxiliary direct classification measurement does little to improve tracking performance.

When the hostile target has distinguishing kinematic features, it is easier to resolve. The hostile target has tighter and more frequent turns. This makes it easier to distinguish the two-target classes, but the local filters associated with the hostile target drag the location estimates away from the actual path. Figure 5.15 shows the two-target, $\mathscr{Y}[k]$-GWE performs about as well as the two-target, $\mathscr{Y}[k]$-GWE in which the friendly and the hostile are more closely matched.

For this specific example, the use of the γ_x-distributed regime model improves performance about as much as having two target classes degrades it. The tracking performance of the GWE is uniformly superior to the nominal EKF in every case studied. Of course there are many combinations of kinematics and measurement quality that we did not investigate. But the various forms of this engagement illustrate the utility of the hybrid problem formulation in target tracking and identification.

Although modeling precision does not lead to proportionate tracking improvement, the fidelity of the regime model and the quality of the regime measurements have a significant effect on the ability of the GWE to identify the target class in an expeditious manner. Any kinematic distinctives and/or a high quality direct measurement of target type yield faster target resolution and more confidence in that resolution. The synergies in tracking and identification should be explored when prompt action must be taken on the basis of a threat.

Finally, the efficacy of a direct, even if noisy, indication of target type, what we called the video sensor, is very clear. Even though the GWE proved able to classify targets from kinematic indications alone, command architectures to support immediate deployment of countermeasures will be more effective if they make timely use of a direct target-type sensor.

Chapter 6
Assurance Regions on a Road Grid

Abstract Most common motion models assume a path in an open, convex region. Suppose, however, we know that the target is constrained to a road grid. The target motion is still in an open region. But the lateral accelerations are much smaller than the longitudinal acceleration. When the target reaches a junction, it may accelerate laterally as it maneuvers through the turn. This change in direction could be included in the location algorithm if we knew precisely where the target was. But this is unknown until after the fact. Thus, the turns come as a surprise to the classifier. In this chapter we show how the GWE can be accommodated to grid motion. But with a hybrid algorithm, intra-junction segments are even easier to navigate than are segments in a tangled environment.

6.1 Introduction

Most of the engagements we have studied thus far involve tracking an agile target moving in an open convex region in the plane; for example, in an open field. We assumed a sensor suite that provided a noisy location measurement every T seconds. We linearized the nonlinear measurement equations and determined a family of local LGM motion/observation models. Within this paradigm, the intensity of the wide band exogenous disturbance was adjusted to account for a variety of exogenous influences both structured and unstructured; e.g., the geometry of the engagement, target type, etc.

We explored the advantage of enhanced sensor architectures. These might include measures of an extended target (LIDAR or video), not just its point location. They might include non-temporal data sets; e.g., motion on a directional road segment. Such augmented algorithms yielded improved tracking, prediction, and classification performance.

In this chapter, we will revisit the problem of tracking and illuminating an agile target. In this engagement, the target moves upon a known road grid. But rather than a continuing on a straight path, the target may or must turn at predetermined road junctions. This stored grid map provides the tracker/illuminator with an important complementary data set.

© Springer International Publishing Switzerland 2016
D.D. Sworder, J.E. Boyd, *Locating, Classifying and Countering Agile Land Vehicles*, DOI 10.1007/978-3-319-19431-8_6

Fig. 6.1 When the target is confined to a road grid, lateral uncertainty is constrained by street width

Figure 6.1 illustrates some of the issues that arise when tracking a grid constrained target. A vehicle is moving on the indicated street map. Suppose the target is proceeding west on the upper (primarily) east-west segment of the map. The target could turn south at a point slightly west of mid-picture. Or it could continue west and turn north (or south) at the west terminus of the segment. The longitudinal motion necessarily is along the street, and the lateral motion is constrained by the street width.

A reductionist modal model for a vehicle traveling on a rectangular grid oriented in the cardinal directions would be to select the direction of motion as the regime state: $S = \{north, \ldots, west\} = \{1, 2, 3, 4\}$. Each regime has an associated kinematic model: the wideband acceleration is much larger longitudinally than laterally. Linking the regime events into a string, the modal state is generated in the usual way; e.g., if $L = 3$, ι is a radix four number of length three. For example, the mode of a target that turned north from west $2T$ seconds ago would be labeled $\phi[k] = (114) : \phi[k]$ is the retrograde segment (north, north, west).

In most of the engagements we have discussed heretofore, handoff conditions were ignored. For example, the agile target moving toward an objective as described in Chap. 5 had a regime change when its turn rate changed. But across such regime events, position and velocity were continuous. To be sure, jerk was discontinuous at a regime transition, but jerk was not a component of the kinematic state.

Engagements with a more complex structure have been mentioned in passing. In one case, a maneuvering target has two motion regimes; constant velocity (CV) and coordinated turn (CT). Suppose that when the vehicle makes a turn (CV↦CT), the vehicle slows because of increased drag. When the target moves again to straight flight (CT↦CV), the speed loss is restored. The kinematic state experiences a predictable change in speed at a regime transition time. This abrupt increment in the kinematic state is delineated with a handoff transformation $\mathbf{H}^+$ that maps $\mathbb{R}^n$ to $\mathbb{R}^n$.

While the handoff transition is a temporal adjustment, there may also be constraints of a non-temporal sort. We will collect these constraints in the data set $\mathbf{C}^{\mathscr{M}}$. For example, in an earlier engagement, a north bound target was confined to a north-south road. In this case the coordinates of this road would be an element in $\mathbf{C}^{\mathscr{M}}$. The target motion is thus limited, and an assurance region need only cover a portion of a narrow road bed.

We will append the *para-measurements*, $\mathbf{H}^+$ and $\mathbf{C}^{\mathscr{M}}$, to the $\{\mathscr{Z}[k]\}$ data set. While the para-measurements can have a complex structure, in this chapter they have a rather simple form. We will consider problems in which handoff depends only on the predecessor and successor regime states at the time of a regime change: if $\iota = (ijl) \mapsto \iota^+ = (pijl)$, $\mathbf{H}^+$ depends only on the pair of modal prefixes (i,p). Similarly, $\mathbf{C}^{\mathscr{M}}$ maps $\mathbb{R}^n$ to $\mathbb{R}^n$ and depends only on the modal prefixes.

We usually think of the presence of handoff conditions as complicating the tracking problem. A random kinematic adjustment at a regime change introduces yet another source of uncertainty. Even in the conventional problem formulation, the assurance regions may be uncomfortably large. An abrupt handoff transformation may make things even worse.

However, para-measurements which restrict the permissible motion space can actually be helpful in target location. A non-temporal relation may interlink $\{x[k]\}$ and $\{\phi[k]\}$ in a manner that restricts the search area of the target—or compresses the assurance region.

To the degree that $\{x[k]\}$ influences $\{\phi[k]\}$, the hypothesis that the regime process is exogenous is contravened. Since the GWE is derived on the basis of a Markov modal model, we will have to adjust the algorithms carefully. For reasons that become clear in the example, we will use the basic hybrid tracking algorithm with adjustments that seem appropriate to the application. These alterations are rather delicate. We will present the performance contrasts in this chapter with side comments on the modified GWE. The detailed development of the grid constrained tracker/illuminator can be found in the appendix.

6.2   The Engagement

6.2.1   Kinematic Model

We wish to track and illuminate a vehicle moving on a road grid. The road network is shown in Fig. 6.2. The target begins at the location *start* in the figure. It is northbound with coordinate $-500E$. It continues north until it comes to a three-way junction at $-200N$. Though it does not have to, the target turns east. At $-350E$, the target comes to another three-way junction. The target must turn here, either to

Fig. 6.2 A local road grid is shown. The full network is shown with the small line width. A target moves north beginning at *start*. The actual target path is shown with an increased line width. Turns are permitted only at the junctions

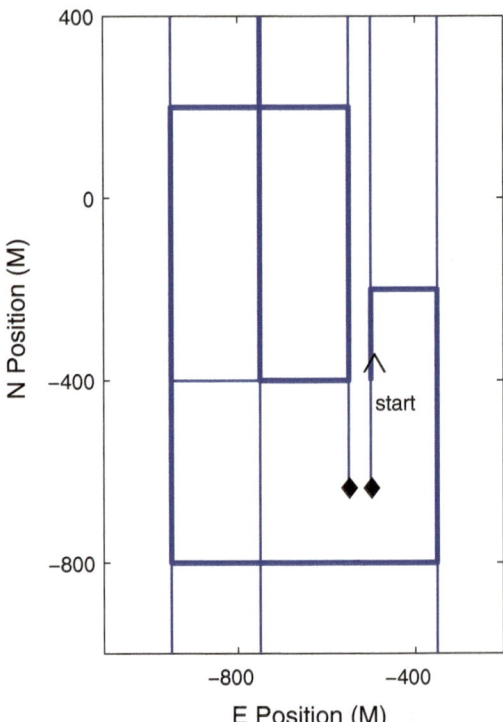

the north or to the south. The target now moves south. And so the target completes the indicated path, exiting the region at $(-950, 400)$.

The terrain is such that the main roads are north-south; e.g., -950E. The region west of -950E and east of -350E is not accessible. The north-south strip between -500E and -550E is difficult to cross. The vehicle must move to the bypass road at -800N to move west of -500E. The extensions of two north-south road segments are shown to dead-end at -650N. There are limited east-west through connectors; e.g., -400N, but these are smaller streets. Because the roads are better in north-south the nominal speeds are higher: the nominal speeds are 20 m/s (44 mph) in north-south and 10 m/s (22 mph) in east-west. The roadbeds are 10 m wide (30 ft). We will refer to the map data set as $\mathcal{M}$.

In the comprehensive kinematic model, each of the local models is tuned to a specific mode. We will represent intra-sample motion with a CV motion model:

$$
\frac{d}{dt}\begin{bmatrix} X \\ Y \\ V_X \\ V_Y \end{bmatrix} = \begin{bmatrix} 0 & 0 & 1 & 0 \\ 0 & 0 & 0 & 1 \\ 0 & 0 & 0 & 0 \\ 0 & 0 & 0 & 0 \end{bmatrix}\begin{bmatrix} X \\ Y \\ V_X \\ V_Y \end{bmatrix} + \begin{bmatrix} 0 & 0 \\ 0 & 0 \\ \sqrt{P^w_{x;\iota}} & 0 \\ 0 & \sqrt{P^w_{y;\iota}} \end{bmatrix}\begin{bmatrix} \dot{w}_X \\ \dot{w}_Y \end{bmatrix}, \tag{6.1}
$$

where $\dot{w}_Y$ is a unit white-noise acceleration, and ι is the modal state.

Equation (6.1) is the conventional kinematic model we have used throughout the book. The identity of the regime enters in the action matrix, P^w. For the most part, the exogenous accelerations are longitudinal with lateral motion restricted by the road width. Hence, the intensity ellipses of P^w are highly eccentric.

Specifically, we will suppose that the target motion is regular and the roads are narrow. To design the tracker, we must complete the CV-model by quantifying the acceleration process. The nominal 1σ-longitudinal velocity increment is: 0.31 m/s^2 in north-south and 0.14 m/s^2 in east-west. The lateral velocity increments are a tenth of the longitudinal: 0.03 m/s^2 in north-south; and 0.01 m/s^2 in east-west. A constant 1σ-acceleration over an interval of 5 s would yield a position increment of about 4 m longitudinally and 0.4 m laterally in north-south. Over a 100 second interval, the 1σ-change in north-south speed due to $\{w_t\}$ would be of the order of 30 m/s.

At a junction, however, the target may (or must) turn; e.g., at the west terminus of the uppermost street in Fig. 6.1, a westbound vehicle must turn. When a turn event occurs, the acceleration has a component perpendicular to the velocity entering the intersection—impulse acceleration is the normative approximation. There is a complementary acceleration in the longitudinal direction as well. Such large, albeit brief, accelerations are difficult to accommodate in a classical LGM model because the impulsive nature of the acceleration violates the usual smoothness conventions. We will assume that U-turns are disallowed in this engagement.

As described, the evolution of the modal state violates one of the fundamental hypotheses underlying the engagement model. For example, the modal event $(i, \iota^-) \mapsto (p, \iota)$ is assumed to be exogenous in the development leading to the **GWE**. But this cannot be true. If the kinematic state of the westbound target is far from the west terminus, it is very unlikely to turn north at the next sample interval: the regime change $\phi = 4 \mapsto \phi = 1$ is prohibited until $\chi[k]$ is just east of the terminus. Thus, the kinematic state imposes restrictions on the modal process in order that the target path be in compliance with the road map, $\mathcal{M}$.

As viewed from the tracker, this encounter is rather complicated. The target encounters several junctions along the path. For example, there is a T-junction at $(-500, -200)$ where the vehicle turns east. There is a 4-way junction at $(-750, -800)$, but the vehicle ignores it. The vehicle also ignores the 4-way junction at $(-750, 200)$ going both east and later going north. As the vehicle moves, it changes direction in an abrupt and unpredictable manner. We will suppose the vehicle exits a turn at the same regime-adjusted speed it enters; i.e., if the nominal road condition suggests a speed doubling after a turn, the target will double its speed if it turns.

6.2.2 Measurements

6.2.2.1 Temporal

The kinematic measurement architecture uses three sensors. The first two are conventional range-bearing sensors with 1σ-quality 95 m in range and 35 mr in bearing—(35 m in cross-range at 1 km). They are located at $(-300, 0)$ and $(-1000, 0)$. The former is well east of the target and in the north region. The latter is only slightly west of the west-most path segment. We would expect this measurement to be quite accurate as the target approaches the turn at 200N. The sensor geometry degrades in the south. We would expect that the west turn at -800N might be hard to isolate—note that the target could have continued south through that junction.

Additionally, there is an acoustic speed sensor. Speed is linked to direction. A high speed points to north-south motion with low speed pointing to east-west. But there is no sub-directional distinction. We will suppose that the raw acoustic measurement is reduced to a statement of a cardinal direction: the measurement is placed in one of the four directional bins. Direction is properly classified only 37.5 % of the time. The acoustic signal becomes the direct regime measurement, $z[k]$.

The kinematic measurements are generated every 5 s. The sensor errors are independent spatially and temporally. As usual, the range-bearing measurements are converted from a polar coordinate system to a common east-north Cartesian coordinate system and stacked as the 4-vector $y[k]$. This along with the acoustic measurements generates the filtration $\{\mathscr{Y}[k]\}$. The change in coordinates pivots about the location estimate, and the coefficients of the $\mathscr{Y}[k]$ model are $\mathscr{G}[k]$-random processes. We will use the $\mathscr{Y}[k]$ designation when the acoustic sensor is not used; e.g., in the EKF, and reserve the $\mathscr{G}[k]$-designation for algorithms using the full data array.

6.2.2.2 Non-temporal

We will suppose that there is a road map showing the permissible locations of the target. The specific map is that shown with small line width in Fig. 6.2. The target moves in the cardinal directions and is constrained by the indicated road junctions. There are junctions at which the target must turn, and there are junctions at which the target may turn.

The map in the figure is an example of a non-temporal para-measurement data set. Despite the data fusion issues introduced by this auxiliary data structure, the map is undeniably useful in so far as it places most of the terrain in the figure off-limits. Only the narrow roads are permissible locations for the target. A turn is a transition event in the untimed modal string. Turns are location specific; i.e., they must be compatible with the map. Thus, there is a tight link between regime transitions and specific locations.

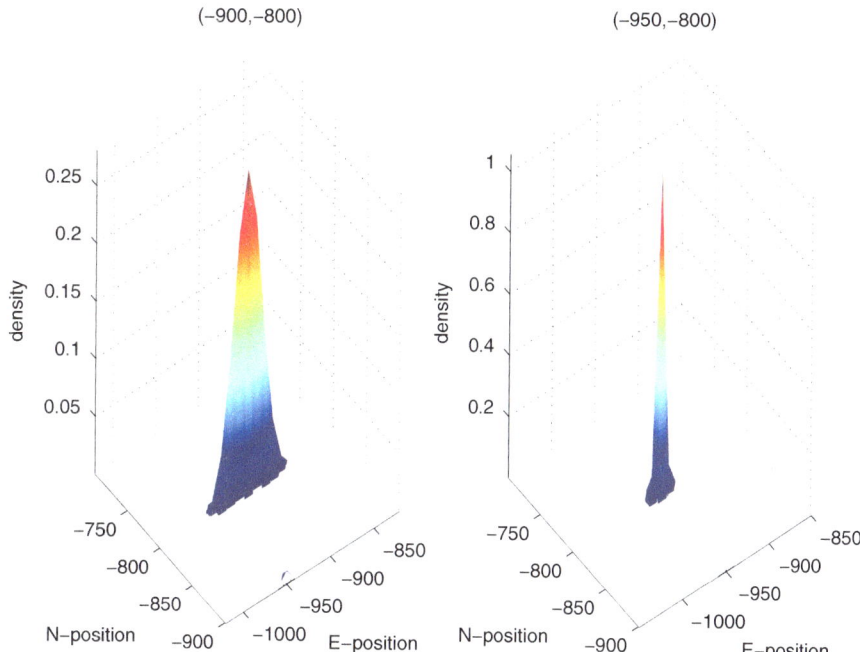

Fig. 6.3 The diffuse conditional density of position preceding a turn becomes more focused after the turn at $(-950, -800)$

To illustrate the advantage of having the map, consider the following situation. The target has been west-bound on the road with coordinate $-800N$ for several successive sample intervals: $\iota = (444)$. During this time the unstructured accelerations are primarily in an east-west direction. The uncertainty tends to grow over time as a function of the size of both P_4^w and P_4^n and reside primarily in east-west. In the left panel of Fig. 6.3, a typical conditional probability density for the target location as it passes $(-900, -800)$ is shown. The density is diffuse and spreads out along the road. The lateral uncertainty is much smaller since the road is narrow. While it may be true that the support of the density covers all of $\mathbb{R}^2$, the spillover onto prohibited terrain (north and south of the road) is minimal.

Now suppose at time $t = kT$, the west-bound target executes a north turn at $(-950, -800)$. The regime shifts from $\phi_t = \mathbf{e}_4$ to $\phi_t = \mathbf{e}_1$. The local model changes abruptly, and the kinematic state must be brought into compliance with a north motion: the east-west velocity is suddenly small. But even more, a north turn can only be executed at the specific location indicated on the map. Thus the east-west uncertainty that has built up over the long west-motion interval is excised. Going forward, the kinematic state is re-initialized with small uncertainty: east-west motion is produced by small lateral accelerations; the north direction is longitudinal.

After the turn, the conditional density in the left panel is replaced with one more of the form given in the right panel of Fig. 6.3. With the target at $(-950, -800)$, the covariance is much smaller, and the assurance regions are much tighter. A location-specific regime change focuses the location estimate and reduces the error that built up during an interval of regime constancy. Thus we see that the para-measurement can be an effective tool in precise target location.

We now have a complete model. The kinematics are delineated by (6.1). The quality of the kinematic and modal observations are described above. The para-measurement data set is specified in Fig. 6.2. Next we must design a tracker that provides a high quality assurance regions for the target.

6.3   The EKF

6.3.1   Nominal EKF

The conventional approach to this problem would be to base tracking upon a suitable EKF. The EKF is a unitary estimator that explicitly acknowledges neither the turns nor the complementary measurements.

The regime transition events are isolated. With no mechanism to distinguish the endemic accelerations from those created by the turn events, we might argue that P^w should just ignore the turns and use the longitudinal uncertainty as the action matrix; $P^w = \max_i P_i^w$, where the maximization is component-wise. Then P^w is diagonal with 1σ-acceleration: 0.31 m/s^2 in north-south and 0.14 m/s^2 in east-west. We will call the corresponding tracker the *nominal* EKF.

A sequence of kinematic measurements compatible with the channel noise matrix was generated. Based upon its internal geometry, the EKF generates the $\mathscr{Y}[k]$-conditional distribution of the kinematic state. The distribution is assumed to be (nearly) Gaussian: $\mathbb{P}(x|\mathscr{Y}[k]) = \mathbf{N}(\hat{x}[k], P_{xx}[k])$. From this, we can deduce the $\mathscr{Y}[k]$-conditional distribution of position. This latter is: $\mathbb{P}(\chi|\mathscr{Y}[k]) = \mathbf{N}(\hat{\chi}[k], P_{\chi\chi}[k])$. The conditional mean, $\hat{\chi}[k]$, is the best estimate of the location of the target. The notional quality of the estimate is given by $P_{\chi\chi}[k]$.

The response of the nominal EKF is shown in the left panel of Fig. 6.4. The feather plot links the true location of the target with $\hat{\chi}[k]$. The ellipse giving the assurance region of probability 50 % is drawn about $\hat{\chi}[k]$. We will call the assurance region $\mathbf{A}_{0.5}$. Using the terminology we favor in this book, $\mathbf{A}_{0.5}$ is the *elliptical error probable* (EEP).

The motion is almost exclusively CV and thus conforms to the LGM template that forms the basis for the EKF—though the observations are not a linear function of x_t. The tracker uses a plausible (even conservative) choice of P^w. Still, the EEP regions usually fail to capture the target: converting a probability into a frequency, the EEP should capture the target half of the time. The errors are particularly egregious where the sensor geometry (GDOP) is bad; e.g., along the stretch -550E, the sensor error ellipses overlap and are perpendicular to the target motion. The

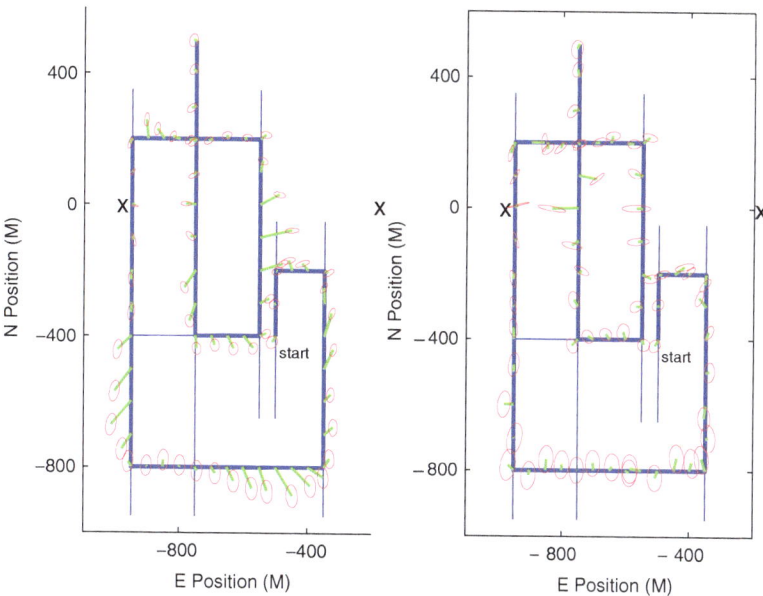

Fig. 6.4 A target follows the bold path beginning at *start*. Two range-bearing sensors are marked with *times symbol* in each figure panel, and they sample position every 5 s. An EKF generates both a location estimate and a tight EEP assurance region centered on the conditional mean. The *left panel* shows the nominal EKF. The *right panel* shows a pseudo-noise augmented EKF processing the same measurement sequence

EKF also lags after a turn; e.g., in the region near $(-400, -800)$, even though the EKF has good estimates of location and speed entering the turn. In this instance, the problem is a low SNR rather than unfavorable sensor geometry.

Figure 6.5 shows a sequence of contour plots of the conditional position density generated by the nominal EKF preceding, simultaneous with, and after a turn event. The target, marked with $\times$ in the left panel, approaches the T-junction from the east traveling at 10 m/s. The velocity has been constant for 500 m. The tracker places the target well. The uncertainty spreads across the road, and the slight velocity bias is to the north. The target reaches the junction in the middle panel. The tracker is accurate with a slight north-west velocity bias. But after the turn, the EKF is slow to increase the north speed. The target is now on the fringe of the notional density—the position error is over 50 m.

The conclusion to be drawn from Figs. 6.4 and 6.5 is not that the EKF is a poor estimator. The location errors are well under than the 100 m range error for the most part. Rather the primary deficiency in the estimator resides in its overconfidence: $\mathbf{A}_{0.5}$ fails to capture the target more often than not. If countermeasures are distributed according to the notional EEP, the target would usually be secure from harm.

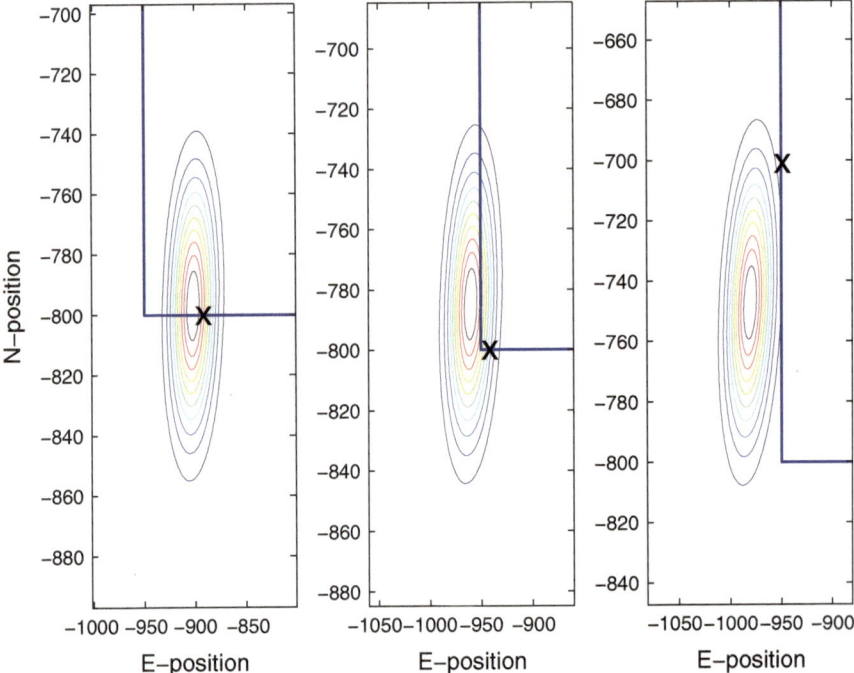

Fig. 6.5 The EKF generates the $\mathscr{Y}[k]$-distribution of target position. A contour plot of the density shows the EKF is accurate approaching the T-junction at $(-950, -800)$. But it drifts into the prohibited region after the turn. The true target location is indicated with a *times symbol* in the figure

6.3.2 Pseudo-Noise Augmentation

The lags in the normative EKF can be reduced by increasing the size of the exogenous acceleration in the kinematic model. That is, let us replace P^w with something bigger: $P^w \mapsto P^w + \Delta P^w$; $\Delta P^w > 0$. This arbitrary increase in P^w is referred to as pseudo-noise augmentation [15, p. 24]. Pseudo-noise is known to increase the bandwidth of the tracker, and the choice of ΔP^w is an engineering judgement.

We argue as follows: In north-south, the normative acceleration is $0.31\,\text{m/s}^2$. When there is a turn, the north-south velocity changes by $20\,\text{m/s}$. These changes take place every $30\,\text{s}$ or so. We could collect these accelerations and conclude that the average north/south acceleration is on the order of $1\,\text{m/s}^2$. Because of their impulsive nature, we will be conservative and increase the acceleration uncertainty by a factor of 9. We will arbitrarily increase the action matrix so that the 1σ-increment in velocity is: $1.26\,\text{m/s}^2$ in east-west, $2.5\,\text{m/s}^2$ in north-south.

Pseudo-noise makes the tracker faster, but it does so at the expense of increasing the area of the EEP. This is clearly seen in the right panel of Fig. 6.4. Processing

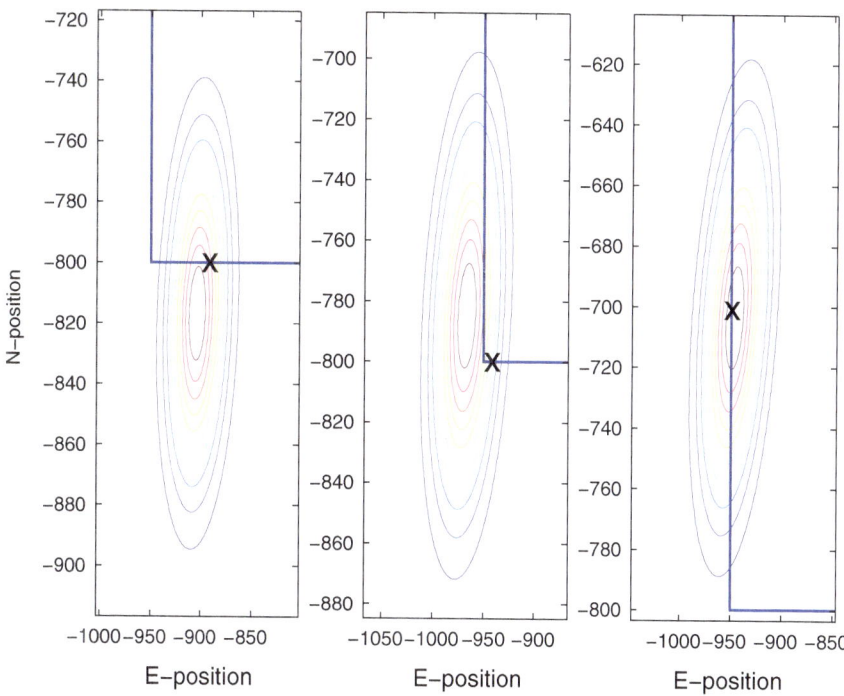

Fig. 6.6 A contour plot of the position density of the augmented EKF approaching the T-junction at $(-950, -800)$ shows a more rapid response but larger error ellipses—see contrast with Fig. 6.5

the same range-bearing sequence that was used in the nominal EKF, the augmented tracker responds more quickly. The location estimates in the right panel are less sensitive to junctions, but the distributions are more diffuse—the EEP is uniformly bigger. The normalized radial error is less than 1σ in low SNR regions. But when the GDOP is high, the standard error can be excessive as seen in the northbound segment at $-750E$.

The position density of the augmented EKF-tracker is shown near the T-junction at $(-950, -800)$ in Fig. 6.6. The density ellipses generated by the augmented tracker are certainly more accurate than those of the normative EKF shown in Fig. 6.5. However, the areas are considerably bigger too.

To illustrate the differences in the two EKF-trackers, look at the performance of each subsequent to the $(-350, -800)$ junction. This junction is a T where the target might have continued south. Instead, the target made an abrupt turn to the west. The left panel of Fig. 6.7 shows the EEP for the normative EKF; the target position is shown with an $\times$. The mean position error is about 100 m and the EEP lies in a prohibited region south of the road.

The right panel of Fig. 6.7 shows the EEP generated by the augmented EKF. This EEP region actually intersects the road and includes the target location on its boundary. But it is larger.

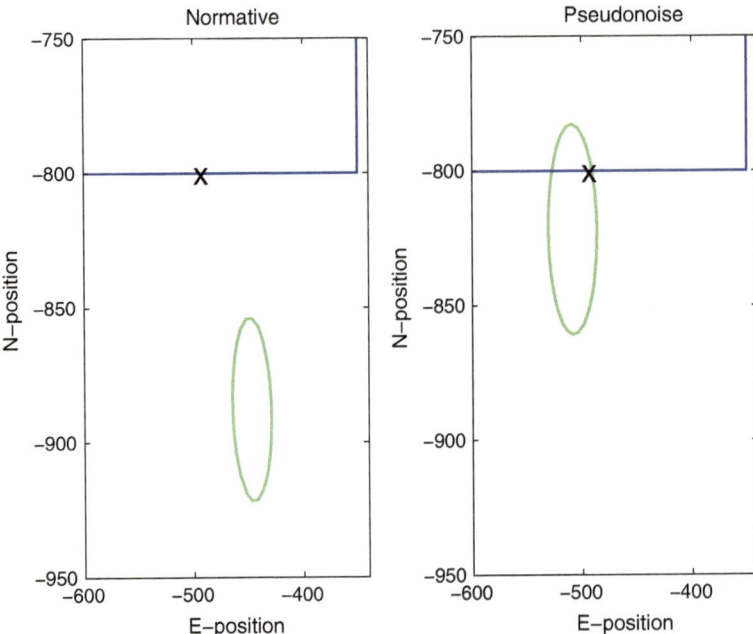

Fig. 6.7 Following the turn at $(-350, -800)$, the nominal EKF lags the target and places the EEP in a prohibited region. Pseudo-noise augmentation improves the response and places the true location within the computed EEP

In this engagement, the broad performance of the augmented EKF is superior to the normative EKF. Moreover, $A_{0.5}$ covers the target most of the time in the former. Unfortunately, the EEP-areas are larger with pseudo-noise augmentation— the density of the countermeasures is reduced with the addition of pseudo-noise.

6.3.3   Tracker Contrasts

The nominal CV-model is a conservative representative of target motion between junctions. The test path has no intra-sample acceleration. The acceleration matrix, P^w, is smaller in the nominal algorithm than it is in the augmented EKF. We say that the update is more model-driven in the nominal EKF than it is in the augmented EKF: the latter is said to be more data-driven. More specifically, the nominal tracker assigns less weight to a single measurement than the augmented tracker does.

But both EKFs ignore the particular character of the turn events. With the memory intrinsic to the EKF, the errors subsequent to a regime change continue well after the event: longer in the nominal tracker than in the augmented version. Figure 6.8 shows the radial error sequence for each tracker. Recall the range-bearing

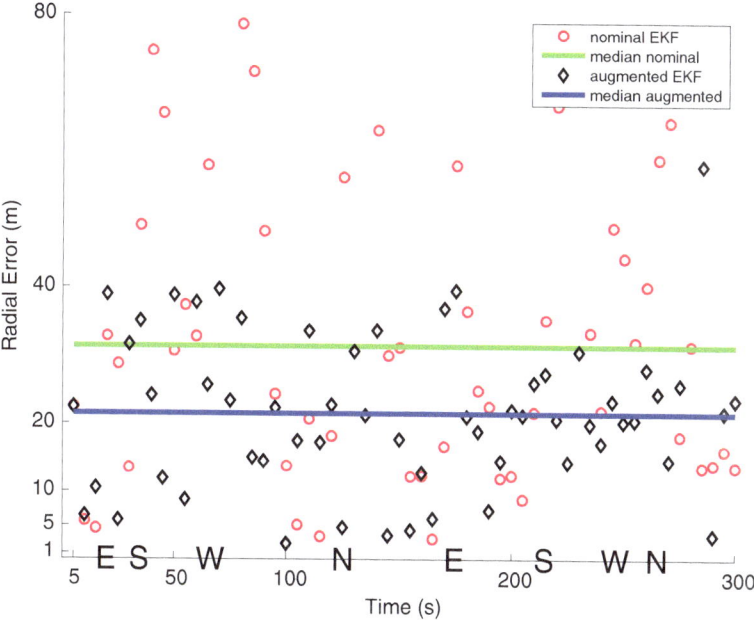

Fig. 6.8 The radial error of the nominal tracker tends to be considerably bigger than that of the augmented tracker. The radial error is shown as a function of time. The turn events are shown on the time axis

sequence is the same for both. The horizontal axis is time. Turn events are shown along the axis with the direction of the turn shown as a place label: the symbol E at $t = 15$ s denotes the first turn to the east in Fig. 6.4. The two trackers begin at the same point, but the errors soon separate themselves. The nominal tracker has significantly larger errors—often between 40 and 80 m. The errors of the augmented filter are seldom more than 40 m. The errors for both trackers are larger when the GDOP is large. Both trackers correct their errors as a turn event recedes into the distant past. But with their memory, neither tracker is able to reach quiescence between turn events.

   While not uniformly inferior to the augmented tracker, the nominal tracker has a significantly larger median tracking error. The median tracking error for the augmented tracker is 21 m, 3 dB less than that of the normative tracker. The largest errors are associated with nominal tracker, nearing 100 m in some cases.

   To understand how the two trackers interpret their estimates, recall some of the properties of the classical Gaussian filter. The Kalman filter generates the $\mathcal{Y}[k]$-conditional distribution of the position: $\mathbb{P}(\chi|\mathcal{Y}[k]) = \mathbf{N}(\hat{\chi}[k], P_{\chi\chi}[k])$. The conditional mean, $\hat{\chi}[k]$, is the best mean-square, point estimate for the target. The quality of the estimate is given by $P_{\chi\chi}[k]$.

   We can simplify the manner in which the position error is displayed by choosing a coordinate system referenced to the uncertainty in the estimate. The normalized

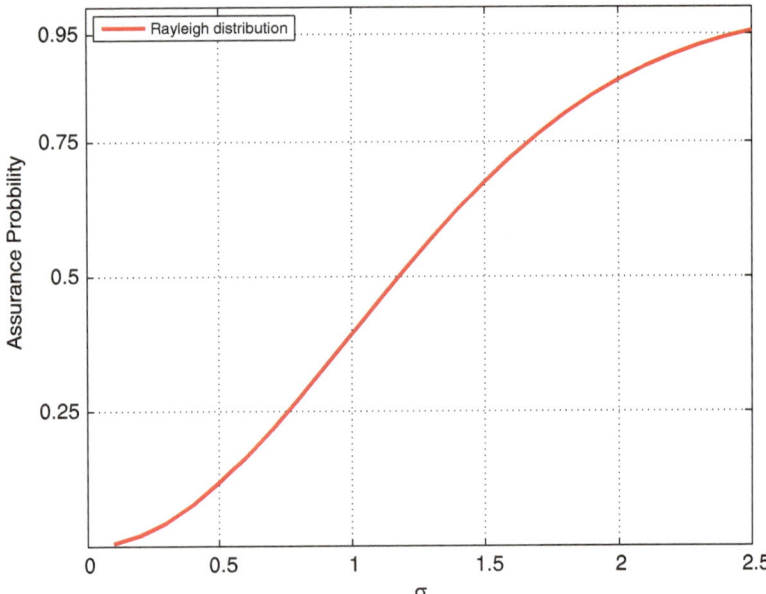

Fig. 6.9 For a Kalman filter, the normalized radial error is Rayleigh distributed. To achieve an 80 % assurance region, a 1.8σ coverage is required

tracking error is a unit Gaussian random 2-vector: $\tilde{\chi}_F[k] = F_{\chi\chi}[k]\tilde{\chi}[k] \sim \mathbf{N}(0, I)$. In this coordinate system, the components of the tracking error are independent with standard deviation equal to one.

Although $\tilde{\chi}_F[k]$ is unit-free, we will refer to the units of $\|\tilde{\chi}_F[k]\|$ as σ—in contrast to $\|\tilde{\chi}[k]\|$ which has units of meters. For example, if $\|\tilde{\chi}_F[k]\| = \|\tilde{\chi}[k]\|_{D_{\chi\chi}} = 2$, we would say that the tracker estimate is off by 2σ. In the coordinate system associated with $P_{\chi\chi}$, the tracking error is displayed relative to its expected size.

The radial error in the modified coordinate system has a Rayleigh distribution with variance one: $\mathbb{P}(\|\tilde{\chi}[k]\|_{D_{\chi\chi}}) \le u) = 1 - \exp(-\frac{1}{2}u^2)$ [10, p. 180]. Figure 6.9 shows a plot of a unit Rayleigh distribution. If we want a tight 95 % assurance region, we would select the region interior to the $\|\tilde{\chi}[k]\|_{D_{\chi\chi}} = 2.5\sigma$ error ellipse. Similarly, the probability that $\|\tilde{\chi}[k]\|_{D_{\chi\chi}} < 1.18\sigma$ is 50 %. It follows that $\chi \in \mathbf{A}_{0.5}$ (or the EEP) is the interior of the $\|\tilde{\chi}[k]\|_{D_{\chi\chi}} = 1.18\sigma$ ellipse centered on $\hat{\chi}$. The normalized radial error seldom exceeds 3σ in a classical tracking problem.

In the LGM problem, the tracker will place the target within $\mathbf{A}_{0.5}$ roughly half of the time. Of course, this is not a proper LGM problem; e.g., the measurements are nonlinear. But we have been conservative in so far as the intensity of the exogenous acceleration in the kinematic model exceeds the actual acceleration except at isolated points. Of more importance, we have simplified the representation of the accelerations at the turn events: ignoring them in the nominal tracker and over-weighting them in the augmented tracker.

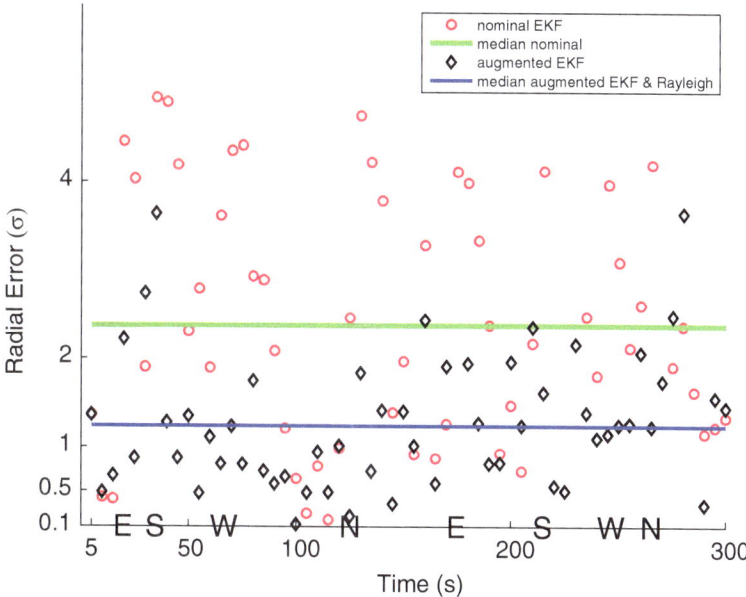

Fig. 6.10 The EEP generated by the nominal tracker is of dubious value. The computed EEP for the augmented tracker is more credible

To test the fidelity of the EEP calculation, Fig. 6.10 shows the normalized error, $\{\|\tilde{\chi}[k]\|_{D_{\chi\chi}}\}$, for both the nominal and the augmented tracker. We see from the figure that the nominal tracker fails to properly calculate its EEP. While $\{\|\tilde{\chi}[k]\|_{D_{\chi\chi}} < 1.18\sigma\}$ should be true 50 % of the time, it is actually true about 20 % of the time. Indeed, the median value of $\{\|\tilde{\chi}[k]\|_{D_{\chi\chi}}\}$ is approximately 2.5σ, and this corresponds to a predicted 95 % assurance region. The largest values of $\|\tilde{\chi}[k]\|_{D_{\chi\chi}}$ occur subsequent to a regime event. The errors decay after an event—sometimes rapidly. But the error path is irregular, and the tracker never has a chance to reach a quiescent state. Conventional wisdom suggests that we can disregard 3σ events. But it is apparent that the nominal tracker makes such errors frequently.

Figure 6.10 also shows the normalized errors in augmented tracker. They are far smaller than those seen in the nominal tracker. In fact the median error is 1.24σ— very close to the 1.18σ predicted from the Rayleigh distribution. The rate of larger errors is in keeping with that predicted: the radial error exceeds 2σ 15 % of the time. Again, the filter does not achieve quiescence—though it approaches it somewhat more rapidly than does the normative tracker.

The augmented tracker is better at computing EEP than is the normative tracker. One could imagine that the tracker could be made less sensitive to turn events by increasing P^w even more. Unfortunately, there is a concomitant penalty associated with pseudo-noise augmentation. As P^w is increased, the EEP ellipses get bigger. The assurance regions are useful only to the degree that they can be used for countermeasure allocation or separation from geographic features. For example,

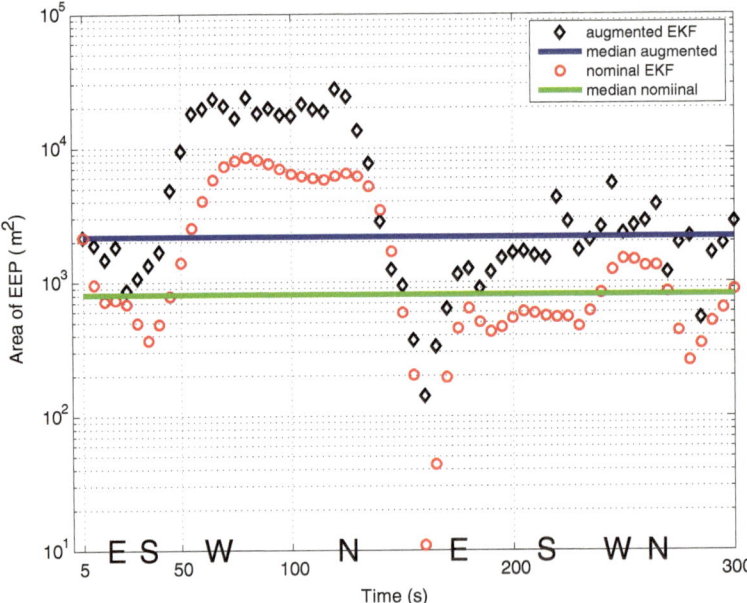

Fig. 6.11 The areas within the EEP ellipses are larger in the augmented tracker than they are in the normative tracker. This is particularly true when the GDOP is high

if we are using $\mathbf{A}_{0.75}$ to achieve p-level separation of the target from some hard boundary, an excessively big $\mathbf{A}_{0.75}$ would make the controller too conservative. Hence, the bigger the assurance region is, the less its value.

If we look at the area of the EEP ellipses, we find that the augmented tracker has EEP areas that are as much as four times that of the nominal algorithm, particularly when the GDOP is big or when the SNR is poor. Figure 6.11 shows the sample function of the area of the computed EEP ellipses for both trackers. The median area of the nominal tracker is a third of that of the augmented tracker. With limited defensive assets, the countermeasure density of the nominal tracker is several times that of the augmented tracker—though unfortunately the EEP of the former often misses the target altogether. It is also true that the segment of the path that has the smallest EEP regions is also a section of the path in which the normalized radial error is large.

It is apparent from this example that a properly designed **EKF**-tracker will follow a multi-regime motion with small error. It achieves this by over-weighting the wideband disturbances to account for the abrupt accelerations associated with regime changes. By increasing P^w, the tracker is made faster, and the influence of regime changes is isolated in the tracker. However, the countermeasure windows increase and reduce the probability of neutralizing a hostile target.

6.4 The Map-Enhanced GWE

6.4.1 Algorithm Overview

6.4.1.1 Introduction

In the illustrative example presented here, the pseudo-noise augmented EKF yields estimates better than the nominal alternative. However, the augmented tracker must generate large assurance regions to achieve this benefit. The EKF is a unimodel algorithm that ignores the factors that distinguish the individual segments of a motion: speed, direction, and compliance with the junction restrictions. It is precisely when the motion regimes have such distinctives that multiple model algorithms become attractive. Hybrid trackers need not average across the kinematic regimes.

The GWE is a more nuanced algorithm than is the EKF. The GWE integrates a statement of direction with the range-bearing measurements that are the inputs to the EKF. Lacking a ground map, the EKF can make little use of direction reports. For example, when the target approaches the T-junction at $(-950, -800)$, the direction is classified as east-west 75 % of the time based upon Doppler. When the target arrives at the junction, it must turn. But the direction is indeterminate: $z[k]$ points north with probability 0.38; $z[k]$ points south with probability 0.38. Hence, the Doppler measurement gives an indication of a regime change without clearly indicating what that change is.

The direction of the regime change is partially resolved by the ground map, $\mathcal{M}$. The map delineates all permissible road segments and junctions. These latter are tabulated in $\mathcal{M}_{cr} = \{\chi_r; r \in N_{cr}\}$. For example, the first element in $\mathcal{M}_{cr}$ might be the statement: $(-450, -200)$ is a north-south T-junction with east access. This would tell the GWE that there was a T-junction oriented north-south with permissible east exit. All junctions are listed in $\mathcal{M}_{cr}$ along with access constraints; e.g., a four-way intersection permits all forward directions. Road segments then join compatible junctions in the cardinal directions.

Let us continue with the engagement described earlier and derive an algorithm that fuses the same range-bearing data with the complementary map information and the Doppler measurements. We are particularly interested in the quality of the assurance regions. We want assurance regions to be both small and predictive of target placement.

In what follows we will outline the way in which the para-measurements are introduced into the GWE-tracker. The basic algorithm has to be adjusted in ways that violate the canonical assumptions on the regime dynamics. These algorithmic adjustments are spelled out in detail in the Appendix. Here we give an overview and the simulation results.

6.4.1.2   Geometric Constraints

Movement on a road grid severely limits the search region for the target vehicle. But it complicates the structure of the hybrid tracker. As derived in earlier chapters, the GWE is composed of $\mathbf{S}^L$ local filters operating in parallel on a common data sequence. In this engagement the regime set will be the cardinal directions, and we will choose $L = 3$ for our memory depth: there are 64 local filters. With a sample interval of 5 s, the explicit memory in the GWE is 15 s.

To extrapolate the kinematic states of the target, the GWE must first identify where it is on the map and determine all junctions accordant with estimated location and velocity. Suppose the target is located at $(-800, -800)$ and is moving west; $\phi[k] = \mathbf{e}_4$. There is only one junction accepting a westbound vehicle from $-800N$: $\chi_5 = (-950, -800)$. Access to the junctions at $\chi_4 = (-750, -800)$ and $\chi_3 = (-350, -800)$ is precluded since they are not congruent with west velocity. Since we are prescient, we know that the vehicle will turn north at χ_5. The GWE-tracker only knows that this turn will ultimately be made, but the forward direction is not known.

To illustrate the alternatives faced by the GWE, let us focus on the ιth local filter with initial condition: mean state $m_\iota- \approx (-800, -800, 0, -10)'$, covariance $P_\iota-$, and modal suffix, $\iota^- = (44)$ (this local tracker has been initialized near the true location and with the true direction). There are several events that are possible on the forward interval. Let us review the possibilities for the local filter, $\iota = i44$:

1. *The target could continue west* $(i = 4) : \iota = (444)$. It might or might not project beyond the χ_5 junction.
2. *The target could turn north* $(i = 1)$ *or south* $(i = 2)$: $\iota = (144)$ or $\iota = (244)$.
3. *The target could turn east* $(i = 3) : \iota = (344)$. This is prohibited.

There are more possibilities at other points along the path; e.g., there could be several junctions accepting the current velocity. But the above illustrates the complexity of the tracking problem. Extrapolation yields a swarm of 64 target locations, with their associated covariances.

6.4.1.3   Kinematic Update

The update of a local filter is based upon the conventional EKF rules. Unfortunately, the diffuse sensor noise may cause the forward estimate to move away from the road grid. To restore map compliance, we use the $\mathbf{C}^{\mathcal{M}}$ transformation to find the point on the map, $\hat{\chi}_\iota^{+F}$, nearest our projected location. We then move the estimate back to the map. This is the first benefit of the map.

Map compliance also requires that the velocity be adjusted. To first order, the velocity must be longitudinal and correspond to the specific regime prefix in ι: This completes the kinematic update. There are several engineering choices that are made to establish the local algorithm; e.g., the initialization at a turn and the intensity of the exogenous lateral accelerations.

6.4.1.4 Modal Update

The GWE delineates the likelihoods of the various modal hypotheses in terms of a (normalized or unnormalized) $\mathscr{G}[k + 1]$-modal probability mass function which we write as $\{\alpha_\iota[k+1]; \iota \in \kappa\}$. The mass function is updated in two steps. First, with the kinematic adjustment based upon $y[k + 1]$, and second, with the regime adjustment based upon $z[k + 1]$.

Let us first look at the kinematic update. The $\iota = ijl$-filter begins with an initial condition $(\alpha_\iota[k], m_\iota[k], P_\iota[k])$. It extrapolates forward under the prefix $i \in S$. At $t = [k + 1]T^-$, it is now in state $(\alpha_\iota[k], m_\iota^-[k + 1], P_\iota^-[k + 1])$. The filter index remains ι, but the local moments move up one sample time.

At time $t = [k + 1]T^-$, the tracker makes a measurement represented by $y[k + 1] = H_\iota x[k + 1] + \sqrt{P_\iota^n} n[k + 1]$: the model relates the raw measurement $y[k + 1]$ to the model state ι through both the geometry of the sensors (H_ι) and the peculiarities of the exogenous noise (P_ι^n). The covariance of the kinematic measurement is labeled P_ι^y.

The channel noise matrix requires a coordinate change from a polar systems to a rectilinear one. The transformation is based upon the local estimate of position; i.e., on $\hat{\chi}_\iota$. Hence, the intensity of the sensor noise, P_ι^n, conveys some information on the modal state; the size of the noise in $y[k + 1]$ could favor a specific groups of modes.

The situation regarding P_ι^- is more subtle. The covariance is intimately linked to the path of the vehicle; e.g., is the target moving east or north? And the current velocity is a function of the full modal string—of which ι is a prefix.

Define

$$E_\iota^y = \Delta \|m_\iota\|_{D_\iota}^2 - \|y[k + 1]\|_{D_\iota^n}^2. \tag{6.2}$$

E_ι^y is a unit-free quantity that is dependent upon the kinematic measurement and the mode. To see more clearly what the ιth update accomplishes, let us focus just on the position components of m_ι and on the upper block of D_ι. The units of the mean state are meters, and the units of the upper elements of D_ι are m^{-2}. So the vector $D_\iota m_\iota$ has units m^{-1}. The term $D_\iota m_\iota$ expands the mean state by emphasizing those components those value is the most sure. The direction in which we have more confidence (say lateral) is favored when multiplied by D_ι over the direction in which we have less confidence (say longitudinal).

When $y[k + 1]$ is observed, both m_ι and D_ι change. The former in the direction of the measurement, and the latter gets bigger to the degree and in the direction that the measurement conveys useful information. This means that $\|m_\iota\|_{D_\iota}^2$ will change as well. When $y[k+1]$ is near the extrapolated state, the associated mode is favored: for a given mode ι, the bigger $\Delta\|m_\iota\|_{D_\iota}^2$ is, the more likely the ιth mode is the correct one. Internally, the GWE computes both d_ι and P_ι and uses the equivalence $\Delta\|d_\iota\|_{P_\iota}^2 = \Delta\|m_\iota\|_{D_\iota}^2$ to compute this term in E_ι^y.

The above comment is for purpose of illustration. The units of m_ι and D_ι contain compounds of both meters and seconds. But the term $\|m_\iota\|_{D_\iota}^2$ emphasizes the kinematic states in which the ιth filter has most confidence. The increments in $\Delta\|m_\iota\|^2_{D_\iota}$ should favor the correct sub-filter: this filter moves a larger multiple of the standard units of information when the kinematic measurement is matched to the filter.

It is also true that the size of the state correction is partially due to the size of the measurement itself. To adjust for this, another unit-free term, $-\|y[k+1]\|^2_{D^n_\iota}$, is included in E^y_ι. This term is bigger when the measurement is in a direction of lower SNR, and it acts to normalize for the size of the raw observation. Because E^y_ι is used to contrast the importance of alternative modes, when D^n_ι is independent of the mode, $\|y[k+1]\|^2_{D^n_\iota}$ is constant across modes, and it will be ignored in computing E^y_ι.

For each of the 64 filters in the GWE, the kinematic update of the modal mass function is

$$\alpha^-_\iota[k+1] = \alpha_\iota[k]|F^y_\iota|\exp(\tfrac{1}{2}E^y_\iota). \tag{6.3}$$

Equation (6.3) updates the $\mathscr{G}[k+1]$-mass function with the $y[k+1]$ data. As with the kinematic state, the local filters are updated in parallel. As written, $\alpha^-_\iota[k+1]$; $\iota \in \kappa$ is unnormalized.

The next step in generating the $\mathscr{G}[k+1]$-modal mass function is the integration of the regime measurement $z[k+1]$. The first issue to be addressed is the proper model for $z[k+1]$. Is the measurement at $t=(k+1)T$ the regime proxy in the kth interval? Or is it the forward regime in the $(k+1)$th interval?

The answer to this depends upon the delay in reducing raw observation data to a classificational statement: at time $t=(k+1)T^-$ the regime is actually p while the current regime proxy is i. With the simple acoustic measurement, we will assume that the regime state is measured with quality matrix $\mathbf{D}$ and without latency. For $i \in \mathbf{S}$,

$$z[k+1] = \mathbf{D}\phi_p[k+1] + \eta[k+1], \tag{6.4}$$

where $\mathbf{D}$ is the discernibility matrix. For $\iota^+ \in \kappa^+$, the update for a regime measurement is

$$\alpha_{\iota^+}[k+1] = \alpha^-_\iota[k+1]\Pi_{pi}z[k+1]'\mathbf{D}_{.p}. \tag{6.5}$$

Equation (6.5) generates a mass function on κ^+—a set S-times the size of κ.

6.4.1.5 Transition Rates

To understand the structure of the regime update equation, observe that the factor Π_{pi} is the probability that the mode makes the transition $(ijl) \mapsto (pijl)$. Since the GWE formalism is based upon the (compromised) premise that $\{\phi[k]\}$ is a Markov process, this probability can be expressed as a function of the regimes; i.e., $\mathbb{P}(i \mapsto p)$.

The proposed transition is then confirmed by a measurement, $z[k + 1]'\mathbf{D}_{.p}$. This factor tends to be higher in the actual direction of travel. For example, at the beginning of the interval at time $t = 115$ s, the target is westbound: $\iota = 444$. It is also westbound at the end of the interval: at $t = 120$ s, $\iota^+ = 4444$. The acoustic sensor will classify the motion as westbound with probability 38 %, eastbound with probability 38 %, northbound with probability 12 %, southbound with probability 12 %. The GWE will weight the elements of $\alpha_\iota^-[k + 1]$ that correspond to east-west at three times the level of the north-south elements.

The nature of the forward regime transition probability matrix is somewhat contrary in this engagement. A premise in the development of the GWE is that $\{\phi[k]\}$ is exogenous: $\Pi_{ip} = \mathbb{P}(\phi[k + 2] = \mathbf{e}_p|\sigma_\phi[k + 1])$ is a function of the ordered pair (pi) alone. In this case, the regime sequence is precursory to the kinematic state sequence. But in the tracking example, the regime transitions and the kinematic state are entwined. To illustrate, consider a slow westbound target at $\chi[k] = (-800, -800)$. The vehicle cannot turn north over the next interval: $444 \nRightarrow 1444$ because it will not reach the junction. We would have to say then that $\Pi_{41} = 0$. Alternatively, if $\chi[k] = (-950, -800)$, a turn must be made: $444 \nRightarrow 4444$ and perhaps $\Pi_{41} = 0.5$. Unfortunately at $t = 115$ s, both $\chi[k] = (-800, -800)$ and $\chi[k] = (-950, -800)$ are possible since the support of the $\mathscr{G}[k + 1]$-conditional distribution of location covers both alternatives—though the former is far more likely.

We will approximate the forward transition probability with a Π matrix that takes into account the binary nature of the motion. Specifically, we will compose Π from a pair of primitive rate matrices: I and $\Pi_\vdash$. The former captures the situation when the target is between junctions, and the latter captures the situation when the vehicle is at a junction.

Between junctions, the regime transitions are not permitted; the target maintains its current direction. This simply means that $\mathbb{P}(p = i) = 1$. At a junction, the situation is not as clearly delineated. If the target encounters a north-south T-junction from the east; e.g., $(-950, -800)$, $\mathbb{P}(p = 1 \text{ or } p = 2) = 0.5$. But the 4-way junction at $(-750, -800)$ would be captured with $\mathbb{P}(p = 1) = \mathbb{P}(p = 2) = \mathbb{P}(p = 4) = 0.33$ if all forward directions are equally likely.

The transition behavior at a junction depends upon the junction type: does a westbound target encounter a 4-way or a T-junction; can the target make a U-turn? In this GWE algorithm we will simplify things and delineate junction behavior with the single matrix $\Pi_\vdash$ corresponding to a 4-way without U-turns. This is not a fully prescriptive choice in this engagement: it makes every junction accessible from every point on the road with compatible east-north coordinates. But it simplifies internal structure of the algorithm. The GWE is not required to find only those junctions accepting a given direction of motion.

In principle, the tracker uses the appropriate rate matrix, Π, at every observation time. Unfortunately, the GWE does not place the vehicle determinately on $\mathscr{M}$; the target will be in a neighborhood of any junction, albeit with small probability in most cases. Hence, the choice between $\Pi = \Pi_\vdash$ and $\Pi = I$ cannot be made with certainty. For example, the westbound, 444-tracker at $(-850, -800)$, projects ahead to a neighborhood $(-900, -800)$. Even though this is well short of $(-950, -800)$, the junction is within the support of the $\mathscr{G}[k]$-distribution of the local filter.

More specifically, consider the 444-filter at time $t = 115^-$ s. The target is moving west and $\hat{\chi} = (-920, -800)$. The west-variance is $281\,\mathrm{m}^2$, that is, $\sigma = 17\,\mathrm{m}$. Looking west, the filter sees the junction at $\hat{\chi} = (-950, -800)$: the junction is 1.7σ away in a west direction. The normative position distribution for this local filter is Gaussian. On this basis, the probability that the target lies west of the road terminus is given by the complementary error function, **erfc**. But $0.5\,\mathrm{erfc}(1.7/\sqrt{2}) = 0.04$: there is a 4 % chance that the target lies in the prohibited zone west of the junction. A plot of $0.5\,\mathrm{erfc}(0.7\sigma)$ is shown in the Appendix. If the GWE places the target at the junction: $\hat{\chi}_\iota = (-950, -800)$, there is still a 50 % probability that the target has not reached the junction. On the other hand, if m_ι is 1σ east of the junction, the probability the target lies at the junction is 15 %.

To capture the ambiguity in location and the concomitant ambiguity in transition matrix, we will use **erfc** to blend the primitives in the GWE, smoothly changing from I to $\Pi_\vdash$ as the target nears a junction.

To quantify the notion of approach to a junction, we use the Mahalanobis distance to measure the target/junction separation. At time $t = (k+1)T$, the local position statistics of target position are $(\hat{\chi}, P_{\chi\chi})$. Suppose that the rth junction is compatible with ι: if $\iota = 444$, and χ_r is west of $\hat{\chi}_\iota$ with the same north coordinate, then χ_r is a possible regime transition point for the target.

The Mahalanobis distance, ϱ, from the target to the rth junction is defined as $\varrho[m_\iota[k+1], \chi_r] = \sqrt{\|\hat{\chi}_\iota[k+1] - \chi_r\|^2_{D_\chi}}$. After a $y[k+1]$ update, let $\varrho(m_\iota[k+1])$ be the distance to the nearest congruent junction. For example, a westbound target, $(\iota = 444)$, would look for the nearest west accessible junction with the same north coordinate. A gauge of the likelihood of encountering the junction during the next sample time is $\mathrm{erfc}(0.7\,\varrho(m_\iota[k+1]))$.

This GWE adjusts the Π matrix as a function of target position. It replaces the true position with the local mean, $\hat{\chi}_\iota$. It assumes the likelihood of making a regime change is dependent upon the longitudinal separation from the nearest junction measured in standard units. The modal transition matrix used in this engagement is

$$\Pi_\iota = \Pi_\vdash \mathrm{erfc}(0.7\,\varrho(m_\iota[k+1]^+)) + I\,\mathrm{erf}(0.7\,\varrho(m_\iota[k+1]^+)). \qquad (6.6)$$

If there are no compatible junctions, we will set $\Pi = \mathbf{I}$.

Equation (6.6) provides a smooth flow from $\mathbf{I}$ to $\Pi_\vdash$ as the target approaches a junction. Properly $\Pi_\vdash$ should be matched to the junction type. This refinement has not been incorporated into the GWE displayed here: $\Pi_\vdash$ corresponds to 4-way junctions without U-turns. This means that χ_ι is the distance to the *nearest* junction even if that junction is inaccessible. This will create anomalous estimates from time to time. A striking example of this will be seen in the following section.

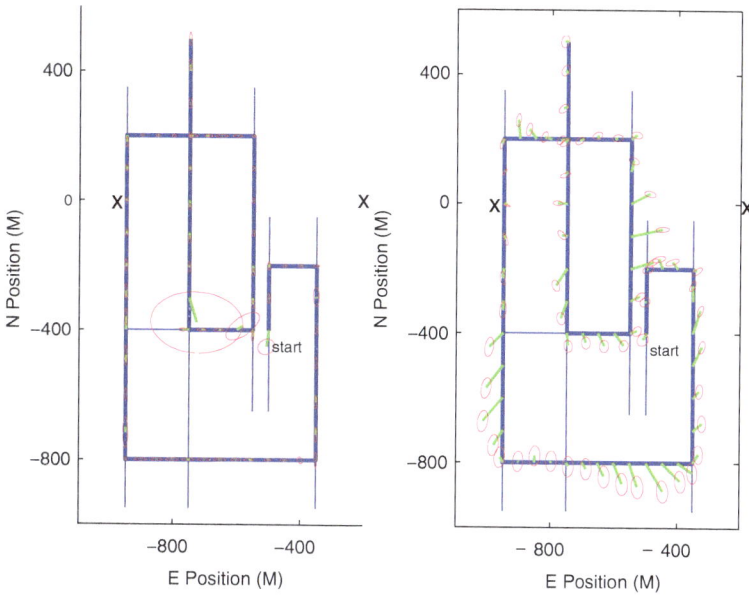

Fig. 6.12 The target follows the path shown. The augmented EKF generates both a location estimate and a tight 50 % assurance region as shown in the *right panel*. The *left panel* shows the EEP generated by the GWE using the projected distribution $\hat{\mathbb{P}}$. In essentially all cases, the EEP for the GWE is so small that it does not display in the figure

6.4.2   The *GWE-Tracker*

6.4.2.1   Tracking Performance

Let us revisit the engagement studied in the earlier sections. The target is moving on a rectilinear grid as shown in Fig. 6.2. Figure 6.12 shows the response of the GWE (left panel) and the augmented EKF (right panel). The range-bearing sequence is the same for both trackers.

Figure 6.12 illustrates the advantage accruing to proper use of the hybrid tracking algorithm. Despite the fact that the projected $\mathscr{G}[k]$-distribution is being to determine the assurance regions, the EEP ellipses are so small as to be invisible on the graph— with two notable exceptions that we will return to later. In Fig. 6.3, we see the $\mathscr{G}[k]$-density going into the T-junction at $(-950, -800)$. The left panel shows the conditions 50 m east of the junction and the right panel shows the conditions at the junction. When the target is at $(-900, -800)$, $\mathbb{P}(\iota = 444|\mathscr{G}[k+1]) \approx 1$: the $\mathscr{G}[k+1]$-distribution is essentially Gaussian and $\mathbb{P} \approx \hat{\mathbb{P}}$. However, when the target is at $(-950, -800)$, the likelihood of the true mode is reduced: $\mathbb{P}(\iota = 444|\mathscr{G}[k+1]) = 0.81$, and a fraction of $\mathbb{P}$ points east. In both cases, the EEP ellipse is hidden beneath the graph of the road displayed in Fig. 6.12: in the latter case because the density is so tight; in the former case because the uncertainty is distributed east-west.

The advantage of the **GWE** is twofold. It knows there the junctions are (they are listed in $\mathcal{M}_{cr}$), and it has some warning before a turn event occurs. But there are situations in which the way the **GWE** integrates the totality of information is not sufficient to resolve the event ambiguity. Consider the tracker state after the target turns west from the $(-550, -400)$ T-junction. In 5 s, the target progresses to the point $(-600, -400)$.

The T-junction is contained within a rather labyrinthine environment. The target is initially moving south: $\phi_\iota = (222)$. The **GWE** location estimate is accurate and EEP is so small it does not show on the graph. As the target approaches the junction, it could continue south, it could turn west, it cannot go east, and the probability of returning north is quite small. The regime measurement is useful for indicating a turn event, but it does not give a direction.

The **GWE** used in this example has an abridged junction logic: all junctions are said to be 4-way. Thus the **GWE** imagines an east turn (prohibited) to be as likely as a west turn. There are no congruent junctions to the east of $(-550, -400)$, and therefore all east-directed local filters must stop at the junction. This causes an accumulation of extrapolations at $(-550, -400)$; i.e., all of the $\iota = 3jl$ filters collect at the junction. The $\mathcal{G}[k+1]$-probability of a west extrapolation is $\mathbb{P}(\phi = \mathbf{e}_4) = 29\%$. With probability $\mathbb{P}(\phi = \mathbf{e}_3) = 23\%$, the east regime is nearly as likely as the west regime.

The $\mathcal{G}[k+1]$-density is shown in Fig. 6.13. The biggest probability mass accumulates near the true location, $(-600, -400)$. However, the probability mass

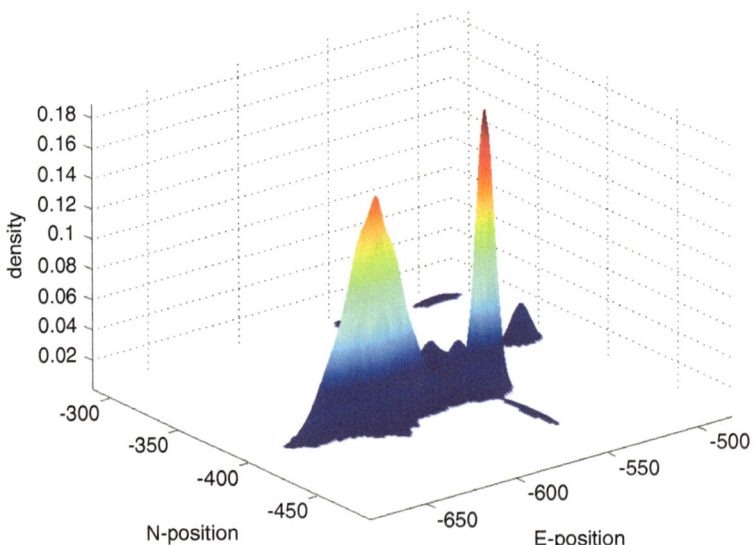

Fig. 6.13 The GWE finds the west turn at $(-550, -400)$ hard to follow. The simplified turn logic leads to an accumulation of probability at the junction despite the fact that the target has moved west to $(-600, -400)$. The $\mathcal{G}[k+1]$-conditional location density is shown

at the junction is also large; indeed, the mode of the distribution is located at the junction rather than $(-600, -400)$.

There are a variety of tracking artifacts exposed in Fig. 6.13. There is a small south component of the distribution which is to be expected since it is permissible for the target to pass through the junction: $\iota^+ = 2222$. There are a couple of north humps that are the result of modes that correspond to a slowed south motion; e.g., $\iota^+ = 2332$.

There is even a component on the road segment at $-500E$. This is prohibited in $\mathscr{M}$ because this would require jumping across the prohibited corridor from $-550E$ to $-500E$. But the GWE logic assumes that all junctions are accessible from every direction. This small probability is an instance of the *frontage road* effect in which the tracker places the estimate on a nearby road segment even though there is no direct connector.

The most prominent artifact in Fig. 6.12 is $\hat{\mathbb{P}}$-EEP ellipse at $(-750, -300)$. The regime measurements are a hindrance when they are wrong, and the motion is in a tangled portion of the path. As the westbound target approaches the 4-way junction at $(-750, -400)$, the regime measurement signals north-south twice before the junction is actually reached. The GWE arrives at the junction uncertain of the direction of extrapolation. The $(-750, -400)$ 4-way has compatible junctions in all directions. This spreads $\mathbb{P}$ as it gives weight to each of the potential regimes.

The fine structure of the $\mathscr{G}[k+1]$-density is shown in Fig. 6.14. The target entered the 4-way junction at $(-750, -400)$ with confidence in the west-bound motion. But, the $\mathscr{G}[k + 1]$-data set is not sufficient to resolve the turn. The 64-term density has 14 distinguishable peaks. The mode of distribution is correct. But all four directions from the junction have representation. The prominent east hump is due to the fact that the GWE is slow to leave the junction at $(-550, -400)$. Because the primary uncertainty is longitudinal, the $\mathscr{G}[k + 1]$-distribution tends to spread to the east.

In contrast with the EKF-trackers, the GWE acknowledges this lack of confidence. The large EEP ellipse simply warns the system that the GWE is confused, and no actions should be taken. Fortunately, the confusion is resolved in the next time step, and at $(-750, -200)$, $\mathbb{P}(\iota = 111) = 99.7\%$. Suffice it to say that the $\hat{\mathbb{P}}$ covariance ellipses are a very conservative measure of the assurance regions.

The fine structure of the $\mathscr{G}[k + 1]$-densities near the two anomalous points on the path is difficult to interpret from the density plots. Figure 6.15 shows contour plots of the densities: post west turn (left panel); along the west bypass (middle panel); post north turn (right panel). To display a large region, the axes are not square with the horizontal axis compressed. As the target moves south, the GWE mistakes the turn type: the map prohibits an east turn, but the GWE is not aware of this and lets probability accrete at the junction (the left panel).

Passing the junction, the GWE rationalizes $\mathbb{P}$, and by the time the target reaches $(-700, -400)$, $\mathbb{P}$ is essentially Gaussian (middle panel). After another 50 m, the target turns north at $(-750, -400)$ (right panel). This is the most difficult stage of the tracking engagement. The $\mathscr{G}[k + 1]$-density is spread along the road in all four directions in a lumpy manner that gives significant weight to several motion hypotheses.

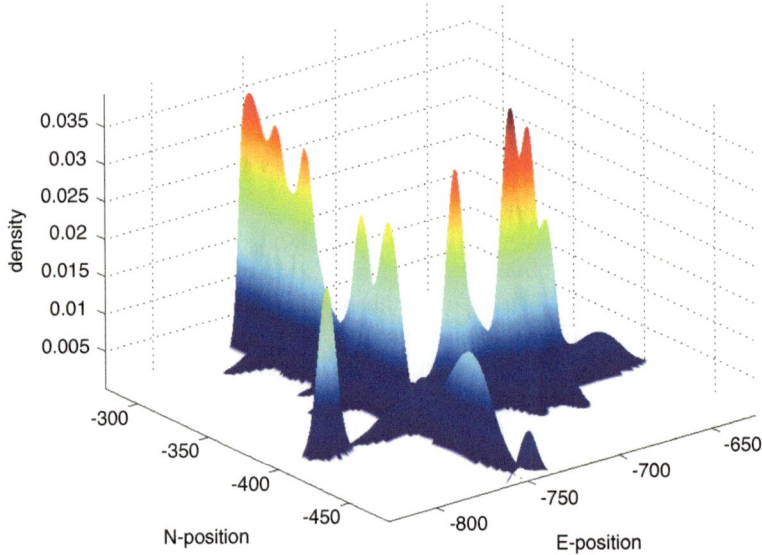

Fig. 6.14 The 4-way junction at $(-750, -400)$ is not well resolved by the **GWE**. The $\mathscr{G}[k+1]$-location density is shown one step beyond the north turn. Significant weight is placed on all of the alternative road segments

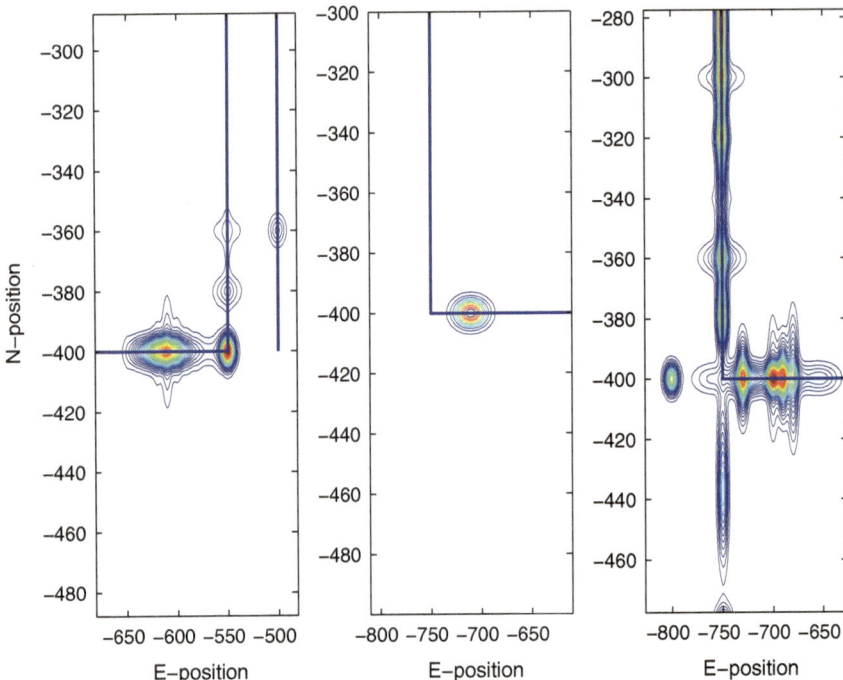

Fig. 6.15 Contour plots of the $\mathscr{G}[k+1]$-location density as the target moves across the -400N bypass segment

The GWE performs well in this engagement. At only two sample points—out of the 61 points total—does the GWE give a poor estimate of location. In contrast to the EKF, this confusion is manifest in the error statistics in $\hat{\mathbb{P}}$. And in each case, the GWE remedies this uncertainty at the following sample time.

6.4.2.2   Contrasting the GWE with the Augmented EKF

The augmented EKF is superior to the nominal EKF because its bandwidth is greater. It achieves its advantage by increasing the size of its notional EEP. The augmented EKF increases its ostensible P^w well beyond the acceleration expected on the straight sections of the path. Thus, the EKF can be said to trade off model accuracy against the quality of the assurance region.

The GWE uses a more nuanced model than do either of the EKFs. It uses a regime measurement to cue it to the forward regime—though this measurement is relatively coarse. Of more advantage is the path map, $\mathcal{M}$. The map identifies singular points on the path where specific events can occur. The GWE needs not average over event strings in the way that the EKFs do.

Despite its obvious benefit, the use of singular points to re-initialize the kinematic estimates has received little attention in the tracking literature. For example, as the target approaches a north turn at $(-950, -800)$ its east uncertainty is large, and the north uncertainty is small. Following the turn, the uncertainty is reduced in both directions; see Fig. 6.3. At the junction, the uncertainty in north and east is of size compatible with the dimension of the intersection. As the target leaves the junction, uncertainty will now build up in the north direction, but it will stay small in the east direction. The turn event compresses the highly eccentric covariance ellipses into a small region at the junction.

A handoff event like a junction is quite useful for reducing the size of the assurance regions coincident with the event. But there are issues that make utilization of such events difficult. We have already observed that the support of $\mathbb{P}(x|\mathcal{G}[k+1])$ is the whole of $\boldsymbol{R}^4$. Because the $\mathcal{G}[k+1]$-distribution is diffuse, even the prohibited region west of -950E is within its support set, albeit there is a small probability of a constraint violation. Indeed, we cannot be sure when or even if we will enter the junction we are using as a reference point. Following entry, we are never completely sure of our exit direction. Hence, we must balance the likelihoods of the various permissible events to arrive at a rational decision regarding target placement.

With these caveats, we can gain insight into the broad advantage of the GWE by contrasting it with the augmented EKF-tracker. The EKF does not acknowledge the multi-regime nature of the engagement nor does it avail itself of a map. Rather, it balances the lateral and longitudinal excitation within a single channel noise matrix. Figure 6.16 shows the radial error sequence in this engagement. Again we note that both trackers see the same range-bearing sequence.

We see that the median tracking error of the GWE is essentially zero—actually 0.1 m. The median error of the augmented EKF is closer to 22 m. The GWE has a

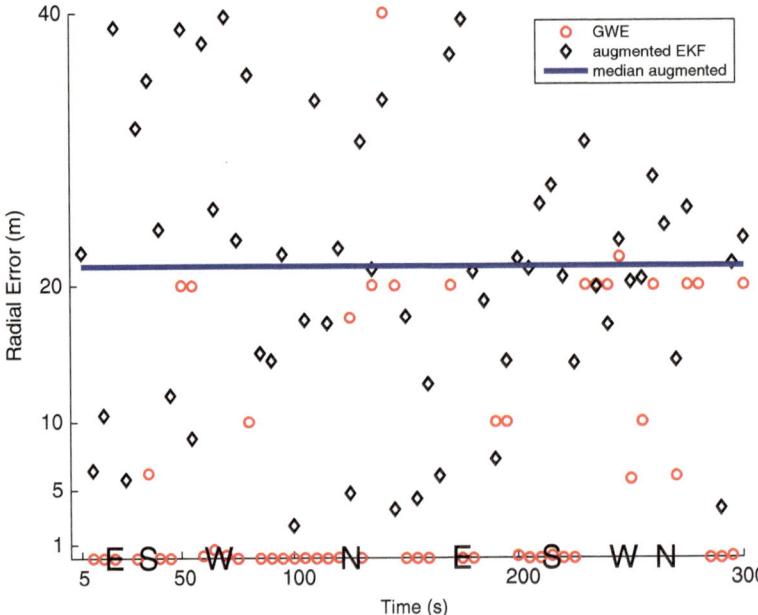

Fig. 6.16 The radial error of the augmented EKF tracker is (almost) uniformly bigger than that of the GWE tracker. The median error is orders of magnitude higher

smaller error at every point on the path—with the exception of the two anomalous points mentioned earlier. When the target is not in close proximity to a turn, the GWE reduces the radial error by two orders of magnitude. When the GWE is uncertain of its estimate, it acknowledges it by expanding its EEP. It then corrects the uncertainty in the next time sample. The granularity of the figure is due to the discrete partitioning of the vehicle path.

To explore the quality of the notional assurance regions, let us revisit the quality of the EEP calculations. We have noted that the median radial error of a Kalman filter is 1.18σ. The augmented EKF is consistent with that expectation; its notional EEP encloses the target position about half the time with a median radial error of 1.24σ. The GWE is much better. Figure 6.17 shows the normalized radial error of the GWE (using the projected distribution $\hat{\mathbb{P}}$) and the augmented EKF. First we note that the median error of the GWE is minute—actually 0.01σ. The radial error exceeds the Rayleigh threshold only 15 % of the time vs. 50 % for the EKF. The normalized error for the GWE is smaller than that of the EKF at almost every sample time. The maximum radial tracking error is 2.6σ for the GWE—contrast 9.7σ in the EKF.

The EEP of the GWE is conservative insofar is its capture frequency is above the Rayleigh threshold. But its usefulness also depends upon its area. A large assurance region is of less predictive quality than is a more compact region. Again we will use the projected distribution, $\hat{\mathbb{P}}$, to delineate the assurance regions. Figure 6.18 shows

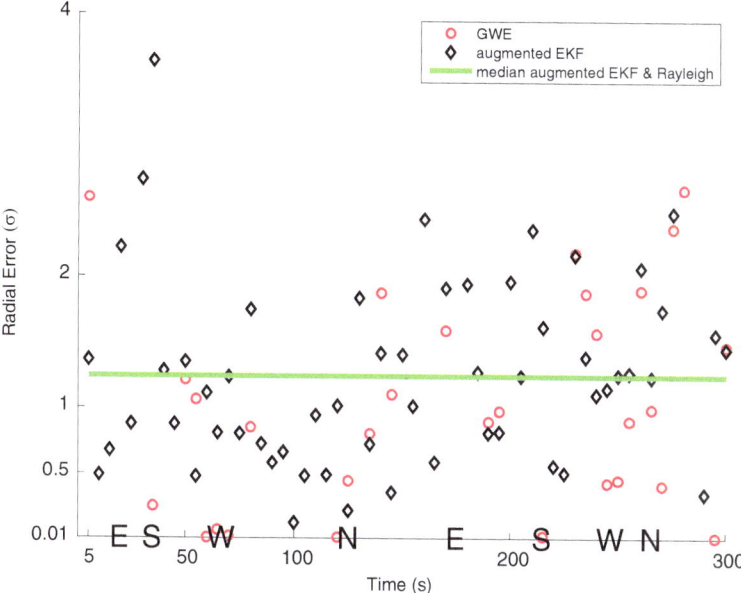

Fig. 6.17 The normalized radial error of the GWE is far smaller than that of the augmented EKF

the areas of the notional EEP for both trackers. After the tracking problem begins, in only two cases does the EKF generate a more compact EEP than does the GWE. Both exceptions are visible in Fig. 6.12. Because of its huge size, the giant GWE assurance region, $4E07$ m^2, is not shown in Fig. 6.18.

One measure of the relative quality of the two trackers is the ratio of the area of the EEPs. The EKF assurance densities are smaller in the low SNR portions of the path. The GWE assurance densities are smaller in neighborhood of a junction. Figure 6.19 shows the area of the EKF-computed area of the EEP normalized by the area of the GWE-computed area of the EEP. The area ratios have a large range, and to plot the values on this engagement on a single graph, we have used a logarithmic scale.

An area ratio less than one indicates that the GWE provides a tighter assurance region than does the EKF. After the initial step in the engagement, the GWE-tracker has a tighter assurance regions at nearly every sample. The two significant exceptions are apparent in Fig. 6.12. All of these calculations are done with the projected distribution,$\hat{\mathbb{P}}$, instead of the fine structure distribution, $\mathbb{P}$. For this reason, we view the size ratios presented in the figure as being conservative. The median improvement in the size of the EEP is about two orders of magnitude.

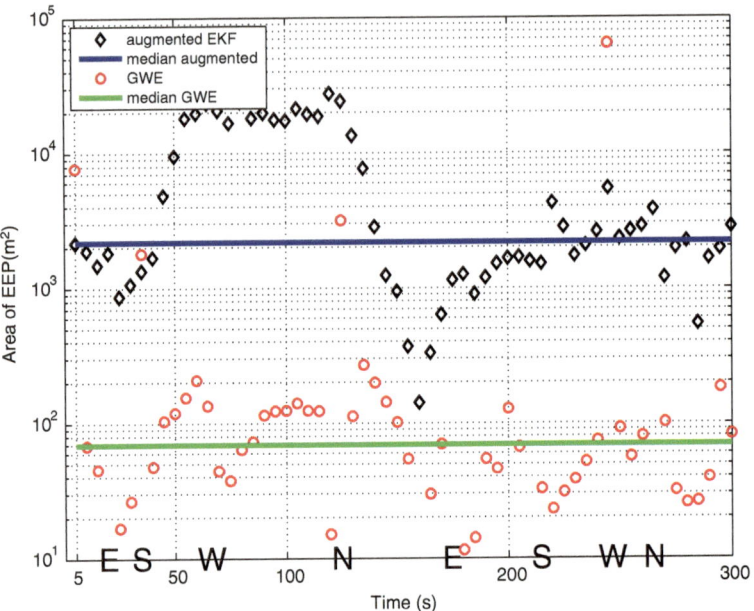

Fig. 6.18 The area of the EEP regions are an order of magnitude larger in the augmented EKF as compared with the GWE

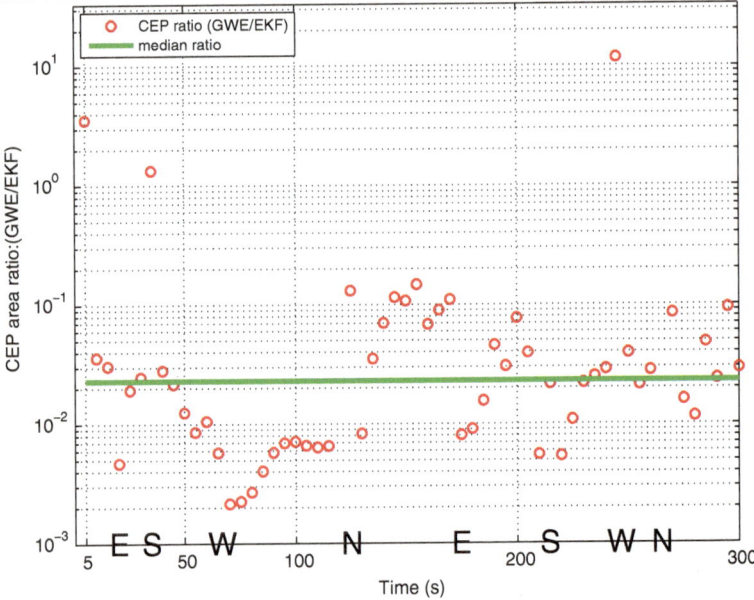

Fig. 6.19 The ratio of the area of the EEP regions favors the GWE by a factor of nearly one hundred

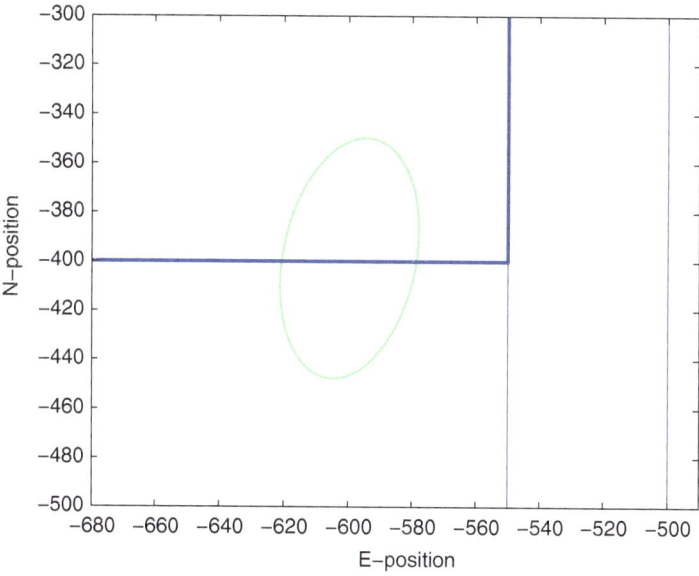

Fig. 6.20 The 75 % assurance region generated by $\hat{\mathbb{P}}$

6.4.2.3 Quality of the Assurance Regions

The notional EEP ellipses are shown in Fig. 6.12 for both the augmented **EKF**-tracker and the **GWE**-tracker. We have used the projected distribution to form the EEP in the latter case. In over 90 % of the samples, $\mathbb{P} \approx \hat{\mathbb{P}}$, and the EEP shown in the left panel is as tight as possible.

There are exceptional cases in which the EEPs in the figure are conservative. This will occur when $\mathbb{P}$ is multimodal. Figures 6.20 and 6.21 illustrate this contrast. The target has been moving south and encounters a T-junction at $(-550, -400)$. Instead of continuing south, the target turns and moves west. There are no compatible junctions to the east, and the **GWE**-tracker accretes probability to the junction itself.

Figure 6.20 shows the 75 % assurance region for target location using the Gaussian approximation $\hat{\mathbb{P}}$. The region is elliptical with axis lengths derived from the Rayleigh distribution. This approximation to $\mathbf{A}_{0.75}$ is simply connected and roughly centered on the target: $\chi = (-600, -400)$. But most of $\mathbf{A}_{0.75}$ lies in the prohibited region off the roadbed. The assurance density could be increased by an order of magnitude if we let $\mathbf{A}_{0.75}$ be the intersection of the covariance ellipse with the 10 m wide roadbed.

For optimal countermeasure allocation, we seek the smallest region in the plane for which we are assured that $\chi \in \mathbf{A}_{0.75}$ with probability 75 %. This would be that region in which $\mathbb{P}$ exceeds the appropriate threshold. Figure 6.21 shows $\mathbf{A}_{0.75}$ derived from $\mathbb{P}$. This computed $\mathbf{A}_{0.75}$ is but half the area shown in Fig. 6.20; 1630 m^2.

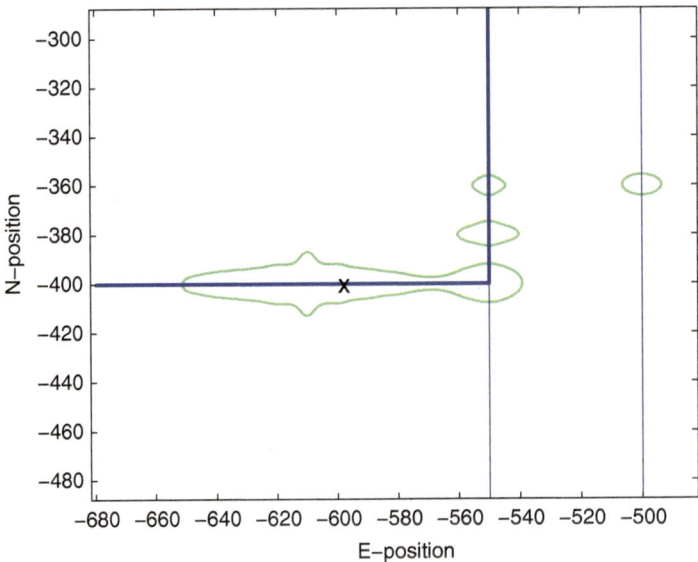

Fig. 6.21 The 75 % assurance region generated by $\mathbb{P}$

The 64-term Gaussian sum, $\mathbb{P}$, displays considerable internal structure. It closely overlaps the road. It preserves a component lying across the separation zone at $-550E$. The intersection of $\mathbf{A}_{0.75}$ with the roadbed would not significantly increase the assurance density as would happen with the $\hat{\mathbb{P}}$ calculation. This $\mathbf{A}_{0.75}$ shows the strong modal peak at the corner—$\mathbf{A}_{0.75}$ from $\hat{\mathbb{P}}$ misses the corner completely and less than 6 % falls along the road.

6.4.2.4 Applications of Assurance Regions

The choice of an assurance region depends in part on the use to which it will be put. If we are going allocate ballistic particles to neutralize the target, we would naturally select a convex cover region; $\hat{\mathbb{P}}$ would be an appropriate choice for the defining distribution. However, smart particles could cover a multiply connected region; $\mathbb{P}$ would be an appropriate choice for the defining distribution. But the assurance region is often much more difficult to find using $\mathbb{P}$. For example, Fig. 6.21 was obtained by a direct search.

The assurance regions deduced from the $\mathscr{G}[k+1]$-distributions are useful in many applications. For example, suppose the path environment were rich in clutter sources. Tracking is improved if a gate is placed about the target estimate, and only those measurements originating from the gated region are accepted by the tracker [27]. A tight assurance region removes more of such artifacts [19, 23].

Conflict avoidance is another situation in which tight computation of A_p is important. We may have a convoy of controlled vehicles. Two such vehicles are not permitted to occupy the same spatial neighborhood: a minimum separation is required from the center of vehicle one to the center of vehicle two. Let the 75 %-level assurance regions of these vehicles be labeled $\mathbf{A}_{0.75}(1)$ and $\mathbf{A}_{0.75}(2)$. Then at level 75 %, an engineer must insist that $\mathbf{A}_{0.75}(1) \cap \mathbf{A}_{0.75}(2) = \phi$ or even that $\|\mathbf{A}_{0.75}(1) - \mathbf{A}_{0.75}(2)\| \geq \delta$ for an appropriate δ.

In many applications, there is a need for target awareness transcending the kinematic mean. For example, suppose $\check{A}$ is a region of the position space, entry into which prohibited. An engineer may wonder how likely it is that the target moves to within a distance δ of $\check{A}$. We must accept the fact that a tracker does not place the target determinately. Indeed, any non-trivial position region is possible, at least with some small probability. We can, therefore, never say with certainty that $\check{A}$ has not been breached.

While we cannot be sure that kinematic state violations will not occur, we can quantify likelihood of such events. For example, from $\mathbb{P}$, we can evaluate $\mathbb{P}(\|\chi[k] - \check{A}\| > \delta)$: the probability that the target is at least δ meters from the prohibited zone. If this probability is high, e.g., 0.99, we would say that we were in position compliance with the domain avoidance constraint at the 99 % level.

6.5 Conclusions

The EKF is a unimodel algorithm that represents motion uncertainties with a (roughly) isotropic, white acceleration. The factors that distinguish the individual modes on a path composed of segments in the cardinal directions do not appear explicitly. For example, velocity and compliance with the junction restrictions do not enter into the acceleration model. It is precisely when the motion modes are idiosyncratic that multiple-model algorithms become attractive. Instead of a single kinematic template, a family of local models is tuned to the distinctives of the target motion; e.g., direction, nominal speed.

The GWE is a time-discrete hybrid estimator that supervises the simultaneous evolution of a family of local filters and associates a state estimate with each. The kinematic disturbance has both structured and unstructured components. The latter is represented by Gaussian white noise. The former, in the role of a random directional process, creates the path seen in the figures.

The engagement studied in this chapter demonstrates the performance advantage of a tracker that integrates para-measurements into location estimates. The notional assurance regions generated by the GWE are of higher quality than those generated by the EKF.

Assurance regions are not unique: any set that contains $\mathbf{A}_p$ is also an assurance region at level p. But compact assurance regions are most useful for reducing the illumination region for a target, for determining minimum separation of like targets in coordinated motion, and other uses.

Unfortunately, the $\mathscr{G}[k + 1]$-distribution generated by the GWE is more complex than that generated by the EKF. But the corresponding $\mathbf{A}_p$ may not be. The assurance region generated by the EKF is the interior of a covariance ellipse. The same will be true of the assurance region generated by the GWE when there is a single term dominance in $\mathbb{P}$. Surprisingly, this single-term dominance is frequent in the applications studied.

During times proximate to a possible regime transition, $\hat{\phi}[k]$ may be more diffuse with several significant components. Fortunately, $\hat{\mathbb{P}}$ flags these cases with a large notional covariance. Were the countermeasures to be apportioned according to $\hat{\mathbb{P}}$, an engineer might decide not to use his allocation when the EEP fails to meet some simple criterion.

One test for the quality of the location estimate is the confidence that the GWE places in the modal estimate. At every time, $\sum_\kappa \hat{\phi}_\iota = 1$. Typically, there will be one mode for which the $\mathscr{G}[k+1]$-probability is approximately one. When this happens, the GWE is also confident of its location estimate: the area of the assurance region is small. Of course, the GWE can be both confident and wrong, but at least the tracking error tends to be small when modal confidence is high.

But there are times when the GWE fails to identify the mode. These are also times at which the area of the assurance region is large and the quality of the tracker placement is suspect. It would be convenient to have a simple flag for these times of ambivalence. A plausible statistic would be the quadratic norm of the modal probabilities: $\|\hat{\phi}_\iota\|$. When the GWE is confident of its modal estimate, $\|\hat{\phi}_\iota\| \approx 1$. If the GWE is completely ambivalent, assessing all modes as equally likely, $\|\hat{\phi}_\iota\| \approx 0.125$.

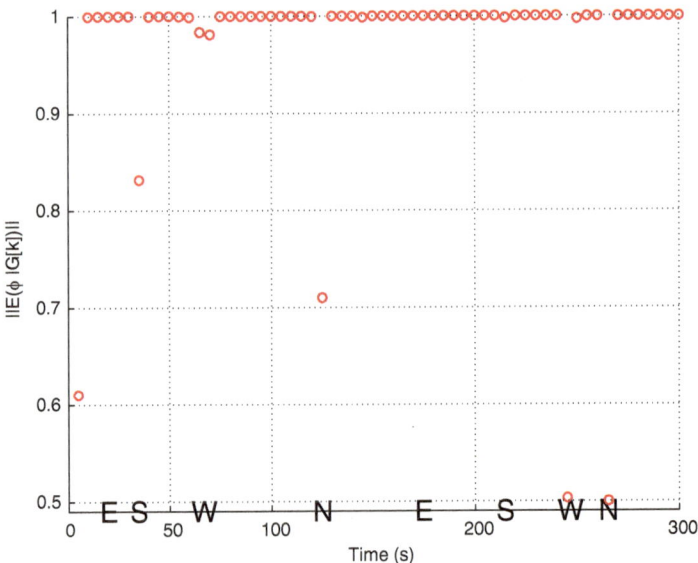

Fig. 6.22 The norm of $\hat{\phi}[k]$ flags times of uncertainty

An even simpler flag would be the regime norm $\|\hat{\phi}_t\|$. At time $t = kT$, $0.5 \leq \|\hat{\phi}_t\| \leq 1$. A plot of $\|\hat{\phi}_t\|$ is shown in Fig. 6.22. The suspect EEP calculations near the turns are manifest in this plot. The two worst estimates are coincident with the smallest values of $\|\hat{\phi}_t\|$. A systems engineer might use a decision logic as follows: Take an action only if $\|\hat{\phi}_t\| \geq 0.75$. This simple rule would avoid those times when the target placement is most suspect.

For a command architecture, the engagements studied have shown the GWE to be much superior to the EKF in efficient deployment of countermeasures. The estimate-quality refinements suggested show promise of even better performance in resource-constrained engagements. More on this topic in the next chapter.

Chapter 7
Action Windows with Resource Limits

Abstract The command algorithm makes resource decisions on the basis of the quality of its location or extrapolation estimates. When resources are expended, they can be distributed in a compact region with a high density. Or they can be distributed across an expanded region. An expanded region not only increases the capture probability but also expends resources at an increased rate. This chapter contrasts the uni-model and multi-model capture regions using the same raw measurement data set. The hybrid algorithm is much more efficient in resource allocation. It employs a confidence index that is used as an enable signal before resources are distributed. Hybrid algorithms improve capture rates with smaller resource demands.

7.1 Introduction

In previous chapters, we have developed algorithms for locating an agile target and predicting its future position. The algorithm provides a notional $\mathcal{G}[k]$-distribution (or $\mathcal{Y}[k]$-distribution) of position now or in the future. From this we can both place the target and determine regions with appropriate capture probability.

For example, consider a hostile target following the road-constrained path as presented in Chap. 6. An EKF-tracker was used to place the target every 5 s. The EKF was viewed as somewhat conservative since the action matrix, P^w, pointed to a bigger wideband acceleration than was actually used in the simulation.

A sample response of the nominal EKF-tracker is shown in Fig. 7.1—see also the left panel of Fig. 6.4. The figure shows the position estimates along with the 50 % assurance ellipses (the EEP regions). The scale has been compressed in north-south so the full path can be displayed.

We observe that during quiescent intervals, e.g., nearing the north turn at $(-950, -800)$, the EKF does well. The error is small and the EEPs encircle the target. Unfortunately, after a turn where the SNR is low; e.g., $(-350, -800)$, or where the GDOP is poor; e.g., $(-550, -50)$, the error is big and the EEP does not properly delineate the uncertainty in the estimate.

© Springer International Publishing Switzerland 2016
D.D. Sworder, J.E. Boyd, *Locating, Classifying and Countering Agile Land Vehicles*, DOI 10.1007/978-3-319-19431-8_7

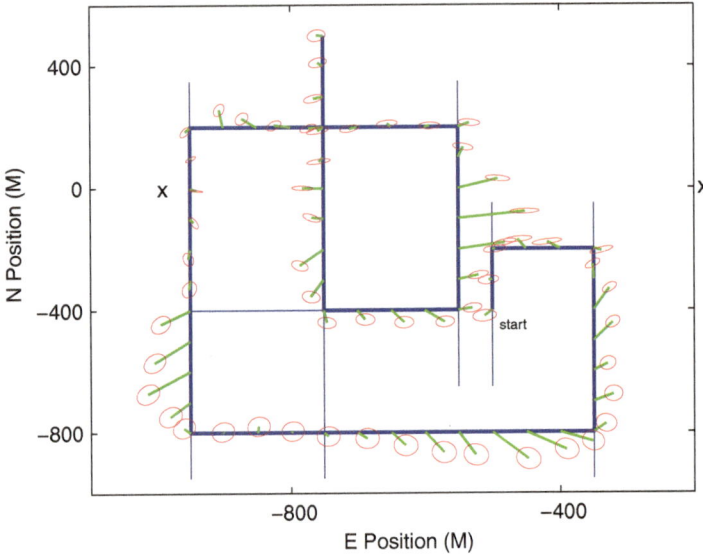

Fig. 7.1 A target of undetermined intent follows the path shown with *bold line* width. The EEP ellipses of an **EKF** are shown every 5 s

We would expect the EEP ellipses to encircle the target about half of the time. It is evident from the figure that this is not the case; in this engagement, the EEP ellipses capture the target at a rate far less than predicted. Consequently, the tracking accuracy of the **EKF**-tracker is not well described by the computed probability distribution.

In Chap. 6 an alternative **EKF**-tracker was proposed. The action matrix in the kinematic model was increased by adding pseudo-noise. This augmented tracker was found to have smaller errors after a turn and an increased bandwidth. The corresponding EEPs were more predictive of location.

Figure 7.2 shows an expanded view of both **EKF**s as the target moves toward the 4-way junction at $(-750, 200)$, both from the west and from the south. The same sequence of kinematic measurements is used in both panels. The computed EEPs of the former are more representative of the actual error than are those of the latter. Unfortunately, the areas of the augmented EEPs are considerably larger than those of the nominal **EKF**.

Let us imagine an engagement in which a target of uncertain intent is detected. The defender must illuminate the target to infer its purpose. The defender will focus on a small area to increase the interrogation accuracy. In another engagement, the defender recognizes a target with hostile intent and must neutralize the target. In the latter case, the defender may use a directed energy weapon (DEW) focused at the current position. Or the weapon may use ballistic particles focused on a future location of the target.

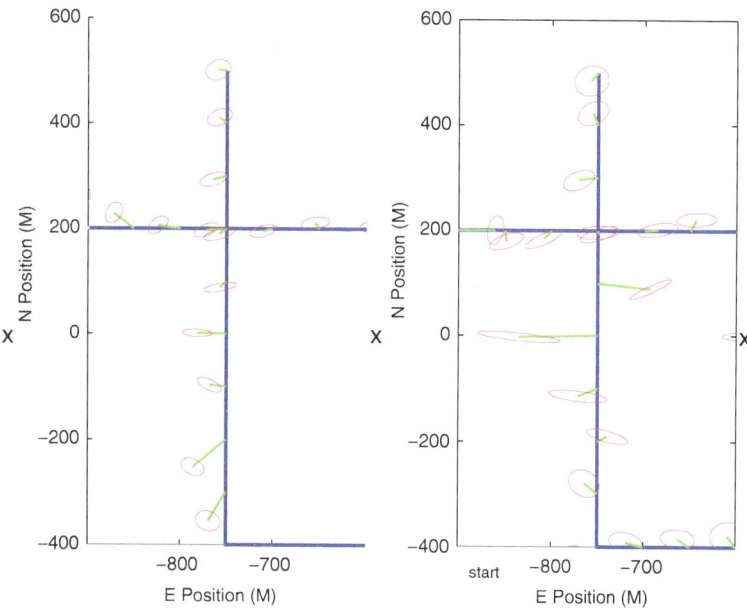

Fig. 7.2 The augmented EKF-tracker has larger EEP ellipses (*right panel*) than does the nominal EKF (*left panel*)

In what follows we will investigate the ability of a defender to place an action window about a target, either concurrently or predictively. If the target is within the action window, it is counted as a *hit*. If the target is outside the action window, it is counted as a *miss*. In either of the above engagements, proper defense requires that we hit the target a specified fraction of the time.

Actually, generating an action window requires energy, either to interrogate the target or to project countermeasures. We will suppose the power required of the action agent at each event time is proportional to the area of the action window. The total energy required of the defensive system is the sum of the power pulses; e.g., the total energy is proportional to the cumulative sum of the areas of the action windows. For the purpose of this exercise, we will measure energy expended in units of square meters.

The design of this defensive system requires a tradeoff of accuracy vs. energy. A tight action window conserves energy; a tight action window misses the target more often. In this chapter, we will not constrain the total energy. But we will contrast the algorithms regarding their ability to use their energy reserves efficaciously.

We will use two of our previous engagements to illustrate the tradeoffs involved: illuminating a target on a road grid (see Chap. 6); neutralizing an agile target with ballistic delay (see Chap. 4). In both engagements we will use the data stream generated for that chapter.

7.2   Action Windows with a Grid Map

7.2.1   The Engagement

7.2.1.1   Review of Kinematics and Sensor Architecture

The underlying engagement model is given in Chap. 6. To recapitulate briefly, the
representation of target motion between turns is CV:

$$
\frac{d}{dt}
\begin{bmatrix} X \\ Y \\ V_X \\ V_Y \end{bmatrix}
=
\begin{bmatrix} 0 & 0 & 1 & 0 \\ 0 & 0 & 0 & 1 \\ 0 & 0 & 0 & 0 \\ 0 & 0 & 0 & 0 \end{bmatrix}
\begin{bmatrix} X \\ Y \\ V_X \\ V_Y \end{bmatrix}
+
\begin{bmatrix} 0 & 0 \\ 0 & 0 \\ \sqrt{P^w_{x;\iota}} & 0 \\ 0 & \sqrt{P^w_{y;\iota}} \end{bmatrix}
\begin{bmatrix} \dot{w}_X \\ \dot{w}_Y \end{bmatrix},
\tag{7.1}
$$

where $\dot{w}_Y$ is a unit white noise acceleration, and ι is the modal state. The path is on
a rectangular road grid oriented in the cardinal directions. The regime state is the
current direction of motion: $i \in (1, \ldots, 4)$. The identity of the mode enters in the
action matrix. Because the exogenous accelerations are primarily longitudinal P^w
ellipses are highly eccentric.

At a junction, the target may (or must) turn. When a turn event occurs, the
exit velocity is perpendicular to the entrance velocity: the nominal acceleration is
infinite. The permissible paths are restricted by the map data set, $\mathcal{M}$. The terrain
is such that the better roads are north-south. Consequently, the nominal north-south
speeds are higher: the 20 m/s in north-south and 10 m/s in east-west. The roadbeds
are 10 m wide (30 ft).

The motion shown in Fig. 7.1 is simple. The target follows the bold path at the
direction-adjusted nominal speed along the center-line of the road. The kinematic
state is sampled every 5 s.

To complete the kinematic model we assume that the wide-band acceleration
is white. The 1σ-longitudinal velocity increment is 0.31 m/s^2 in north-south and
0.14 m/s^2 in east-west. The lateral velocity increments are a tenth of the longitudinal. A constant 1σ-acceleration over one sample interval would yield a position
increment of about 4 m longitudinally and 0.4 m laterally in north-south.

The tracker utilizes three kinematic sensors. The first two are conventional range-
bearing sensors with 1σ quality 95 m in range and 35 mr in bearing. They are located
at $(-300, 0)$ and $(-1000, 0)$. They provide a measurement of position every sample
time. The sensor errors are independent spatially and temporally. Additionally, there
is an acoustic speed sensor. Since speed is linked to direction, the raw acoustic
measurement is reduced to a statement of direction: the acoustic measurement is
placed into one of the velocity bins. The direction measurement, $z[k]$, is properly
classified only 38 % of the time with the errors placed symmetrically.

7.2.1.2 Notional Statistics of the Action Windows

In the conventional LGM tracking problem, the Kalman filter generates the conditional distribution of the target position. At time $t = kT$, this distribution is Gaussian: $\mathbb{P}(\chi|\mathscr{Y}[k]) = \mathbf{N}(\hat{\chi}[k], P_{\chi\chi}[k])$. The conditional mean, $\hat{\chi}[k]$, is the best mean-square, point estimate of location. The covariance of the location error is $P_{\chi\chi}[k]$.

The error in the Kalman-tracker, $\tilde{x}[k] = x[k] - \hat{x}[k]$, is a process of zero-mean independent Gaussian 2-vectors. Earlier we defined the normalized position error to be: $\tilde{\chi}_F[k] = \sqrt{D_{\chi\chi}[k]}\tilde{\chi}[k]$. Then $\{\tilde{\chi}_F[k]\}$ is a unit Gaussian white sequence: $\tilde{\chi}_F[k] = F_{\chi\chi}[k]\tilde{\chi}[k] \sim \mathbf{N}(0, I)$. In the coordinate system of normalized errors, the components of the tracking error are independent spatially and temporally with standard deviation equal to one.

We refer to the units of $\|\tilde{\chi}_F[k]\|$ as standard deviations. For example, if $\|\tilde{\chi}_F[k]\| = \|\tilde{\chi}[k]\|_{D_{\chi\chi}} = 2$, we would say that the tracker estimate is off by 2σ.

The radial error in the normalized coordinate system has a Rayleigh distribution with variance one: $\mathbb{P}(\|\tilde{\chi}[k]\|_{D_{\chi\chi}} \leq u) = 1 - \exp(-\frac{1}{2}u^2)$. Figure 6.9 shows a plot of a unit Rayleigh distribution. The EEP window captures the target 50 % of the time. To achieve this capture rate, we would choose the cover region to be the interior of the centered covariance ellipse with $\|\tilde{\chi}[k]\|_{D_{\chi\chi}} < 1.18\sigma$. If we needed to illuminate the target 95 % of the time, we would cover the $\|\tilde{\chi}[k]\|_{D_{\chi\chi}} = 2.5\sigma$ ellipse centered on $\hat{\chi}$.

Because $\{\tilde{\chi}_F[k]\}$ is white, the hit sequence is an independent binary random process. If we use the EEP to guide the action agent, the probability that the target is missed twice in succession is 25 %. The likelihood of four successive misses is only 6 %; an improbable event in this encounter of about 300 s. Alternatively, if we use a 2.5σ cover region, the probability of a two-miss string beginning at a specific point is only 0.25 %; a rare event.

Unfortunately, to reduce the frequency of miss events by increasing the multiple of σ requires a significant increase in the energy capacity at the action center. The power for a 95 % cover region is over six times as much as that required for an EEP window. The total energy grows quickly with any reduction in the probability-of-miss for strings: a two-miss string at 0.25 % takes thrice the energy of a four-miss string at 6 %.

In the following sections, we will look at the miss-rate vs. energy tradeoff. We would like to have an action algorithm that illuminates and interrogates the target at least half of the time with no miss strings longer than three. And we would like to do this with minimal illumination energy. We will see that we can achieve both aims with a hybrid tracking algorithm.

7.2.2 *EKF-Illuminator*

7.2.2.1 The Nominal EKF-Illuminator

In earlier chapters, we observed that the conventional approach to this interrogation problem would to base the placement of the illumination window upon an EKF-tracker that replaces the accelerations at the turns with a white exogenous forcing term. In Chap. 6 we proposed that P^w should simply use the longitudinal uncertainty to size the action matrix; $P^w = \max_i P_i^w$, where the maximization is component-wise: P^w is diagonal with 1σ-acceleration $0.31\,\text{m/s}^2$ in north-south and $0.14\,\text{m/s}^2$ in east-west.

In that chapter we generated a sequence of kinematic measurements compatible with the range-bearing uncertainty. From this we deduced the $\mathscr{Y}[k]$-distribution of position. The response of the nominal EKF is shown in Fig. 7.1. Despite the fact that the motion is almost exclusively CV, and that the tracker uses a conservative action matrix, the EEP regions fail to capture the target as often as expected.

More specifically, the upper panel in Fig. 7.3 shows the cumulative number of hits as this engagement evolves. We would have predicted about 30 hits on the interval since the EPP should cover the target half the time. Instead we find that we have only 12 hits in 300 s: the hit-rate is only 20 %. Moreover, we see long strings in which we

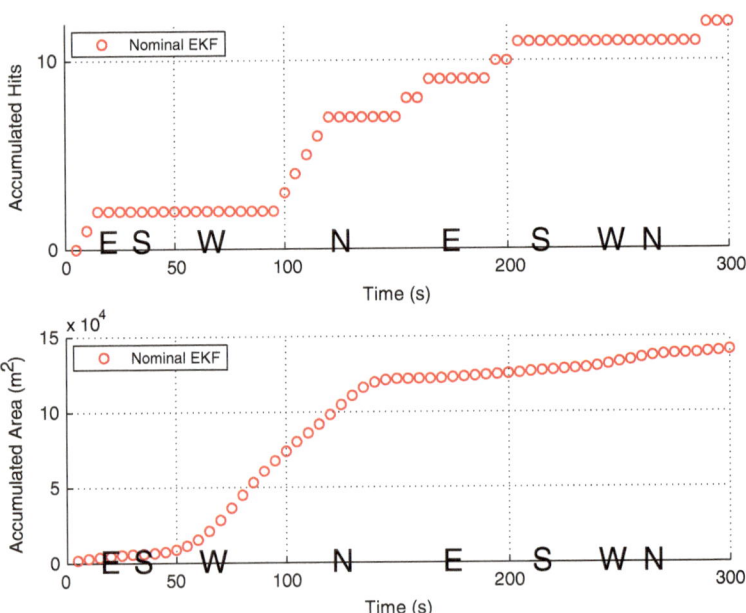

Fig. 7.3 The cumulative number of target illuminations and the associated energy are shown for the nominal EKF. The times of turns are shown across the abscissa

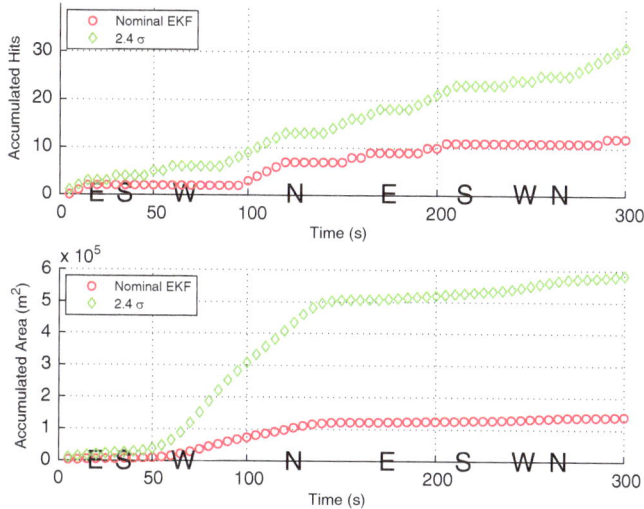

Fig. 7.4 With an assurance region defined by the 2.4σ error ellipse, the nominal EKF achieves a 50 % cover rate

fail to cover the target. For example, as the target makes the south-to-west-to-north passage circa -200N, the target is not illuminated for nearly one hundred seconds. Strings of this length are unacceptable and are not predicted by the EKF.

The energy used in the engagement is proportional to the cumulative sum of the cover areas. The energy process is shown as the lower panel in Fig. 7.3. The units are m^2, and we should think of the shape of the figure as giving insight into the rate of energy expenditure. The total illumination energy expended in this engagement is almost 1.5E5. Most of that is required when the target moves along the west bypass at -800N. In this region, the SNR of the sensors is low and the EEP ellipses large.

The nominal tracker seems unable to properly place the interrogation window. Of course, we might reframe the problem as follows: How big must the assurance regions be in order that we illuminate the target half of the time?

Figure 7.4 shows the cover rate when the 95 % error ellipse is used to determine the cover region. Unfortunately, instead of a cover rate of 95 %, we achieved only a 50 % rate. The energy cost in this new architecture is considerable. The total cover area of the 2.5σ EKF-tracker is now $6E5$, a growth in energy expended by more than a factor of four.

Even though the miss-rate is about 50 %, there are still long strings in which the target is not illuminated. One string of 30 s has a notional probability of 1.6E-6 %. But it occurs during this engagement. The normative EKF-tracker is not adequate for directing the interrogation window.

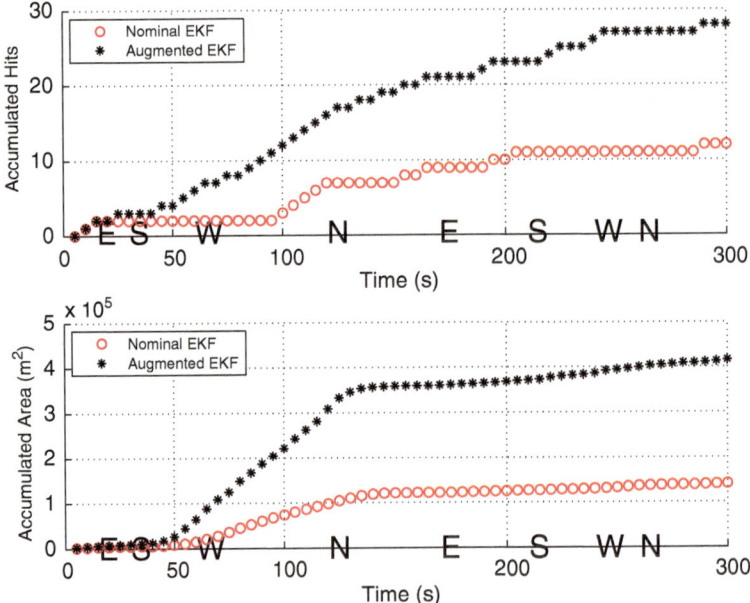

Fig. 7.5 The augmented EKF-tracker achieves the specified hit-rate, but the energy required is significantly increased. There are excessive miss-strings

7.2.2.2   The Augmented **EKF**-Illuminator

In Chap. 6, we proposed increasing the tracker bandwidth by increasing the size of the exogenous acceleration in the kinematic model; i.e., we replaced P^w with: $P^w + \Delta P^w$; $\Delta P^w > 0$. The choice of ΔP^w is an engineering judgement. For reasons outlined in that chapter we increased the acceleration intensity by a factor of 9: the 1σ-increment in velocity is $1.26\,\text{m/s}^2$ in east-west, and $2.5\,\text{m/s}^2$ in north-south.

Pseudo-noise augmentation makes the tracker faster, but it does so at the expense of increasing the area of the EEP. This is clearly seen in the right panel of Fig. 6.4. Processing the same kinematic measurements that were used in the nominal EKF, the augmented tracker responds more quickly. The location estimates in the right panel are less sensitive to junctions, but the distributions are more diffuse.

Figure 7.5 shows the cover sequence generated by the augmented EKF. The augmented EEP regions actually enclose the target about half of the time. Thus one could argue that the computed EEP is predictive of the illumination probability. Of course, the tracker still has strings in which the target is missed: the string of length nine seen in the figure has a notional probability 0.2 % of occurrence.

Unfortunately, the areas of the assurance regions of the augmented EKF-tracker are rather large: the total expended energy using the augmented tracker is 71 % of that required by the normative tracker using 95 % cover regions.

7.2.2.3 EKF Contrasts

As pointed out earlier, the kinematic model used in the nominal EKF has a conservative representation of target motion during most of the engagement. The speed is actually constant along any road segment. Since P^w is smaller in the nominal than it is in the augmented tracker, we might contend that the nominal EKF-tracker uses a more accurate kinematic model than does the augmented tracker. However, the influence of regime transitions is broad, thus weakening that argument.

Both EKF-trackers ignore the turn events and the speed changes following these events. Thus, they depend on a kinematic model that blurs the idiosyncratic character of the motion. The augmented EKF-does a better job of recovering from changes in direction. But when used to direct an illuminator toward the target, the augmented tracker is not much better than the nominal tracker. With a slight reduction in expended energy, the augmented tracker suffers a slight increase in miss-rate. For a given energy budget, the miss-strings in the augmented tracker are longer, and this is viewed as a disadvantage.

On the other hand, the augmented EKF has a clearer sense of how the uncertainties in engagement are reflected in the cover regions. The nominal EKF-tracker achieves about the same miss-rate with about the same energy expended if multi-sigma action windows are used. But the assurance regions that are far bigger than necessary. Thus, it is the lack of conformity of the predicted quality to the actual quality of the illuminator that weighs against the nominal EKF.

7.2.3 GWE-Illuminator

7.2.3.1 GWE Overview

In this engagement, the nominal and the pseudo-noise augmented EKF have comparable miss-rates when they use comparable amounts of energy. Chapter 6 presents an alternative, the GWE, that has much better performance. The GWE is a hybrid algorithm that maintains 64 individual estimators: the modal string is of length three. The GWE uses the same set of range-bearing measurements that is used by the EKF. But it fuses the concurrent speed measurement with the range-bearing measurement. Furthermore, the GWE avails itself of a map of the street grid. Integrating these disparate data sets, the GWE-tracker proves to be superior to either of the EKF-trackers.

Figure 7.6 shows an expanded comparison of the response of the GWE and the augmented EKF again near the 4-way junction at $(-750, 200)$. The $\mathscr{Y}[k]$-distribution of the EKF is nominally Gaussian. The EEP ellipses of the GWE shown in the figure are based upon the projected distribution, $\hat{\mathbb{P}}$, and this distribution is nominally Gaussian as well.

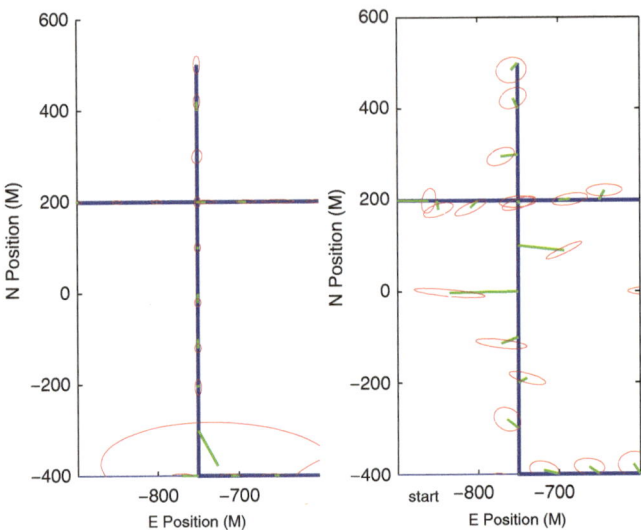

Fig. 7.6 With one major exception, the EEP ellipses of the GWE-tracker are far smaller than those of the augmented EKF

The 4-way junction shown in the figure is difficult for the GWE to navigate. The target makes a north turn at the bottom of the figure. The target then approaches the junction—which it will ignore—with parallel roads both to the west and to the east with the same nominal speed as the −750E link.

Both trackers in the figure use the same $\{\mathscr{Y}[k]\}$ sequence, and the GWE-tracker is clearly superior. With the exception of the sample at $(-750, -300)$, the error is small and the EEP ellipses cover the path. This is in contrast to the augmented EKF which fails in both respects. Unfortunately, the EEP associated with the anomalous event at the north turn is so large that its area overwhelms all of the other error ellipses combined.

7.2.3.2   Modal Confusion

In contrast to the EKF, the GWE acknowledges the hybrid nature of the engagement. As part of the tracking algorithm, the GWE computes a $\mathscr{G}[k + 1]$-estimate of the modal state, $\hat{\phi}_\iota[k + 1]$. If the observation sequence is of good quality, there is usually one $\iota \in \kappa$ for which $\hat{\phi}_\iota[k + 1] \approx 1$. The GWE-tracker emphasizes this local model in extrapolation and computes its uncertainty accordingly.

If the GWE is confident in its identification of the mode at time $t = kT$, the quadratic norm of $\hat{\phi}[k]$ is about one: $\|\hat{\phi}[k]\| \approx 1$. On the other hand, if the tracker is nonplussed, $\|\hat{\phi}[k]\| \ll 1$. Indeed, if all modes are equally likely, $\|\hat{\phi}[k]\|$ can be as small as 0.125.

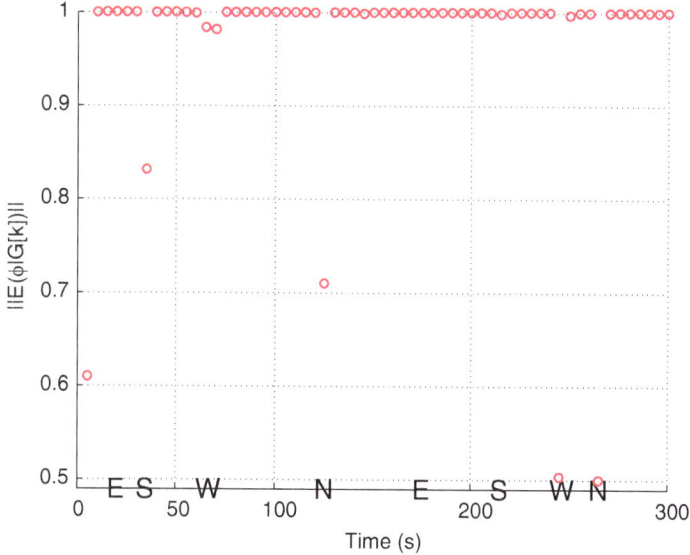

Fig. 7.7   The norm of $\hat{\phi}[k]$ flags times of modal confusion

The norm of $\hat{\phi}[k]$ is an easily computed measure of the modal uncertainty as the target moves along the path. However, it suffers from the fact that past regimes are weighted as heavily in $\|\hat{\phi}[k]\|$ as is the current regime. In what follows, we will use quadratic norm of the regime probabilities, $\hat{\phi}_i; i \in \mathbf{S}$, to gauge the ambiguity. Of course, one could use a weighted quadratic norm on $\hat{\phi}_\iota; \iota \in \boldsymbol{\kappa}$ to achieve a blending of the current versus the recent modal past. But this is not explored here.

With four regimes, $\|\hat{\phi}[k]\|$ ranges between one half and one. Figure 7.7 shows $\{\|\hat{\phi}[k]\|\}$ as computed for this engagement (see Fig. 6.22). The **GWE** is certain of its regime in almost all instances. It resolves the initial condition in one sample.

There are two other occasions where the **GWE** is undecided as to the current regime. Both instances are in convoluted portions of the path, and both are resolved in one time step. After the north turn at −950E, the **GWE** is unsure whether the target is moving north or still moving west into the turn. These regimes are determined to be about equally likely and $\|\hat{\phi}[k]\| \approx 0.7$. This ambiguity is clarified at the next sample time.

Earlier, following the south turn at −350E, an anomalous event occurred in the simulation. The acoustic sensor failed to measure the speed. The **GWE**-algorithm interpreted that to mean that all forward directions should be given equal weight. This is not the best interpretation of a failed measurement. Despite this, the **GWE** was not corrected in Chap. 6 because the tracking error was small, and a single sensor failure was judged to be a problem of slight concern. But as we will see, this artifact has important implications now. Using the current tracking logic, the tracker assigns 80 % confidence to north after the failure but retains a 20 % probability of continued west motion.

There are two additional points on the path that the **GWE** finds unresolvable. The west turn at $(-550, -400)$ and the north turn at $(-750, -400)$ cause the certainty index fall to about one half, the theoretical minimum. In both cases, the turns are difficult to isolate, and in both cases there is resolution in 5 s.

7.2.3.3 Illumination Rates

The sample performance of the map-enhanced **GWE**-tracker is presented in Chap. 6. The **GWE** fuses a noisy speed measurement and road map with the position measurements. From this data sequence, the regime proxy is usually identified with confidence. We propose the following architecture for directing the illuminator: If $\|\hat{\phi}[k]\| \geq \lambda$, we illuminate the EEP; if $\|\hat{\phi}[k]\| < \lambda$, we do not illuminate at all. Clearly, in the latter case, the target is missed. But we avoid using the energy that would be expended on EEP regions where there is large uncertainty in target position and concomitant loss of focus.

The choice of λ is an engineering judgement. If we make λ high, we will accumulate a lot of misses because we will not activate the illuminator. On the other hand, if we set λ low, we will expend a lot of energy covering extended assurance regions.

In this engagement we have set $\lambda = 0.7$. The performance of the **GWE**-illuminator architecture is shown in Fig. 7.8. In the figure, we have contrasted the

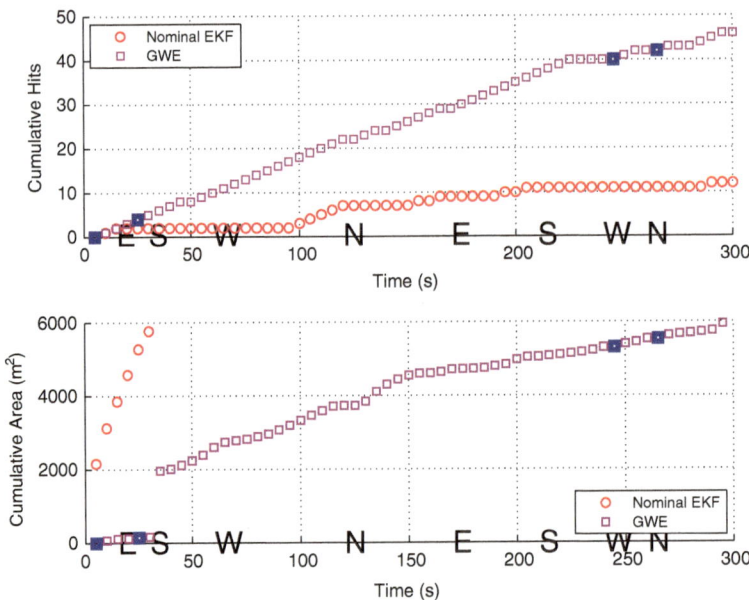

Fig. 7.8 The *upper panel* shows the miss-rate using the EEP regions. The miss-rate is far less than predicted for the **GWE**. The illumination energy required by the **GWE** is orders of magnitude less than that of the nominal **EKF**

GWE with the normative EKF-illuminator with a 50 % cover region. The times at which the GWE flags the regime state as uncertain are shown in the figure with the icon ∎.

To put the figure in context, recall that the predicted miss-rate for the EEP illuminator is 50 % for either algorithm. But the GWE-illuminator actually only misses the target 23 % of the time as contrasted with the 80 % miss-rate of the nominal EKF. This is despite the fact that the GWE selects the cover region based upon the projected distribution instead of the fine-structure of the $\mathscr{G}[k]$-distribution of position.

The GWE experiences one string of five misses (notional probability 3 %) and one string of length three (notional probability 12 %). There are four *failure-to-illuminate* flags. These two strings and four flags cover essentially all of the misses. The normative EKF is seen to be inferior to the GWE with a low hit-rate and long strings of misses.

The reduced miss-rate is achieved in the GWE with far less expenditure of illumination energy. The lower panel in Fig. 7.8 shows the cumulative energy expenditure for both the EKF and the GWE. The nominal EKF has the smallest cover regions of the EKF-trackers. But still its cover area is nearly 1.5E5 square meters. The GWE-tracker uses only 6E3 square meters to reduce the miss rate by a factor of four. In the figure the cover area of the EKF is seen to grow off of the chart within 50 s.

The energy expenditure of the GWE-tracker grows roughly linearly in time except for the noticeable break after the first south turn. It was noted earlier that the GWE found the regime hard to classify at this point because of a failure of the acoustic sensor. But $\|\hat{\phi}[k]\| \approx 0.8$. Further, the GWE captured the target at this time, albeit with a large EEP. Unfortunately, the energy expended at this single point was about a third of the total energy used by the GWE in the engagement.

After the first north turn, the GWE was again uncertain about the direction of travel. The regime norm was close to the boundary: $\|\hat{\phi}[k]\| \approx 0.7$. An illumination was commanded. The energy expended at this time was not so large as in the earlier event, though perhaps because of this, the illumination was unsuccessful.

7.2.3.4 Conclusions

The GWE-architecture is clearly superior to either form of the EKF-architecture. The simple regime-norm logic seems to work reasonably well here. But a weighted modal-norm might be better. The aberrant failure of the acoustic sensor to provide a regime measurement was not handled properly by the GWE-tracker. That logic should be upgraded in the future.

7.3   Ballistic Fire Control

7.3.1   The Engagement

7.3.1.1   Review of the Kinematics and Sensor Architecture

In the previous section, we studied the ability of a tracker-based illuminator to capture a moving target while keeping the requisite illumination energy small. The GWE-illuminator had a lower miss-rate than predicted on the basis of our reductionist model, and it had a much tighter focus than did either of the EKFs. The GWE-illuminator is clearly the algorithm of choice in that engagement.

Let us now look at a different kind of encounter. A hostile target is moving in the plane, but it is not confined to a rigid grid. In contrast to the previous engagement, the kinematic state is continuous. The underlying motion is delineated by the coordinated turn model; see (3.1). There are discontinuities in acceleration but not in velocity. The detailed parameterization of the kinematic model is given in Chap. 3 and will not be repeated.

The nominal EKF-tracker would again be the conventional choice for target placement. The EKF uses the *centered* kinematic model; i.e., CV, with three sensors generating the range-bearing measurements. Figure 7.9 shows the response of the EKF-tracker: see Fig. 3.1. It is good on long CV-sections of the path. It is particularly good on the initial south-east leg where the kinematic model matches the actual path. However, the EKF lags following the turns to an unacceptable degree. A detailed discussion of the EKF is presented in Chap. 3.

In this section, we want to explore a variant of the engagement of Chap. 3. A hostile target is following the path shown in the figure. The range-bearing measurements are those used earlier.

We must design a defensive architecture which will neutralize the target. We have a device which projects ballistic particles over a cover area which contains the predicted target location. The prediction time is equal to the transit time of the particles—firing is immediate after an enabling command is given. If the target is within the region covered by the particles, the target will be destroyed with probability 80 %. We must choose a specific firing protocol; e.g., does the weapon fire every sample time; every other sample; contingent sample times; etc. We must be assured with a confidence of 95 % that the target will be missed no more than ten times over the engagement interval.

Firing every sample time increases the likelihood that the target is neutralized. But the particles are a limited resource. We must select an algorithm which hits the target with a high frequency while minimizing the countermeasure assets consumed. We will measure the number of particles expended by the total area of the cover regions selected. In the next section, we will determine the quality of an enabling logic based upon the EKF. Later we will contrast the EKF with a hybrid fire control.

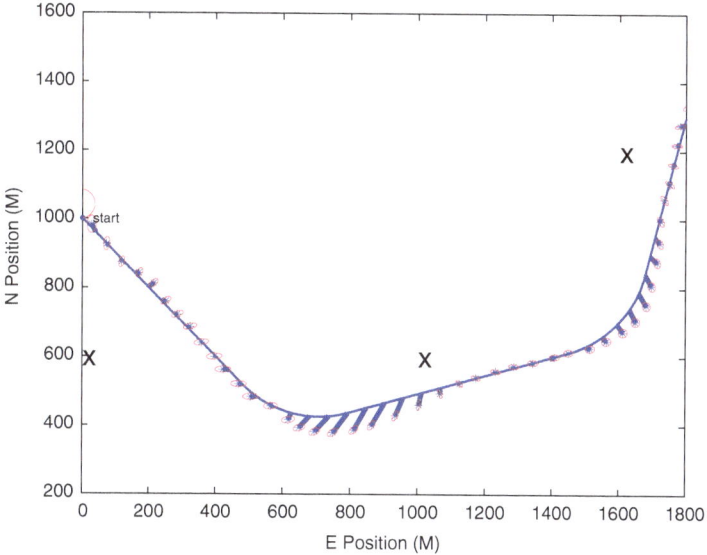

Fig. 7.9 A hostile target is moving along an evasive path in the plane. The EEP ellipses shown are generated by an **EKF**

7.3.2 *EKF Fire Control*

7.3.2.1 Notional Statistics of the Action Windows

The minimum rate of particle expenditure is achieved when we blanket an appropriate error ellipse generated by the **EKF**-predictor. In this section, we will investigate a simplified version the fire control problem in which the prediction time is fixed: either 2 s or 4 s. Actually, the prediction time would depend on the weapon-target geometry with a longer prediction time when firing range is long.

Further, we will suppose the hit-probability is independent of where the target is within the cover area. This ignores the non-uniform coverage of any pre-specified area. Another simplifying assumption is that the cover sequence is an independent-identically distributed binary process. This is clearly not the case for the encounter shown in Fig. 7.9. We make this assumption to size the problem. If the hit sequence is an iid binary process, the number of misses is binomially distributed. A direct calculation shows that we need an 80 % cover probability to achieve a 95 % assurance of no more than ten misses in the encounter.

To achieve an 80 % cover probability, Fig. 6.9. shows that we need to blanket the 1.8σ error ellipse. Under the various hypotheses, the **EKF** fire control should miss no more than ten times, and it should use the minimum number of particles. In the next section, we will contrast the performance achieved with that anticipated. The two- and four-second **EKF**-predictors are presented in Chap. 4, and we will use the data sets generated there in the performance curves that follow.

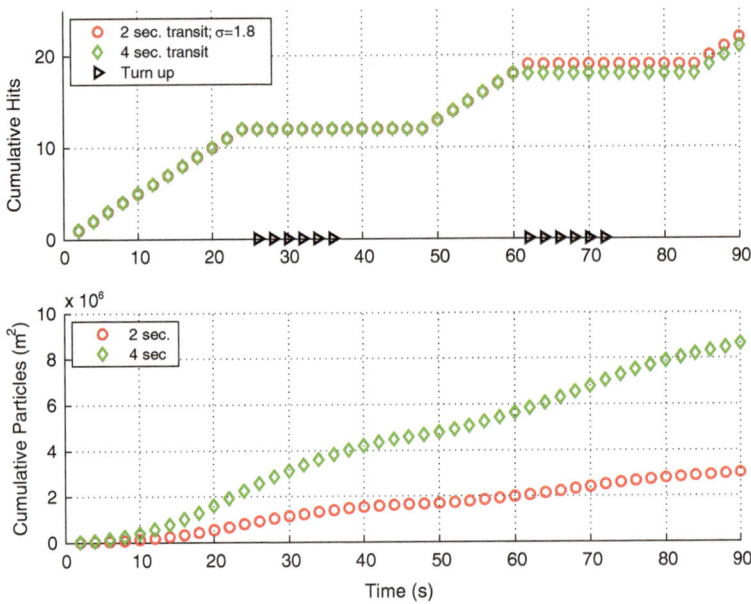

Fig. 7.10 The *upper panel* shows the cumulative hits on a hostile target using the **EKF** fire control architecture. The *lower panel* shows the inventory of defensive assets required with a uniform firing strategy. The times of a turn are shown in the figure

7.3.2.2   Hit Rate

The fire-control algorithm is designed to hit the target at least 36 times during a 90-second engagement. Figure 7.10 shows the performance of the **EKF** fire control for both the two-second and the four-second predictors using a 1.8σ cover region. The hit rates for the two architectures are quite close. Unfortunately, both fail to meet the system specification: the two-second fire control misses 23 times and the four-second fire control misses 24 times. In both cases, the miss-rate is over twice that specified.

From the upper panel of the figure we see that the miss sequence fails the iid hypothesis upon which the choice of the cover region is made. The miss process is highly correlated with few misses during quiescent operation and long strings of misses subsequent to a turn. This behavior is not surprising. The kinematic model underlying the **EKF** fire control is conservative over most of the engagement. After resolving a regime event, the fire control algorithm seldom misses.

While the performance of the nominal **EKF**-architecture is not totally unexpected, it must be acknowledged that it fails to meet the system specification. We can reduce the miss-rate by using larger error ellipses for the cover region. Figure 7.11 shows the performance of the **EKF** fire control when we use the 6σ error ellipses. The two-second predictor misses only 11 times and the four-second predictor misses

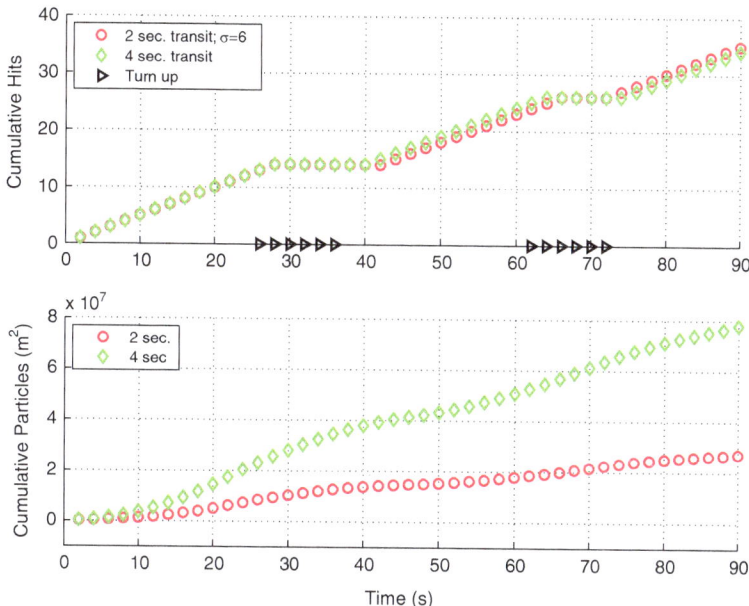

Fig. 7.11 The EKF fire control using 6σ error ellipses has far fewer misses but uses far more particles

only 12 times. Of course, the 6σ ellipses should never miss on an engagement this short. Again the miss strings of the two predictors mimic each other. The 6σ miss strings are much shorter than occur using the 1.8σ cover.

The improved performance in Fig. 7.11 is achieved with a significant increase in the required number of particles. In either of the two figures, the two-second predictor requires about 35 % of the particle reservoir needed for the four-second predictor. The error ellipses for the latter are bigger for the same value of σ. But the 6σ fire control requires 11 times the particle inventory utilized in the 1.8σ cover. This profligate use of defensive assets is not suggestive of good system design.

7.3.3 *GWE Fire Control*

7.3.3.1 $\mathscr{Y}[k]$-Architecture

The EKF-architecture meets the hit-rate specification only when the cover regions are enlarged to a disproportionate degree. The notional cover probability of the tracker-predictor is set at 80 % to be assured that the total number of misses be less than ten. But the 80 % cover failed to hit as often as predicted. Actually, a 6σ cover (essentially 100 % sequential hit-rate) is required to achieve the hit-rate specification.

It is natural to ask if performance would be improved with a hybrid fire control algorithm. In the tracking problem of a target confined to a road course, the GWE-illuminator was much better than the EKF. However, the GWE-tracker availed itself of both a speed measurement and a data set showing the road configuration. These para-measurements proved to be quite valuable. The turn events were used to reset the kinematic state. The turn events also helped the GWE to isolate the regime state.

In this engagement, the advantage of the hybrid architecture is not so evident. There is no map constraint; the target can move at will in the plane. Thus, there are no regime transition-induced kinematic resets. Instead, a regime transition just adds to the location uncertainty that has built up from the start of the engagement. Lacking an additional para-measurement, it is not clear that the GWE fire control provides any advantage over the EKF.

In Chap. 4 we studied the prediction problem using the GWE. Both the two- and the four-second predictors were presented there. The kinematic measurement sequence was that used in the EKF-predictors. The GWE was found to be superior to the EKF regarding target placement. But the 1σ-error ellipses of the former were bigger than those of the latter.

The GWE-illuminator architecture of the earlier section benefitted from its ability to measure modal uncertainty as the engagement evolved. In this way, it avoided using energy when the situation was inauspicious. The GWE achieved a high hit rate with less energy than did its EKF counterparts by not attempting to illuminate the target when the distribution of position is diffuse.

The $\mathcal{Y}[k]$-GWE does not have any para-measurements. This is reflected in Fig. 7.12, a plot of the confidence process, $\|\hat{\phi}[k]\|$. There are only three possible regimes in this engagement: $i \in \{1, 2, 3\}$. Hence, $\|\hat{\phi}[k]\| \geq 0.58$. From an initial value near that minimum, $\{\|\hat{\phi}[k]\|\}$ increases on the initial CV-section of the engagement. When a turn occurs, the uncertainty increases as the GWE attempts to resolve its mode. During most of the encounter, the $\{\|\hat{\phi}[k]\|\}$ process is desultory without a clear indication of motion condition. The general character of the uncertainty process does not inspire confidence for its use as a gauge for a fire-enabling logic.

7.3.3.2 $\mathcal{Y}[k]$-Hit Rate

Since the $\{\|\hat{\phi}[k]\|\}$ process in Fig. 7.12 indicates that the GWE is not able to separate the modes very well, we will use the simple firing protocol we used in the EKF architecture: fire on every sample. The upper panel of Fig. 7.13 shows a plot of the number of hits along the path. The lower panel shows the cumulative number particles expended in units of m^2. This figure should be contrasted with Figs. 7.10 and 7.11

Despite its limited information, the GWE fire-control misses only five times. This is well within the system specification and is fewer than the 6σ EKF fire control. There are no long strings of misses. This is true for both the two- and the four-second flight times.

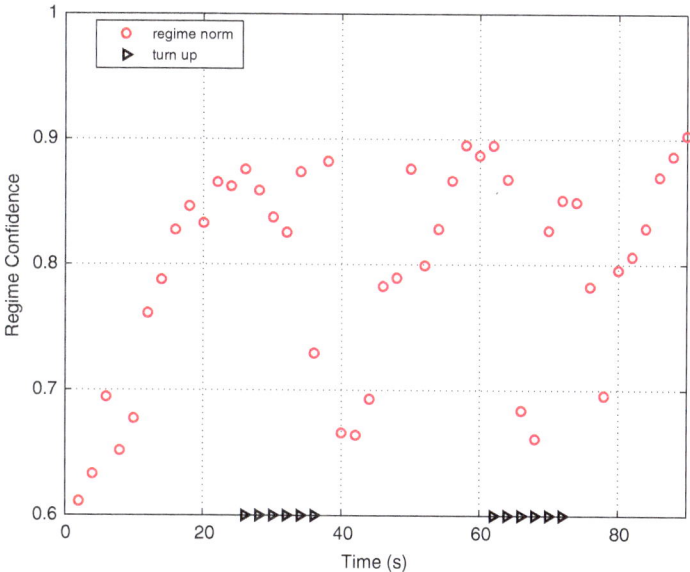

Fig. 7.12   From the $\{\mathscr{Y}[k]\}$ measurements, modal uncertainty is unavoidable

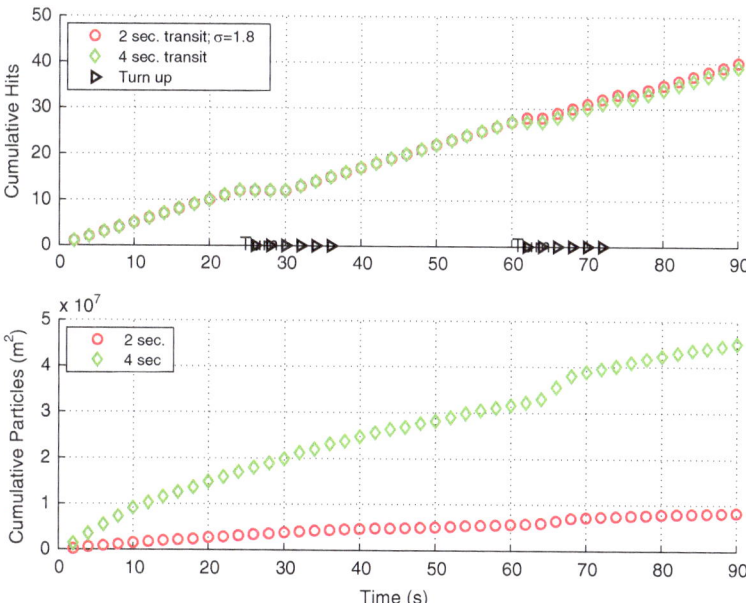

Fig. 7.13   The *upper panel* shows the cumulative hits on a hostile target using the $\mathscr{Y}[k]$-GWE fire control architecture. The *lower panel* shows the inventory of defensive assets required with a uniform firing strategy

The **GWE** reduction in miss-rate is achieved with fewer particles. Only the 6σ **EKF** architecture meets the hit-rate specification, and it uses three times the particles for a two-second particle flight time. If the flight time is 4 s, the **EKF** uses nearly twice the number of particles required by the **GWE**. Further, the 6σ **EKF** misses 10 % more often. Therefore, even without auxiliary measurements, the **GWE** architecture is superior to that of the **EKF**.

7.3.3.3 $\mathscr{G}[k]$-Hit Rate

The $\mathscr{Y}[k]$-**GWE** fire control is more effective than is the **EKF** fire control despite using the exact same sequence of measurements. The improvement is achieved without any clear assessment of the regime. The enabling protocol is simply to fire at every measurement time.

Suppose we now include an auxiliary para-measurement. In the architecture described in Chap. 4, we have a forward regime measurement that is correct 80 % of the time.

In Chap. 4 we generated a $\mathscr{G}[k] = (\mathscr{Y}[k] \vee \mathscr{Z}[k])$ data set and derived a $\mathscr{G}[k]$-**GWE** predictor. As would be expected, the $\mathscr{G}[k]$-confidence process is a more assured indicator of the current regime than is the $\mathscr{Y}[k]$-version. This is reflected in Fig. 7.14. The confidence process $\|\hat{\phi}[k]\|$ tends to stay above 0.9 over the engagement. After the initialization, $\|\hat{\phi}[k]\|$ falls much below 0.75 only twice. The first time is after a turn and the second is near the end of the engagement. In both

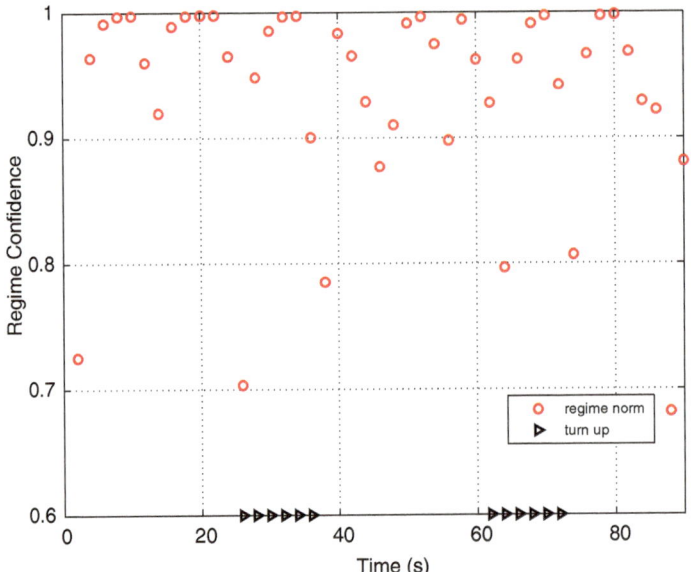

Fig. 7.14 From the $\{\mathscr{G}[k]\}$ data structure, the regime is easier to isolate

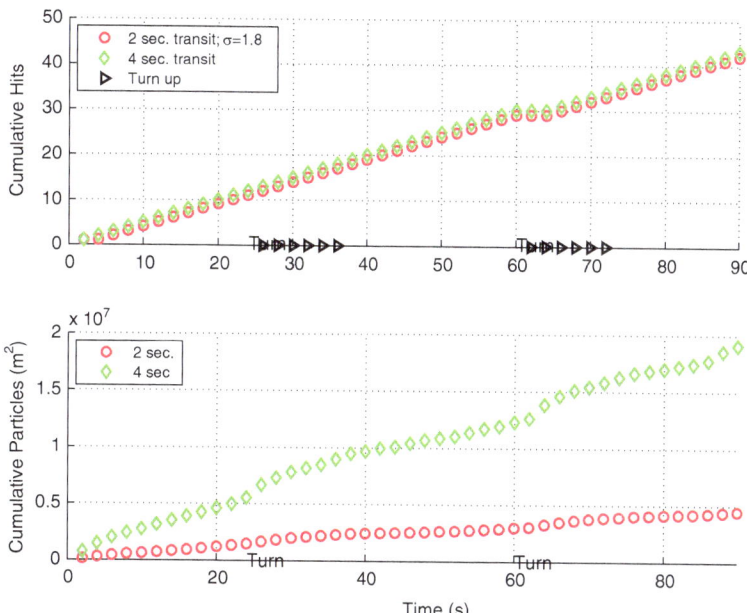

Fig. 7.15 The *upper panel* shows the accumulated hits on a hostile target using the 1.8σ ellipses of the $\mathscr{G}[k]$-GWE predictor and a fire-every-measurement protocol. The *lower panel* shows the inventory of defensive assets required for this uniform firing strategy

cases, the regime uncertainty is resolved in 2 s. This is in contrast to the $\mathscr{Y}[k]$-GWE that spends a third of the time below 0.75.

The $\{\|\hat{\phi}[k]\|\}$ process is useful in identifying regime uncertainty. But it says little about where the target is located. Suppose we use the uniform firing protocol proposed in the $\mathscr{Y}[k]$ fire control and cover the 1.8σ assurance regions every sample time. Figure 7.15 shows that the number of misses is only three—only two in the four-second predictor. But more importantly, the number of particles required in the $\mathscr{G}[k]$-architecture is reduced by more than half when compared with the $\mathscr{Y}[k]$-architecture. This makes the $\mathscr{G}[k]$ fire control the clear algorithm of choice if the relevant sensor suite is available.

The $\mathscr{G}[k]$ fire control with always-on enabling easily exceeds the miss-rate specification. We could improve performance by trading a few additional misses for a reduction in the required number of particles. Suppose we use the following enabling strategy: If $\|\hat{\phi}[k]\| \geq 0.75$, we fire on the 1.8σ ellipse; if $\|\hat{\phi}[k]\| < 0.75$, we do not fire. Clearly, we can only miss more often with this firing logic. Actually, with this enabling rule, we have three more misses in the four-second predictor (Fig. 7.16). Despite the uncertainty in the regime classification, the target was within in the 1.8σ ellipse in each of the excluded samples.

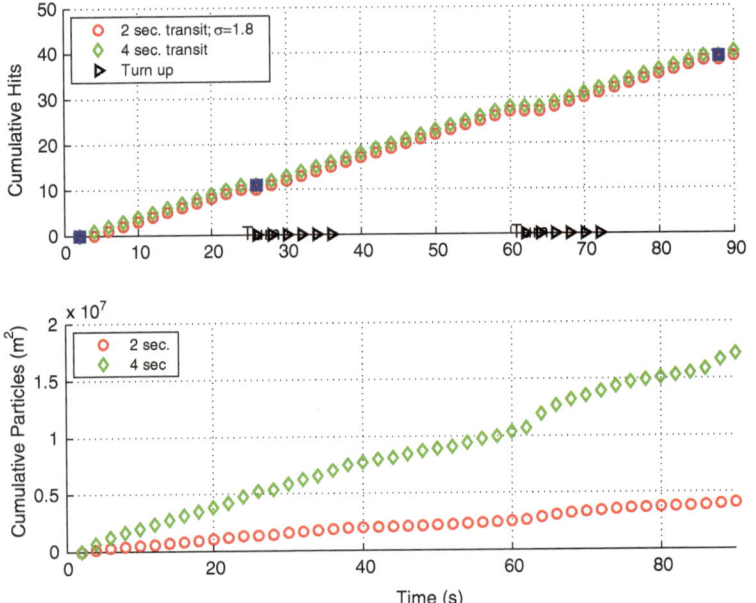

Fig. 7.16 The *upper panel* shows the accumulated hits on a hostile target using the 1.8σ ellipses of the $\mathscr{G}[k]$-GWE predictor and a fire-when-confident protocol. The *lower panel* shows the inventory of defensive assets required for this non-uniform firing strategy

But even with this non-uniform firing strategy, we easily meet the miss-rate specification. And the number of particles used is reduced by about 25 %. The reduction in defensive assets is certainly worth the few additional misses.

7.4   Conclusions

This chapter explores the utility of a resource allocation architecture based upon a hybrid tracking-prediction algorithm. The hybrid estimators are of higher dimension than their EKF counterparts. But they are much more accurate, both in target placement and in the fidelity of their notional assurance regions. The reduction in defensive assets using a hybrid tracker is of an order of magnitude.

The performance improvement associated with a hybrid system is a function of the sophistication of the sensor suite that generates the tracking data. When the basic range-bearing measurements are augmented with para-measurements, performance is enhanced. This is true even in cases where the para-measurement is only indirectly related to location. This enhancement stems more from a reduction in the size of the assurance regions than it does from a reduction of the raw radial tracking error.

The hybrid system architectures proposed here allocate resources according to the projected distribution. This distribution is simple to compute and to interpret. But the assurance regions generated thereby tend to be larger than necessary. The fire-when-confident strategy saved a significant number of particles in the engagement studied. But it also failed to hit the target as often as the fire-every-sample strategy. The regime confidence process fails at exactly those points where the fine structure of the true $\mathscr{G}[k]$-distribution (or $\mathscr{Y}[k]$-distribution) of target placement would be most useful in determining the assurance regions. Logic based upon the $\mathscr{G}[k]$-distribution would be much more difficult to implement. But in applications where the size of the particle (or energy) inventory is decisive, a more nuanced approach to computing the assurance regions would be warranted.

Chapter 8
Serendipitous Events in Tracking and Classification

Abstract Real-world tracking applications encounter circumstances not well modeled in previous chapters. The location sensors might be partially or completely obscured rendering the target invisible. With some knowledge of the terrain and perhaps the placement of well-located features such as buildings, bridges, and tunnels, the tracker can be modified to generate more faithful assurance regions. Isolated temporal events are hard to incorporate into conventional command algorithms. For example, if a hostile target enters a tunnel, the kinematic measurements are interrupted and performance degrades during the period the target is hidden. But what if the location of the tunnel entrance were known from a road map? Ingress could be used to excise certain state errors. Or the map might restrict egress to a few known locations with a useful motion constraint. Ingress and egress are serendipitous events that can be used in a hybrid-tracking algorithm to improve the tracking quality. In this chapter, a particular engagement is chosen to illustrate the utility of event inclusion. The following analysis is quite detailed and engagement specific. The encounter is chosen to illustrate the advantage of integrating random but isolated events into the computation of assurance regions and not to form a definitive template for broad command algorithms. The reader should reflect on the algorithmic adjustments made here, and see how similar adjustments might be made in similar applications.

8.1 System Specifications

An uncooperative target makes use of abrupt maneuvers to make placement and classification difficult. After moving for a period in one motion regime, the target may suddenly change to another. Since a tracker has some of the characteristics of a low-pass filter, the location estimates may fail to respond to changed conditions as quickly as we would like.

Not only does the target placement tend to lag changes in regime, but the assurance regions generated with a reductionist motion model will fail to account for the increased uncertainty generated by such unanticipated events. In Chap. 2 we show examples in which a conventional EKF-tracker is far too optimistic regarding its own performance when unforeseen regime changes arise.

© Springer International Publishing Switzerland 2016
D.D. Sworder, J.E. Boyd, *Locating, Classifying and Countering Agile Land Vehicles*, DOI 10.1007/978-3-319-19431-8_8

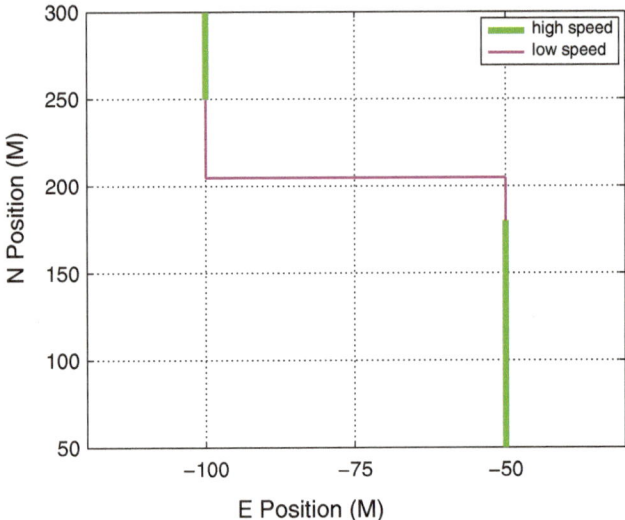

Fig. 8.1 A portion of the path of an uncooperative target is shown. Road segments of reduced speed are marked by *color* and *line* width

To illustrate this tracker conceit, consider the following engagement. A target of undetermined intent is detected moving north from position $(-50, -50)$. It has a speed of 20 m/s (45 mph) along a road at -50E. After continuing along the road for a time, the road narrows, and the target speed drops to 10 m/s. As it continues north, it comes to a 4-way intersection at $(-50, -205)$, and the target turns west. The road segments in east-west are worse than those in north-south and the speed of the target is reduced to 5 m/s. The target continues west until the road turns north at -100E, and the speed returns to 10 m/s. Finally, the target exits this region of reduced speed at $(-100, 251)$ and returns to a north speed of 20 m/s at -100E.

A portion of the target path is shown in Fig. 8.1. Road segments with reduced speed are marked. The geometry is similar to that of Fig. 2.6, but in that engagement the target is northbound at constant speed throughout.

There are three synchronized transmitter-receivers that provide noisy range measurements to a central processor every half second. They are located at

$$(X_r, Y_r) = [(-100, 0), (-20, 0), (-20, 50)],$$

and the standard range error is 2 m.

At the beginning of the engagement there is one transmitter west of the path and two to the east. The sensors are all to the south of the road junctions. In Sect. 2.4.1, this sensor suite was called the *south array*.

Because of the variable topography of the road, the target may be visible to the sensors on certain parts of its path, but not on others. We will suppose that if the target is visible to any sensor, it is visible to all, and conversely. If the target cannot be seen, the classifier is said to be *blind* as long as obscuration exists. If the target can be seen, it is said to be *visible*.

There is a region of 30 m width along the north and east edges of the obscured region $\mathbb{R}^O$ where the target is visible to the classifier. We will call this region $\mathbb{R}^C$. When the tracker determines the target is within this region, it provides an accessibility signal to the classifier, and the target is illuminated and labeled. If captured within the illumination window, it is inspected and classified without error. If, however, the target is not within the window, classification fails. These restrictions hold as long as the target is within the interrogation region—a period of a second and a half at the nominal target speed. If the target is not captured during this interval, the target is classified as hostile, and countermeasures are ordered.

We must derive an algorithm which will classify the target as hostile or benign and initiate intercept as appropriate. The architecture must meet the following specifications:

1. With confidence greater than 80 %, the tracker will properly classify the target within the second and a half during which it is accessible.
2. An illumination window can have an area no greater than 20 m^2 with smaller windows viewed favorably.

To contrast the capabilities of different algorithms, we will look carefully at several related engagements. Although the specific results are sample function dependent, the analysis of a single sample path will illustrate the differences between the various algorithms. We are particularly interested in the influence of obscurations and map information on classification accuracy.

8.2 The Visible Target

8.2.1 The Engagement Model

When first detected, the kinematic state of the target is northbound on a road at -50E: $x[0] = [-50, -50, 0, 20]'$. The classifier utilizes a sequence of range measurements to create a data sequence, $\{\mathscr{Y}[k]\}$, that is transformed into an estimate of location. When the target leaves the low speed segment of the path, the classifier will place a query window about the estimated target location. Target intent will be determined and passed to an action center for response. We will suppose that if the query window covers the target, intent is determined with probability one. Thus, proper action depends on our ability to locate the target with high probability during the interrogation interval.

We will use a model-based tracking algorithm as we have done before. Specifically, we will assume the intra-junction kinematic model is CV:

$$\frac{d}{dt}\begin{bmatrix} X \\ Y \\ V_X \\ V_Y \end{bmatrix} = \begin{bmatrix} 0 & 0 & 1 & 0 \\ 0 & 0 & 0 & 1 \\ 0 & 0 & 0 & 0 \\ 0 & 0 & 0 & 0 \end{bmatrix}\begin{bmatrix} X \\ Y \\ V_X \\ V_Y \end{bmatrix} + \begin{bmatrix} 0 & 0 \\ 0 & 0 \\ \sqrt{P^w_{x;\iota}} & 0 \\ 0 & \sqrt{P^w_{y;\iota}} \end{bmatrix}\begin{bmatrix} \dot{w}_X \\ \dot{w}_Y \end{bmatrix}, \qquad (8.1)$$

where $\dot{w}_X$ and $\dot{w}_Y$ are unit white-noise accelerations. Because the motion is primarily northbound, the exogenous accelerations are larger northbound as well. The intensity of the acceleration is 6.3 m/s^2 in north-south and 4.5 m/s^2 in east-west.

As the target moves along the path, range measurements are taken every half second. The raw measurements are replaced with a vector of east-north location measurements:

$$y[k] = H[k]x[k] + \sqrt{P^n}[k]n[k]. \qquad (8.2)$$

In the stacked model, $H = [H_\chi \quad 0]$ where H_χ consists of the pairs of transmitter-to-target directional derivatives. This change in the measurement coordinate system from range to Euclidean is discussed in Sect. 2.4. The coordinate transformation in (8.2) is only approximate, and the coefficient matrices are random since they depend upon the estimated target location.

Suppose initially that the classifier is not aware of the road path and is unsure where the engagement began, though it is confident that the target is northbound. The initial state estimate is poor: $\hat{x}[0]' = [-14.2; -32.0; 0.4; 21.0]$. The initial tracker placement is well north of the target and east of the road. The velocity estimate is relatively good. The initial uncertainty assessment is optimistic: $P_{xx}(0) = \text{diag}([100, 100, 0.1, 0.1])$.

The basic CV motion model is complete. A complication that did not arise in Chap. 2 is the introduction of road junctions. Here, the target turns and the speed changes. Our CV model of the exogenous acceleration does not well represent such abrupt changes.

8.2.2 Omnipresent EKF

8.2.2.1 Tracker

The conventional approach to this problem would be to design an EKF-tracker that follows the target until it reaches the interrogation region north of 250N. We would then illuminate the target at the three inclusion times using an assurance region of area no greater than 20 m/s^2. For example, if the notional EEP ellipses had area less than 20 m^2, we could be assured of identifying target intent with probability 83 % after three samples—the tracking errors in the EKF are (nominally) Gaussian and iid with probability of inclusion equal to 50 %.

Figure 8.2 shows the response of the EKF beginning at $t = 0.5$. The sensor array and the starting point are shown. The geometry is such that the tracking error is

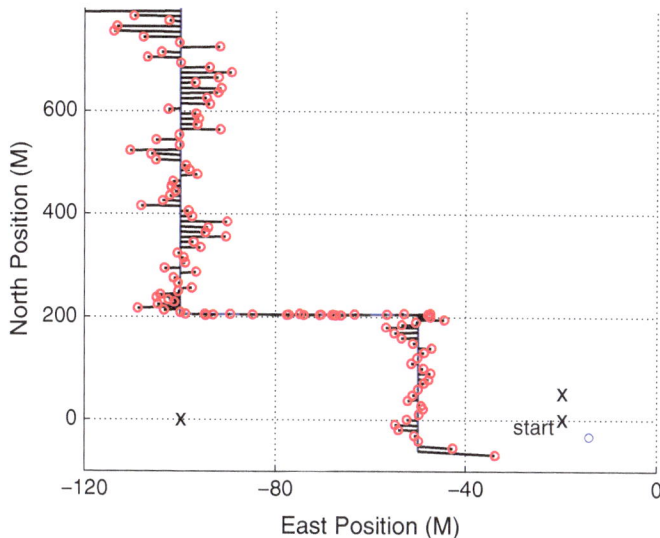

Fig. 8.2 An EKF-tracker is used to follow a target moving along the path shown in Fig. 8.1. The south-array range sensors are shown with *diamond*. The starting point is shown with *open circle*. The tracking error is larger than the range error. Note the scale is not true: the axes in this plot (and in similar performance plots that follow) are unequal to make the cross-track error more visible

primarily in east-west. As the target moves north, the error grows and exceeds the 2m range error. We will call this **EKF** *omnipresent*; the classifier receives the full set of range measurements. However, the **EKF** does not know anything specific about the possible paths. That would be called *omniscient*.

The performance of the **EKF**-classifier is not as good as that on the north path shown in Fig. 2.6. The range data set is noisier, and the kinematic model less faithful. After quickly reducing the large initialization error, the tracking error increases as the target moves north. The sensor geometry is good in the south, and the tracking is accurate even in the presence of the large P^n. But as the target moves north of 200N, GDOP grows rapidly as does the tracking error.

To examine the response of the **EKF**-tracker in more detail, Fig. 8.3 shows an expanded view of three segments of the engagement. Initially, the EEP ellipses are small and capture the target about half of the time—though when they err, the error is surprisingly large. In the segment including the west turn (the middle panel), the tracker is confused by the speed change but accommodates reasonably well to the change in direction with a 5 m north bias. The right panel shows the putative illumination windows north of 250N. The EEP regions are optimistic with the target near the boundary in most cases. But the areas of the EEP ellipses are considerably larger than in the earlier segments of the path.

The EEP ellipses should cover the target about half of the time. Figure 8.4 shows the tracking error in units of notional standard deviation on the segment

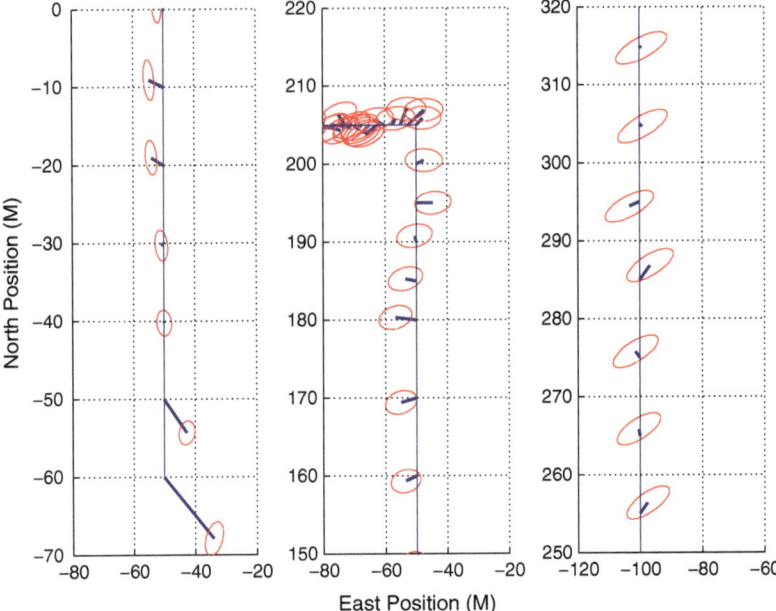

Fig. 8.3 The EEP ellipses are small at the beginning of the encounter. North of the interrogation region, they are bigger but capture the target most of the time

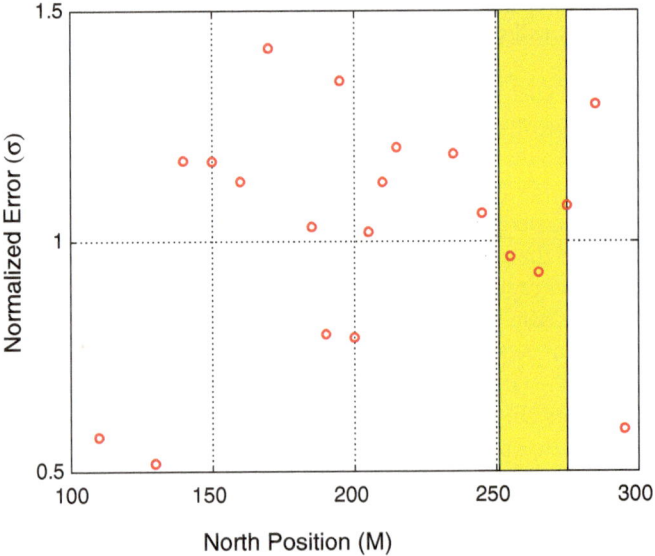

Fig. 8.4 The omnipresent EKF-tracker is used to follow a target moving along the path shown in Fig. 8.1. The EEP ellipses give good coverage in the interrogation interval shown *yellow*

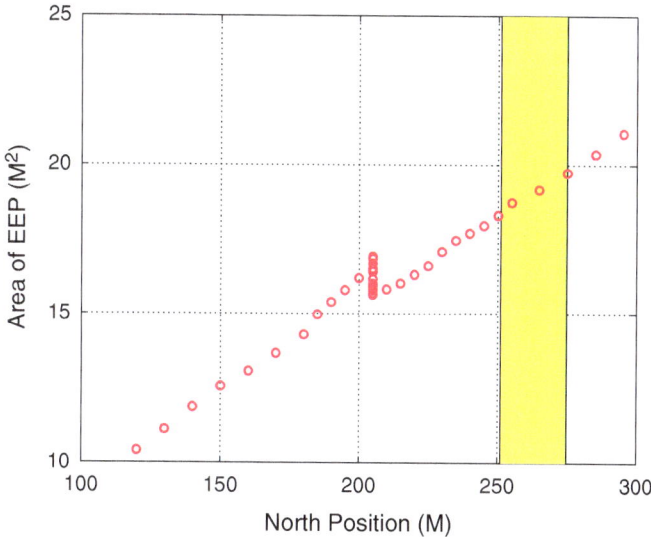

Fig. 8.5 The area of the EEP regions generated by the omnipresent **EKF** grow as the south-array sensor geometry degrades. The area of each of the 50 % assurance regions meets the 20 m² specification within the interrogation interval

(100, 300)N—the EEP ellipses cover the 1.2σ error region. The west segment at 205N is not included since there are overlapping estimates with that coordinate. In general, the EEP ellipses are faithful indicators of the 50 % assurance region. The interrogation interval is shown shaded in the figure. On this sample function, the target is illuminated every sample time.

Figure 8.5 shows the area of the EEP ellipses on the same path segment. At 150N, the area is about 13 m². This grows to nearly 20 m² as the target traverses the interrogation interval. The artifact on the graph at 205N is due to the fact that the EKF-tracker computes several distinct assurance regions with that coordinate.

The omnipresent **EKF** tracker meets the system specifications despite its lack of specificity concerning the path followed. The area of the illumination windows is less than 20 m², and the EEP ellipses cover the target three successive times in $\mathbb{R}^C$—we only required one capture. The omnipresent **EKF** is simple to implement and would be a suitable algorithm for this problem.

8.3 The Veiled Target

8.3.1 The Engagement

Consider now a modified engagement in which as the target moves at reduced speed, it is not visible to the range sensors. This could be due to a barrier as the target passes though a tunnel, a geologic artifact that obstructs the sensor line-of-sight, or similar obstructions. The classifiers treat this condition as a sequence of missed observations.

Figure 8.6 shows the comprehensive road geometry. The shaded area will be called the obscured or veiled region and labeled $\mathbb{R}^O$. The west turn at 205N occurs at a 4-way junction. At the junction, the target will move in one of the three equally likely forward directions.

The target motion is now unpredictable in a more global sense. For example, had the target gone directly though the intersection, it would have continued at a speed of 10 m/s and entered the interrogation region at $(-50, 251)$. This would take about 7 s. Or the target could have turned east and proceeded at a speed of 5 m/s. If it followed this path, it would enter the interrogation region at $(150, 205)$. In fact, the target follows the path in Fig. 8.1 and exits the shaded region $(-100, 251)$ in about 17 s.

Because the target is not visible to the south-array sensors in the shaded region of the figure, the tracking problem is complicated. The tracker must extrapolate across $\mathbb{R}^O$ and interrogate the target when it exits. We will say the tracker lacking measurements across $\mathbb{R}^O$ is "blinded."

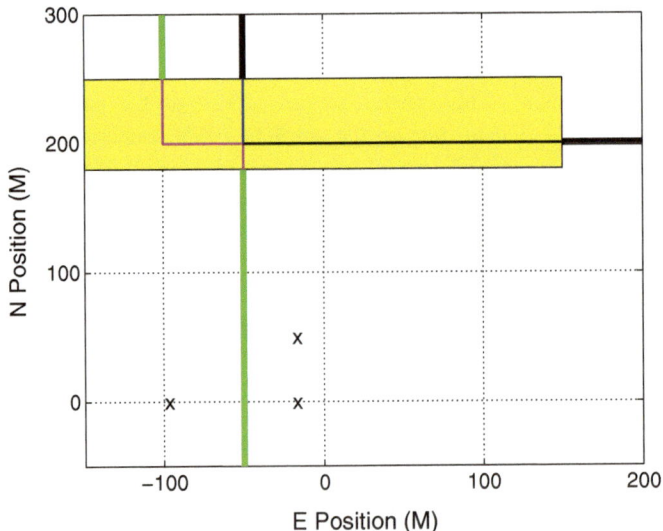

Fig. 8.6 The road grid for the engagement shows a 4-way intersection at $(-50, 205)$. The three forward extensions are shown. The actual path is marked *red-green*. The target is not visible to the south-array sensors in the shaded region

Let us now develop a classification algorithm that meets the performance specifications using an abridged range data set: no range measurements within $\mathbb{R}^O$. This is going to be difficult since the $\mathbb{R}^C$ is contiguous with $\mathbb{R}^O$. But the small set of permissible exit locations is known, and we will use this to our advantage.

8.3.2 Blinded EKF

The most obvious approach to this problem is to again use an EKF to place the target. The EEP ellipses can then be used as illumination windows when the target enters $\mathbb{R}^C$. The range-data is that used in the previous section, reduced by the measurements taken during the passage though the veiled region—though the $\mathscr{Y}[k]$-data set is individualized to this engagement. Tracking south of 180N is necessarily identical for the blind and the omnipresent EKF—same data and same algorithm.

Within $\mathbb{R}^O$, the EKF uses dead reckoning to place the target. But target speed and direction at entry are not known with precision—the estimated speed at entry to $\mathbb{R}^O$ is actually 2.5 % high, and the direction is to the north-west instead of due north. Further, target speed is reduced by 50 % at 180N and the EKF has no way of knowing this. Consequently, the blind EKF extrapolates far to the north-west at a speed that is far too high. This is clearly evident in Fig. 8.7. The EKF estimate tracks well beyond $\mathbb{R}^C$ while the target is actually within $\mathbb{R}^O$—a violation of the

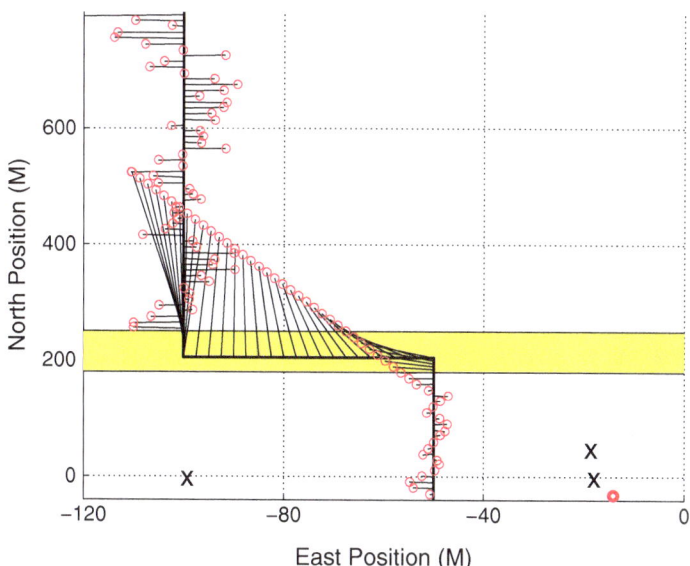

Fig. 8.7 The blinded EKF loses the target as it enters the veiled region. The tracking errors increase during passage toward the interrogation region

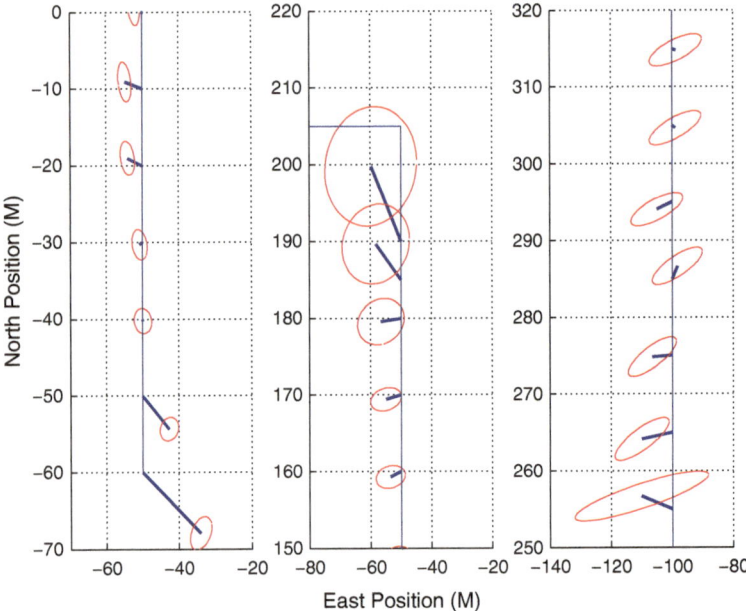

Fig. 8.8 The EEP ellipses for the blinded **EKF**-tracker are larger than those of the omnipresent **EKF**-tracker

formal constraints on the tracking system. Lacking an enable command, the **EKF** does not try to illuminate the target

Once the target exits $\mathbb{R}^O$, the **EKF** corrects itself quickly. Preceding exit from the shaded area, the tracking errors are big and the EEP areas gigantic. However, the location ambiguity is quickly resolved as the target moves north of 250N. Indeed, the blinded **EKF**-tracker matches the performance of the omnipresent **EKF** north of 300N: if the algorithm is stable, the estimate is insensitive to the missing measurements.

Figure 8.8 gives an expanded view of the tracking performance of the blinded **EKF**. The left panel shows the initial phase of the engagement. The left panel of this figure matches the left panel of Fig. 8.3. The south portion of the middle panel matches the omnipresent **EKF**. But when the target enters the veiled region of the engagement at 180N, the target slows. The blinded **EKF** extrapolates forward to the north-west (an error in direction) and strongly to the north (an error in the speed). The areas of the the EEP ellipses grow very quickly.

The EEP ellipses grow so large during passage though $\mathbb{R}^O$ that they cannot be displayed in the figure. But when the target exits, the gross errors are quickly corrected. The right panel shows the segment north of 250N. Contrasting Fig. 8.3 with Fig. 8.8, it is clear that cloaking requires a larger illumination area over the 1.5 s capture interval. Unfortunately, these larger EEP ellipses fail to capture even once.

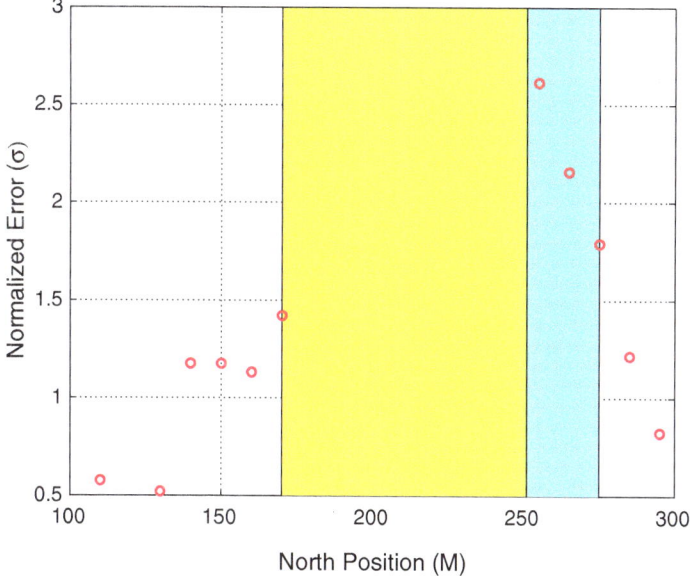

Fig. 8.9 The blinded EKF-tracker keeps the error below 1.5σ until the target enters the veiled region. But the tracking error is several σ at egress

Figure 8.9 shows the size of the tracking error in normalized units. Over the capture interval the error averages over 2σ and is never less than 1.5σ. The blinded EKF-tracker using the notional EEP ellipses fails to classify the target in this engagement.

The blinded EKF fails to capture the target even though the EEP ellipses are significantly larger in the illumination region. Figure 8.10 shows the sequence of EEP areas. Going into $\mathbb{R}^O$, the blind EKF matches the nominal EKF. But entering $\mathbb{R}^C$, the EEP area has grown to thrice that permitted. At the last classification attempt, the EEP area is about $20\,\text{m}^2$, but unfortunately still on the high side. The blinded EKF-classifier can be said to fail in this engagement: the illumination areas exceed specifications, and there are no target captures.

8.3.3 *GWE*

8.3.3.1 The Modal State

Now let us look at a hybrid algorithm for this application. We will drop the appellation "blinded" since we will only consider the GWE in the veiled engagement. The GWE has a separated state space: a modal state, ι, that delineates the motion condition, and a kinematic state, x_t, composed of the conventional kinematic variables. The latter is well described in the previous section. But the former requires some thought.

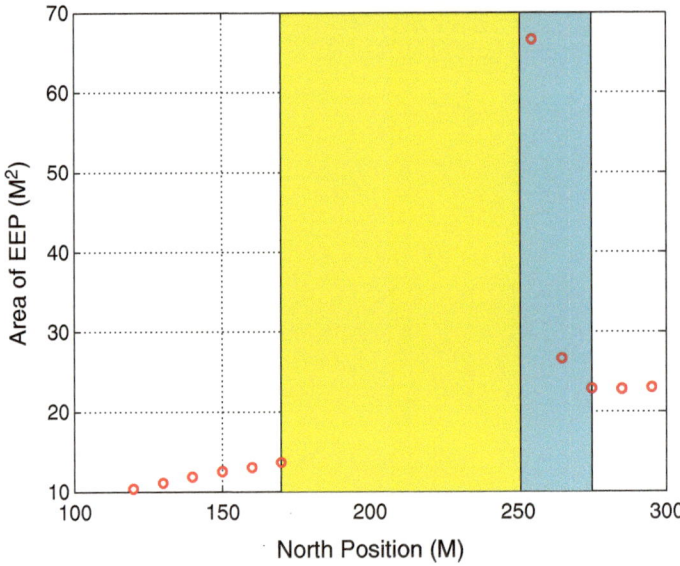

Fig. 8.10 The area of the EEP ellipses of the blind tracker are excessive. After the target leaves the veiled region, the EEP areas exceed $20\,\mathrm{m}^2$ over the interrogation interval

At the beginning of this engagement, the **GWE** has little more information on target motion than does the **EKF**. It has a map, $\mathscr{M}$, of $\mathbb{R}^O$. But it assumes the same lack of motion constraints outside the veiled region as the **EKF** does. The south-most (and north-most) kinematic model is CV. Let us label the south regime, $i = 1$.

Recall that the modal state in the **GWE** is a number (or word or unit vector) composed of a retrograde sequence of regimes. This engagement is delineated by evolution of several non-communicating chains. In the south region, the regime is always $i = 1$. Hence, the mode is of the form $\iota = 111\ldots$. The mode is a string of ones whose length is equal to the temporal sample number; e.g., at $t = 3$, $\iota = 111111$. Let us simplify the modal notation by saying that $\iota = (1; \tau)$ where τ is a temporal pointer that yields the period of the modal string. With our time step of $0.5\mathrm{s}$, $\iota = 111111 = (1; 3)$. This differs from the normative syntax, but we will use this convention because it is concise and evocative.

When the target enters $\mathbb{R}^O$ the range measurements are interrupted. The **GWE** interprets this to mean that the target has now entered the veiled region. The stored map $\mathscr{M}$ shows the entrance point to $\mathbb{R}^O$ to be $(-50, 180)$. The **GWE** knows that the entry event occurred before sample time $t = 12.5$. The passage into $\mathbb{R}^O$ is a serendipitous event which discharges certain position and direction errors that have accumulated since the beginning of the engagement: at $t = 12.5$, the regime changes to extrapolation.

Within $\mathbb{R}^O$, three paths are possible. We will list them: $i = 2$ is west exit; $i = 3$ is north exit; $i = 4$ is east exit. Thus, at the time the sensors are obscured, $(j = 1) \mapsto (i = 2)$ or $(i = 3)$ or $(i = 4)$. Note that the regime state is a path instead of a direction. Because we are doing kinematic extrapolation, we need not break down the directions within the path segment.

The modal state preceding entry is a string of ones. But at $t = 12.5$, the modal state makes the transition $(\iota = 1\ldots) \mapsto (\iota^+ = 21\ldots)$ or $(\iota^+ = 31\ldots)$ or $(\iota^+ = 41\ldots)$. For all ι^+, $\hat{\chi}[25] = [-50 - 180]'$ and the direction of motion is north.

The mode following entry is again a string of length twice the elapsed time in $\mathbb{R}^O$. Thus, the west path would be written $\iota = (2; \tau)$ where τ is the time since entry. For example, if $\iota = (3; 1)$, then at $t = 13.5$ the target is northbound on the road at $-50E$. Note that the time south of $\mathbb{R}^O$ is a $\mathcal{G}[k]$-stopping time.

The **GWE** maintains three estimators as the target crosses the veiled region. The estimators have a common initial condition. The position is that at ingress to $\mathbb{R}^O$; the east speed is essentially zero; the north speed is the speed estimate at entry. The uncertainty in east location and velocity is quite small. The uncertainty in north location and speed is larger because the time of entry is not known precisely, and the entry speed is that estimated by the **EKF**. We will postulate: $\hat{x}[t = 12.5] = [-50, 180, 0, 10.3]'$; $P_{xx} = \text{diag}[0.01, 0.34, 0.01, 6.5]$.

8.3.3.2 Motion Across the Veiled Region

Initially, the three paths in $\mathbb{R}^O$ are equally likely: the likelihood the target will exit to the east is equal to the likelihood the target will exit north. The **GWE**-classifier can use the annotated path map $\mathcal{M}$ of $\mathbb{R}^O$; see Fig. 8.6. The **GWE** begins with a slightly high speed estimate. It will extrapolate forward using the path map and the speed estimate until the target exits from $\mathbb{R}^O$. Lacking range measurements, the **GWE** uses the three map-appropriate LGM models without update. For example, the $\iota = (3; \tau)$ extrapolation uses a CV model with tight restrictions in east-west and more relaxed restrictions in north-south.

Within $\mathbb{R}^O$, the **GWE** generates a three-term $\mathcal{Y}[k]$-distribution of the kinematic states of the target. For example, the $\iota = (3; \tau)$ **GWE** extrapolates essentially due north. Figure 8.11 shows the distribution of north location along the $-50E$ road segment after 17 s in $\mathbb{R}^O$ (and 29.5 s into the engagement) under the $\iota = (3; 17)$ hypothesis.

From Fig. 8.11, the mean north position is nearly 500N: exit occurs at 251N. While the probability that the target is in $\mathbb{R}^O$ is positive after 17 s, it is unlikely to be there. The density in the veiled interval is shown shaded in the figure. It is hard to see since $\mathbb{R}^O$ is several standard deviations south of $\hat{\chi}$. Of course, there are no range measurements at $t = 29.5$, so, if one espouses the $\iota = (3; 17)$ hypothesis, one must believe that the target is, in fact, within the tiny shaded interval.

Another hypothesis is that the target follows the west path: $\iota = (2; \tau)$. Figure 8.12 shows the distribution of north location along the $-100E$ road segment under the $\iota = (2; 17)$ ($t = 29.5$) hypothesis. The mean position, $\hat{\chi}$, is about 260N and the north boundary of $\mathbb{R}^O$ is about 250N. The speed at entry is high and the mean north location is in $\mathbb{R}^C$. No range measurements were taken at $t = 29.5$ because the actual position was $(-100, 250)$.

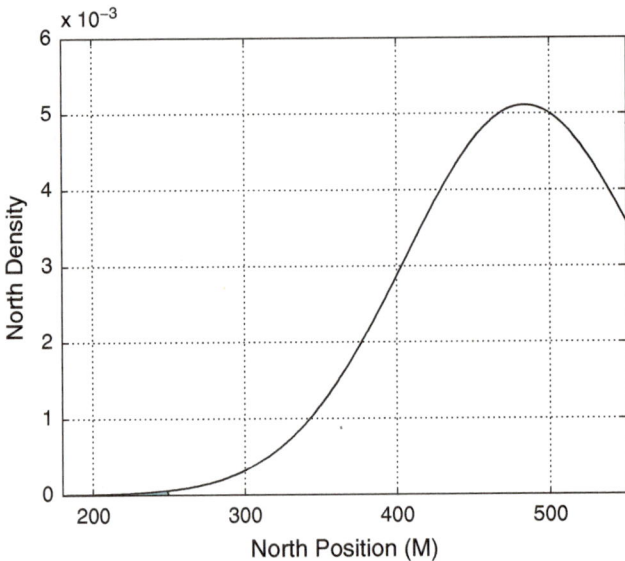

Fig. 8.11 Probability distribution of the predicted target N–S position. After 17 s moving west within $\mathbb{R}^O$, the distribution is heavily weighted to the north. The interval in $\mathbb{R}^O$ along the $-50E$ road segment is shown *shaded* in the *lower left corner* of the figure—it is very hard to see with this scale

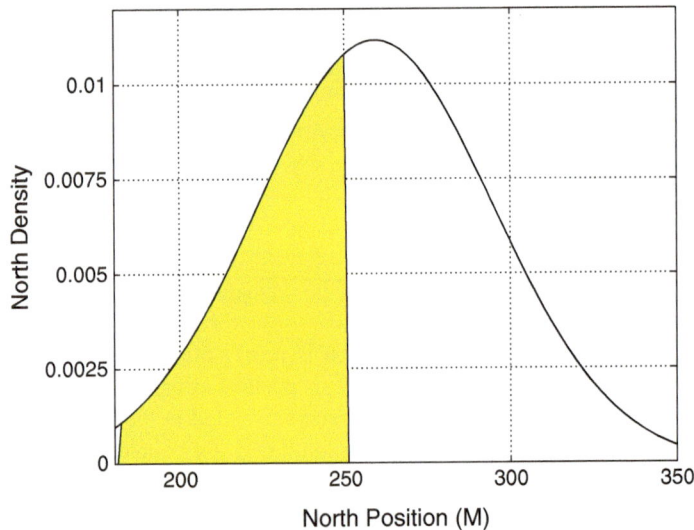

Fig. 8.12 After 17 s moving from entry along the west path, the probability distribution of the predicted target position along the $-100E$ road segment is evenly balanced between $\mathbb{R}^O$ and the north region. The veiled interval is again shown *shaded* in the figure

The third hypothesis in $\mathbb{R}^O$ is $\iota = (4, \tau)$. In this mode, the target enters $\mathbb{R}^C$ from the west. The target is too slow to make this a viable option. So this element of κ will be rejected when the first range measurement is made at $t = 30$.

8.3.3.3   Handoff to Classification

At egress, the GWE must be reinitialized. The two plausible exit points are $(-100, 251)$ if $\iota = (2, 17.5)$, and $(-50, 251)$ if $\iota = (3, 17.5)$. Both exits are to the north and both have the same speed. In neither case is the GWE-classifier aware of the north road segments in $\mathbb{R}^C$. For either case, the north-most kinematics use the CV model used in the south.

Despite the fact that the target model is the same for both exit alternatives, motion into $\mathbb{R}^C$ requires two new regime states: $i = 5$ for a west exit; $i = 6$ for a north exit. We would use $i = 7$ for an east exit, but we have excluded that hypothesis for times greater than $t = 30$. This expansion of the regime space may seem counterintuitive since the motion outside of $\mathbb{R}^O$ is CV in all instances. But the GWE regime label points to separated kinematics, either by motion equation or by initial condition. It is this latter to which the exit mode points: $\iota = 52 \ldots$ or $\iota = 63 \ldots$.

The GWE updates $\hat{\phi}[k]$ at $t = 30$. The three motion modes were equally likely at $t = 12.5$. But $\iota^- = 44 \ldots$ can be eliminated as soon as the range measurements are restored. It is apparent from the figures that $\iota^- = 22 \ldots$ is much more likely than $\iota^- = 33 \ldots$ in so far as the probability that the target is in $\mathbb{R}^O$ is far higher under $\iota^- = 22 \ldots$. We will assign probability one to the former and continue with the GWE. Since the chains do not communicate, it must be that $\hat{\phi}[k] = e_5$ for all $k > 59$.

With the probability of $\iota = 52 \ldots$ essentially one, the target leaves $\mathbb{R}^O$ at $t = 29.5^+$ from the west exit. We will assume that the exit velocity is northbound and the mean speed is that at entry to $\mathbb{R}^O$. We will complete the GWE by setting $P_{xx}[t = 29.5] = P_{xx}[t = 12.5]$.

Figure 8.13 shows the response of the GWE—contrast this with Fig. 8.7. In the earlier figure, the EKF developed very large errors as it dead-reckoned target motion in the obscured region. The GWE, however, recognizes the loss of measurement data and avoids the wild target excursion indicated by the EKF. Only the path segments where the target is visible to the sensors are displayed. From initiation of the engagement to entrance to $\mathbb{R}^O$, the GWE matches both of the EKFs.

In $\mathbb{R}^O$, the $\mathscr{Y}[k]$-distribution of the target position is a three-term Gaussian sum. But once the target exits $\mathbb{R}^O$, the distribution coalesces into a single Gaussian. The event *target egress* is the regime measurement that compresses the state distribution.

At exit, the GWE provides better tracking performance. Figure 8.14 contrasts the EEP ellipses for the blind EKF with those for the GWE. The EEP areas of the GWE (right panel) are smaller and more faithful than are the corresponding ellipses for the blind EKF (left panel). But in neither case do the EEP ellipses cover the target half of the time.

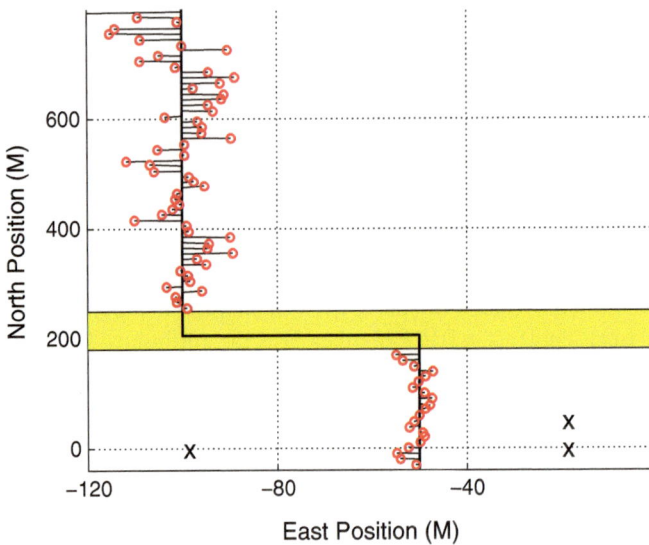

Fig. 8.13 The GWE-tracker reinitializes the placement of the target as it enters $\mathbb{R}^O$. The tracking error at exit is far smaller than that of the blind EKF

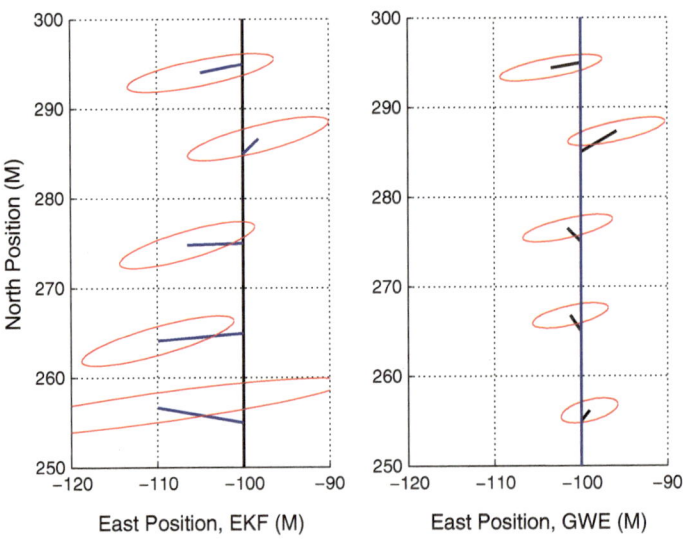

Fig. 8.14 The *left panel* shows the EEP ellipses for the blind EKF-tracker at exit from $\mathbb{R}^O$. The *right panel* shows the EEP ellipses for the GWE-tracker using the same measurement data set. The former are much bigger than the latter

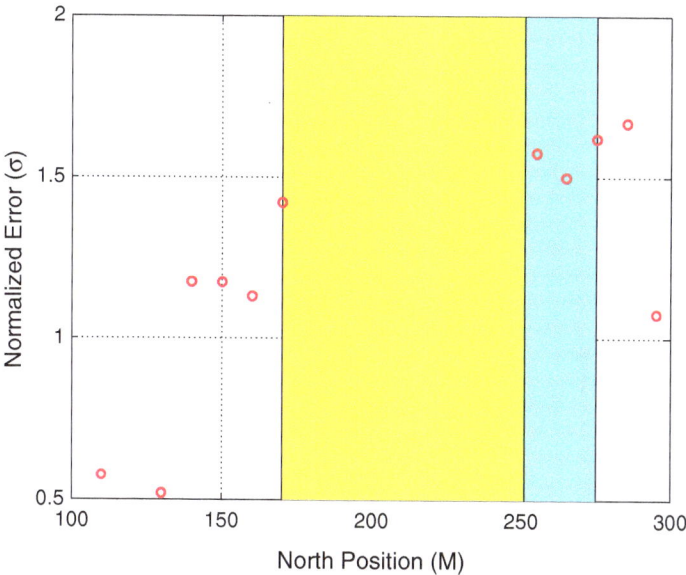

Fig. 8.15 The GWE-tracker keeps the error to about 1.2σ until the target enters $\mathbb{R}^O$. But the errors in the illumination region exceed the EEP limits

Figure 8.15 shows the normalized errors of the **GWE**-classifier. South of $\mathbb{R}^O$, the EEP ellipses capture the target at a satisfactory rate. Upon entering the interrogation region, the EEP ellipses no longer suffice. The standard error is about 1.6σ whereas 1.2σ is required. Thus, the target will escape interrogation with the EEP-logic.

Classifying the target based upon illumination of the EEP ellipses does not meet the system specifications. But this failure is caused by the notional accuracy of the **GWE**-tracker. The **GWE** identifies the exit point of $\mathbb{R}^O$. The **GWE** is not certain about the exit speed, but knows the east coordinate at egress with probability essentially one. Figure 8.16 shows the areas of the **GWE** EEP ellipses. The first range measurement in the illumination is at 255N. The tracker allows the target to stray from the road a bit. The areas of the EEP ellipses are between 5 and $10\,\mathrm{m}^2$. Even if we use an ellipse of size 1.6σ, the illumination areas are less than the $20\,\mathrm{m}^2$ permitted. Indeed, with the 1.6σ regions, the confidence that the target will be classified is notionally 98 %: actually, the target is illuminated every time.

8.3.4 Algorithm Contrasts

In this engagement, the target follows a path along which it is hidden for an interval of time. For example, the target might move along the base of a canyon and not be visible to the sensors. When thus obscured, target placement becomes a dead reckoning problem as studied in Chap. 4: find the $\mathscr{Y}[k]$-conditional distribution of $\chi[k+r]$ with $r > 0$.

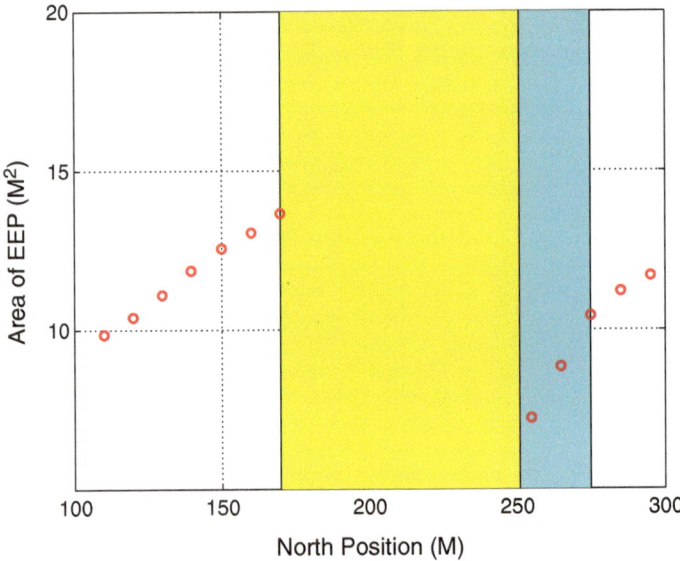

Fig. 8.16 The areas of the EEP ellipses of the blind tracker are very small. After the target leaves the shrouded region, the EEP areas barely exceed $10\,\mathrm{m}^2$ over the interrogation interval

Sensor blinding is a part of many realistic scenarios. The blinding could be partial; e.g., smoke or partial obscuration by interfering objects in the near-field. Or it could be total; e.g., the target is hidden behind a wall or building. But in either case, the obscuration event gives useful information if the local motion is well described.

In this section we pose a classification problem in which a target enters a region inaccessible to the sensors. The target must be located using the noisy sensor measurements that remain and be classified according to its intent. The sensor measurements are of moderate quality, and the illumination interval is short. The blind **EKF**-classifier is not able to meet the system specifications. Either the illuminator must be made more sensitive or the illumination interval must be made longer.

The **GWE**-classifier has only one advantage over the **EKF**-classifier. It has the annotated map $\mathscr{M}$ of the veiled region. We have shown that in this engagement, the **GWE** can use the range data set found lacking by the **EKF** to properly classify the target over the short interrogation period.

As the target passes though $\mathbb{R}^O$, neither the **GWE** nor the **EKF** has a clear sense of location. The flag, *target egress*, is useful for the **GWE**. Note that both the **EKF** and the **GWE** receive the same $y[60]$ range measurements. But only the **GWE** uses this event to localize the exit path. The **EKF** simply adjusts the $\mathscr{Y}[k]$-location estimate to accommodate the new range information.

Since the range data set is common to the algorithms, as the target moves away from $\mathbb{R}^O$, the location estimates converge. In the north region of the motion space, the placement errors are essentially the same for the blind and the sighted algorithms—the benefit of the serendipitous event decays with time. Of course, this is a very simple problem: the modal chains are disjoint and modal mixing in the estimates is not required. The GWE selects the exit point based almost exclusively on the elapsed time in $\mathbb{R}^O$. Consequently, much of the sophisticated logic within the GWE is not exercised.

8.4 The Veiled Target with More Ambiguous Para-Measurements

8.4.1 The Engagement

In the previous section, we studied an encounter in which an agile target must be classified as hostile or benign based upon an image captured in an interrogation region, $\mathbb{R}^C$. Within $\mathbb{R}^C$, a small illumination window is placed about the target and intent determined. We seek to keep the window small to minimize energy consumption and to reduce image processing complexity. But the target moves through $\mathbb{R}^C$ quickly, and we want to be assured of a small probability of both Type 1 and Type 2 errors.

The classification problem is complicated by the fact that the sensors are obscured preceding entry to $\mathbb{R}^C$. In the encounter above, we found a hybrid algorithm was able to classify the target and meet the relevant system specifications. A GWE-tracker located the target well, and an expanded EEP ellipse captured the target during the 1.5 s interrogation interval. The EKF alternative failed to find the target even once.

The blind EKF and the GWE trackers both use dead-reckoning to extrapolate across $\mathbb{R}^O$. Both begin with the same ingress conditions and use the same south range data set. Both extrapolate for the 17 s that the target is hidden, and resume updates when the target reappears. But only the GWE can locate the target with confidence.

The GWE utilizes the annotated map $\mathscr{M}$ of $\mathbb{R}^O$. The map shows the possible paths within $\mathbb{R}^O$ along with the condition of the interior road segments. For example, the GWE knows that target speed is reduced by 50 % at ingress. The GWE knows that east-west segments are even more restrictive, with a nominal safe speed only a quarter of the entry speed. The GWE maintains three distinct extrapolators, one for each interior path. At egress, the GWE identifies the appropriate exit location and the direction of motion. The speed at exit is taken to be that at entry as is the uncertainty in the speed. Thus, we have a nearly linear H^+ map—not quite linear since speed is not a linear function of velocity at ingress.

At exit, the GWE merges the location estimates of the three extrapolators. The GWE actually has a simple form. The trackers make an accurate estimate of the entry conditions into $\mathbb{R}^O$. During passage though $\mathbb{R}^O$, the three permissible modes are equally likely. The correct mode is identified at egress with essentially zero error. Thus, the classifier is simply an EKF reinitialized to match the west path. The GWE-illumination windows cover the target at every sample time with size well within specifications.

The GWE-classifier is so accurate because the exit times are well delineated by the entry conditions. While there are three paths, the north path traverses $\mathbb{R}^O$ too quickly; the east path traverses $\mathbb{R}^O$ too slowly; and the west path traverses $\mathbb{R}^O$ at roughly the right speed. Even though each path retains a positive probability of truth, north and east are so unlikely as to be ignored.

Let us consider a modified engagement with the same specifications. The sensor geometry is the same: three synchronized range sensors with period $T = 0.5$ s. But now the standard range error is six times larger: $\sigma = 12$ m. This makes the entry to $\mathbb{R}^O$ harder to resolve without a map. And the speed estimate at ingress is also degraded.

Once in $\mathbb{R}^O$ the motion hypotheses are expanded. The motions are such that the mean time to exit is the same for all three exits. Thus, the time of the first accessibility signal is no longer a definitive indication of location.

One advantage the classifier has in this engagement is that $\mathcal{M}$ conveys some information about motion through $\mathbb{R}^C$. This will be discussed in more detail within the context of the GWE.

8.4.2   Blinded EKF

An EKF-classifier was presented in the previous section. It uses a simple CV model for the kinematics. The sensor noise in this engagement is six times more intense than we saw earlier. This makes the transformation of the polar measurement into a Euclidian coordinate system more problematic. The performance of the blinded EKF is shown in Fig. 8.17. The tracker is clearly not adequate for this task. In the south, the GDOP is relatively good, and some estimates are within 20 m or so of the target. When the target enters $\mathbb{R}^O$, the EKF tacks off to the northeast—the full array of estimates are not shown in the figure because the error is so large.

As the target moves to the north of $\mathbb{R}^O$—and into $\mathbb{R}^C$—the EKF is not able to locate the target or the road. The velocity is mainly north with a big lateral spread. As the target moves farther north, the velocity estimate is more east-west. Again some of the tracking errors are too large to be displayed in the figure.

As we did in Fig. 8.8 we can display the EEP ellipses for the EKF as it enters and as it exits $\mathbb{R}^O$. The left panel in Fig. 8.18 shows the EEP ellipses as the encounter begins. The tracking errors are on the order of 20 m and the target is in no case covered by the ellipse. In the center panel, the target enters the veiled region. The EKF does capture the target twice with the large diameter EEP ellipses. However, after entrance to $\mathbb{R}^O$, the error grows more rapidly than does the radius of the EEP ellipse, and capture fails.

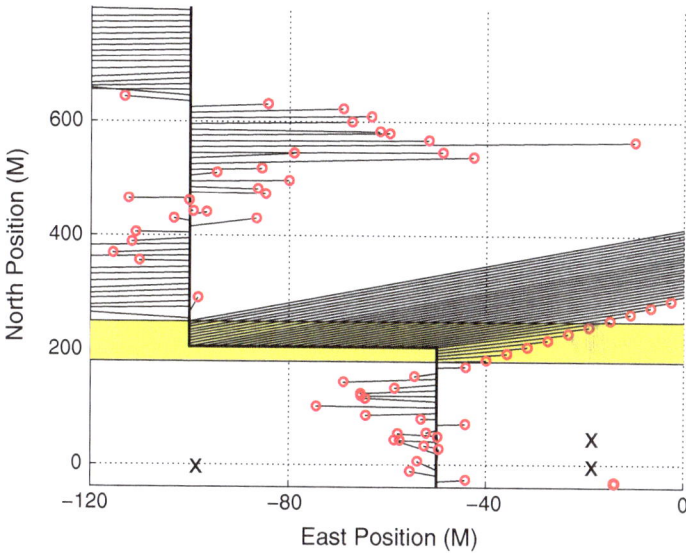

Fig. 8.17 The blinded **EKF** has very large errors when the range noise increases to $\sigma = 12$ m. The location estimates are shown with a *open circle*

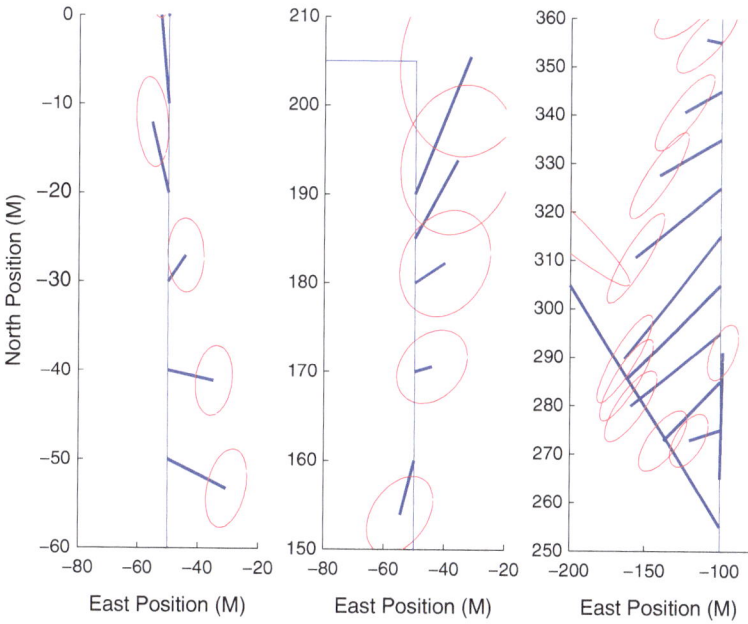

Fig. 8.18 The EEP ellipses of the **EKF**-classifier are excessively big and are of low quality

The error and the EEP ellipses grow quickly in $\mathbb{R}^O$. Suffice it to say that the EKF loses its way in the veiled region: $\hat{\chi}$ moves well beyond $\mathbb{R}^O$. This has some important implications. The range data set is converted to a Euclidian form using $\hat{\chi}[k]$ as a pivot. But $\hat{\chi}[k]$ is so far from $\chi[k]$ that the reforming of the measurements is of dubious accuracy.

We see the difficulty in capturing the target in the right panel of Fig. 8.18. The first update after egress has an error of a couple of 100 m despite the fact that the range error is of order 12 m, and there are three independent sensors. The tracking error is quickly reduced to less than 100 m, but the EEP ellipses are far too small to illuminate the target in the interrogation region.

The EKF is a simple recursive algorithm that fuses a sequence of range measurements into a plausible estimate of target location. However, obscuring the target for a period causes tracking quality to degrade significantly. The EKF fails in this application: the tracking error is too big and the notional assurance regions are larger than specified.

8.4.3   GWE

8.4.3.1   Map Annotation in $\mathbb{R}^O$

Now, let us relax some of the motion constraints in the veiled region and see if the GWE retains any advantage. The regions $\mathbb{R}^O$ and $\mathbb{R}^C$ are as before, and the path followed by the target is the same. But the annotation of the $\mathbb{R}^O$ map $\mathcal{M}$ is not as tight in this encounter. While the paths remain the same, the road conditions and the behavior of the target are less precisely specified.

The GWE was quite effective in the earlier encounter because it was able to isolate the mode at egress with very high confidence. The entry point to $\mathbb{R}^O$ was known. The conditions of the road segments within $\mathbb{R}^O$ were that north-south roads support a speed of half of the entry speed; east-west roads support speeds reduced by half again. From $\hat{v}[k]$ at entry we estimated the entry speed. The initial direction in $\mathbb{R}^O$ is north. The position at entry is defined by the map—though temporal sampling leaves additional residual north uncertainty. From these constraints, $\mathbf{H}^+$ at entry is determined. Dead reckoning is then used to extrapolate across $\mathbb{R}^O$.

In this engagement there is a single entry point to $\mathbb{R}^O$, and there are three possible points of egress. In the previous section they were labeled $\iota^+ = 52\ldots, \iota^+ = 63\ldots,$ and $\iota^+ = 74\ldots$: exit-west, exit-north, exit-east. We know that at $t = 30$, the target enters $\mathbb{R}^C$—the target spends the interval $[12.5, 29.5]$ within $\mathbb{R}^O$. Although the time preceding egress is not a $\mathscr{G}[k]$-stopping time, the entry time into $\mathbb{R}^C$ is.

Now suppose the environment across $\mathbb{R}^O$ is more ambiguous. Let us first look at the west path which is that realized by the target. Denote the entry speed by $\bar{v}$. From the map the direction is north. As it was before, north speed $= 0.5\bar{v}$ until junction; east speed $= 0.25\bar{v}$ until corner: north speed $= 0.5\bar{v}$ until exit from $\mathbb{R}^O$. We called this string $\iota = (2; \tau)$ within $\mathbb{R}^O$ and $\iota^+ = (5; \tau)$ within $\mathbb{R}^C$. This modal structure differs in no essential way from the previous engagement.

The second alternative is the north-exit path. Here $i = 3$ and the speed is $0.5\bar{v}$. But to confuse the defender, the target may stop for an unknown period within $\mathbb{R}^O$. The mode $\iota = (3; \tau)$ consists of north motion at $0.5\bar{v}$, and an independent stop period that is uniformly distributed with mean 8.5 s. For example, the target may enter $\mathbb{R}^O$ at north speed $0.5\bar{v}$, stop at the junction for 5 s, return to the previous speed, and continue north until exit. As before, we will call the exit string $\iota^+ = (6; \tau)$ within $\mathbb{R}^C$.

The final possibility is that of an east exit. Here, the target proceeds to the junction as before and then turns east. But the road conditions in this region of $\mathbb{R}^O$ are not well known. The map annotation suggests two possible east speeds. The east segment may be as it was earlier with slow east speed; $0.25\bar{v}$. Or it may be that the road condition from the junction is quite good, and the east speed is that at entry; $\bar{v}$. We will call the former mode $\iota = (4; \tau)$ and the latter $\iota = (8; \tau)$. The map annotation gives both speeds and assigns probability 0.58 to the former. We will denote the exit mode in the low speed case $\iota^+ = 74 \ldots$ and the high speed case $\iota^+ = 98 \ldots$.

Upon leaving $\mathbb{R}^O$, the kinematic model is again CV. But in this engagement, the map indicates that the motions are strongly directional. For example, a north exit implies that the target moves narrowly north on a road of width about 9 m; an east exit implies the target moves narrowly east. In both cases, the initial speed is the speed at ingress to $\mathbb{R}^O$. Over time, the target may drift from the direction at egress, but over $\mathbb{R}^C$ we have reason to expect the lateral motion to be tightly constrained.

The interrogation region is as it was earlier. The classifier will seek to illuminate the target at exit from $\mathbb{R}^O$ to either the north or to the east. The classification must be accomplished with probability of over 80 % and illumination area less than $20\,\text{m}^2$. We will not require the capture region to be convex—or even simply connected.

8.4.3.2 The Kinematic Model

The motion model south of $\mathbb{R}^O$ is that we used in the previous work. It is CV with a drift to the north and higher exogenous acceleration in north-south. In this encounter, the sensor noise is increased, and the location estimates are poor when south of $\mathbb{R}^O$.

At ingress to $\mathbb{R}^O$, the east-west uncertainty is excised. As we extrapolate across the veiled region, we will suppose that the entry speed is the north component of velocity at entry: $\bar{v} = \hat{v}_2$. In the sample engagement, the speed preceding entry is 24.7 m/s with $\bar{v} = 23$ m/s. Both are significantly higher than the actual 20 m/s entry speed.

The range noise is large, and there is an increase in the kinematic error and in P_{xx}. At ingress, H^+ excises the location error to large degree along with the uncertainty in east-west velocity. We are, however, left with significant uncertainty in north speed. Specifically, the standard error in north speed is 3.7 m/s. The realized speed error is thus about one standard deviation.

The road conditions within $\mathbb{R}^O$ are not precisely delineated. Let the speed within $\mathbb{R}^O$ be labeled $\nu[k]$ where k is the number of samples in $\mathbb{R}^O$. Then the mean speed, $\bar{v}$ will be assumed to follow the rules given earlier: $\bar{v}[k] = 0.5\bar{v}$ up to the junction; $\bar{v}[k] = 0.25\bar{v}$ west from the junction; $\bar{v}[k] = 0.5\bar{v}$ north from the junction if moving; $\bar{v}[k] = 0.25\bar{v}$ east from the junction if moving slowly; $\bar{v}[k] = \bar{v}$ east from the junction if moving fast.

In contrast with the earlier engagement, the GWE sees more variability in the road conditions. We will assume that the speed variance in $\mathbb{R}^O$ grows proportionally to the estimated speed and to the square of the elapsed time. For example, under $\iota = 3\ldots$, the increment in P_{vv} is $0.5\bar{v}(\Delta t)^2$ when moving and zero when stopped. The speed covariance on the other paths grows similarly.

8.4.3.3 Motion Across $\mathbb{R}^O$

In this engagement, the time within $\mathbb{R}^O$ is actually 17 s along the west-exit path. The target enters $\mathbb{R}^O$ before $t = 12.5$ and exits before $t = 30$. Because of the high entry speed estimate, the GWE expects the traverse time under $\iota = 2\ldots$ to be somewhat less than 17 s. The GWE completes the extrapolation using the rule that if $\hat{\chi}$ extends beyond $\mathbb{R}^O$ during the veiled period, the reckoned speed and direction is that preceding exit.

Let us look at the kinematic estimate just preceding egress under $\iota = 2\ldots$. With the given initial speed estimate, the estimated north position is $\hat{\chi}_2 = 278N$. This leads to a peculiar situation. At $t = 29.5$, the target is known to be in $\mathbb{R}^O$ since the target is not visible to the sensors. But the mean estimate is near the north boundary of $\mathbb{R}^C$. Of course, the GWE acknowledges the uncertainty in target placement. The entry speed is uncertain. The speed within $\mathbb{R}^O$ is uncertain too. All modes begin with a standard deviation in position of 2 m, and the standard error grows formulaically. After 17 s, the standard deviation in north position grows to 72 m along the west path. The actual position is 250N and the position error along $-100E$ is 28 m; or 0.39σ. Thus, the GWE is predictive of the egress conditions.

Figure 8.19 shows the Gaussian location density along $-100E$ under hypothesis $\iota = 2\ldots$ at $t = 29.5$. The veiled region is shown shaded. The figure contains a variety of impermissible artifacts; e.g., the target is not permitted to make a U turn and exit to the south at $-100E$. But the totality of such motions is viewed as unlikely and will be ignored. Under the hypothesis that $\iota = 2\ldots$, the north location density at the point preceding egress is $d\mathbb{P}(\chi_2 = 250) = 0.0052$.

Next look at the extrapolation along the north path. This path has a constant speed with a random halt. The halt interval could be very short, and the $\iota = 3\ldots$ GWE would move well north of $\mathbb{R}^C$—to 378N perhaps. Or the halt could be long, and the target might remain near the entrance to $\mathbb{R}^O$. For the conditions of this engagement, the estimated north position is $\hat{\chi}_2 = 278$ m. The standard deviation in north position is 94.5 m. Under the hypothesis that $\iota = 3\ldots$, the north location density at the point preceding egress is $d\mathbb{P}(\chi_2 = 250) = 0.0040$. The $\iota = 3\ldots$ north placement is essentially the same as that of the $\iota = 2\ldots$ algorithm. But the latter has a lower uncertainty at egress, and the likelihood of being near the exit from $\mathbb{R}^O$ is significantly bigger.

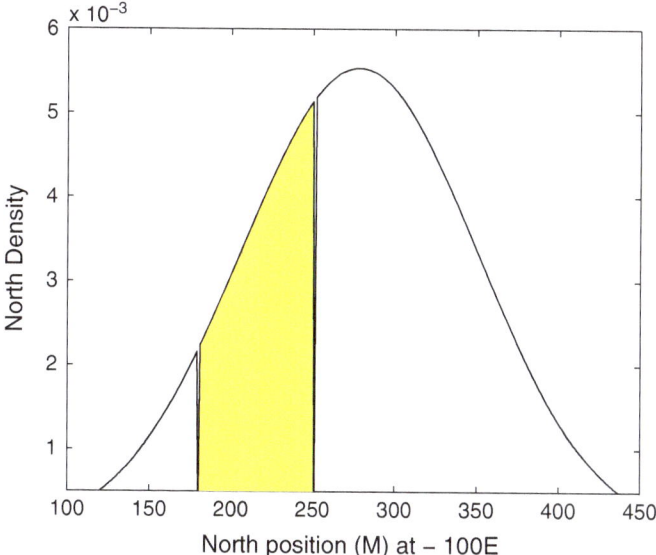

Fig. 8.19 The north position density at $t = 29.5$ of the west-exit GWE. The extrapolation places the target at the north boundary of $\mathbb{R}^C$. The $\mathbb{R}^O$ region is shown *shaded*

The north-exit density is not Gaussian. It is roughly the convolution of a uniform density and a Gaussian density with variable variance. Figure 8.20 shows both the density and a Gaussian approximation to the density—called the umbrella density in the figure. The umbrella density has thinner tails and a somewhat higher mode than does the exact density. It also has a 2.5 % higher probability of being at the exit of the veiled region. Still, the two densities are so close that we will use the umbrella (projected) density in what follows to delineate the pre-egress conditions.

The situation when the target turns east from the junction is more complex. If the eastbound road conditions are such that the target speed is reduced, the estimated east position at $t = 29.5$ is $\hat{\chi}_1 = 35$ m. However, if the road conditions are such that the target moves rapidly, the estimated east position is $\hat{\chi}_1 = 291$ m. In the former case, the target placement is well within $\mathbb{R}^O$, while in the latter case, the target placement is well east of $\mathbb{R}^C$. The standard error is 72 m under $\iota = 4\ldots$ and 84 m under $\iota = 8\ldots$. The uncertainty grows with speed, and the speed under $\iota = 8\ldots$ is higher by a factor of four.

The target placement in the east direction is the composite of two Gaussian distributions with widely separated means. The slow motion is somewhat more likely than is the fast motion. Figure 8.21 shows composite east density at $t = 29.5$. The modal distinctiveness is clear from the figure. Although the mean position is on the east boundary of $\mathbb{R}^C$, the likelihood that the target is at the east egress point is small: $d\mathbb{P}(\chi_1 = 150) = 0.0016$ if $\iota = 4\ldots$ and $d\mathbb{P}(\chi_1 = 150) = 0.0011$ if $\iota = 8\ldots$. The Gaussian approximation to the east location is also shown

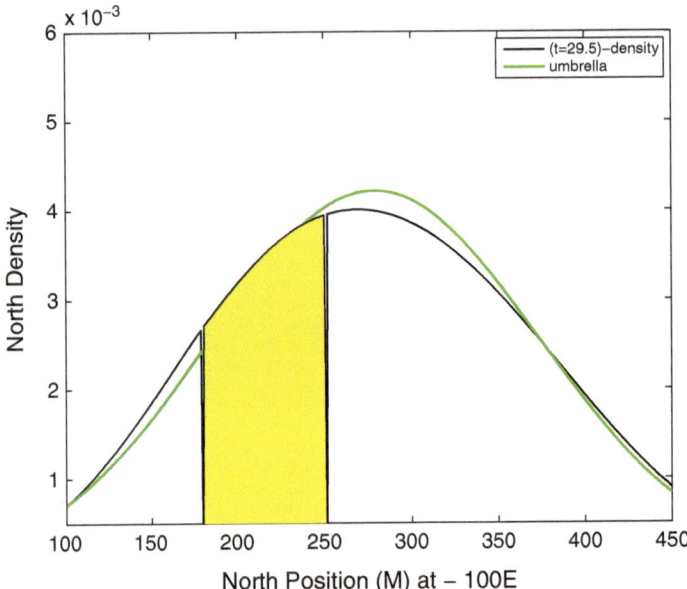

Fig. 8.20 The position density at $t = 29.5$ of the north-exit GWE. The extrapolation places the target at the north boundary of $\mathbb{R}^C$. The $\mathbb{R}^O$ region is shown shaded. Also shown is a Gaussian density matching the first two moments of the location density

in the figure. In contrast to the situation with the north exit, the projected distribution is not a good approximation to the position distribution. For this reason, we will continue to maintain the separate modes, $\iota = 4 \ldots$ and $\iota = 8 \ldots$.

Initially, all of the motion hypotheses within $\mathbb{R}^O$ are assumed to be equally likely. The exit time, $t = 30$, is a $\mathscr{G}[k]$-stopping time. Looking back from $t = 30$, the time of exit favors the west. Let us use the densities of the alternative modes at the egress point as an unnormalized modal density for the modes at $t = 29.5^+$. Then over the four permissible modes in $\mathbb{R}^O$, we would write: $\hat{\phi}(29.5)^+ = [0.44; 0.34; 0.13; 0.09]$.

Figure 8.22 shows the notional location density generated by the GWE. The composite density has four distinct local maxima. The west-exit is the strongest followed by the north-exit. Both are in $\mathbb{R}^C$ for the most part, and the road width is about a meter. The two east-exit maxima are weaker with only a part of the $\iota = 8 \ldots$ hypothesis shown because of its extent to the east. The probability that the target is within $\mathbb{R}^O$ is actually less than 50 %.

8.4.3.4 Handoff to $\mathbb{R}^C$

At $t = 30$, the target exits $\mathbb{R}^O$ and begins the traverse of $\mathbb{R}^C$. The event *target egress* is the relevant modal measurement. The exit time is a $\mathscr{Y}(30)$-stopping time, and the antecedent placement is adapted to $\mathscr{Y}(30)$.

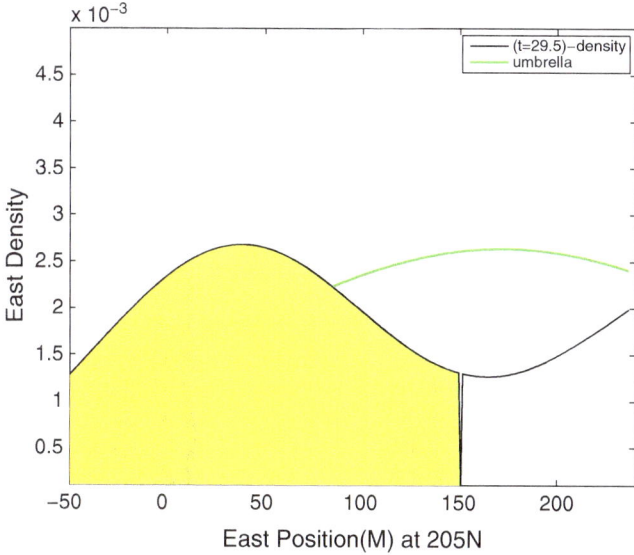

Fig. 8.21 The position density at $t = 29.5$ of the east-exit **GWE**. The density is bimodal with a strong likelihood that the target is well within $\mathbb{R}^O$, and a strong likelihood that the target is east of $\mathbb{R}^C$. The $\mathbb{R}^O$ region along the 205N path is shown *shaded*. Also shown is a Gaussian density matching the first two moments of the location density

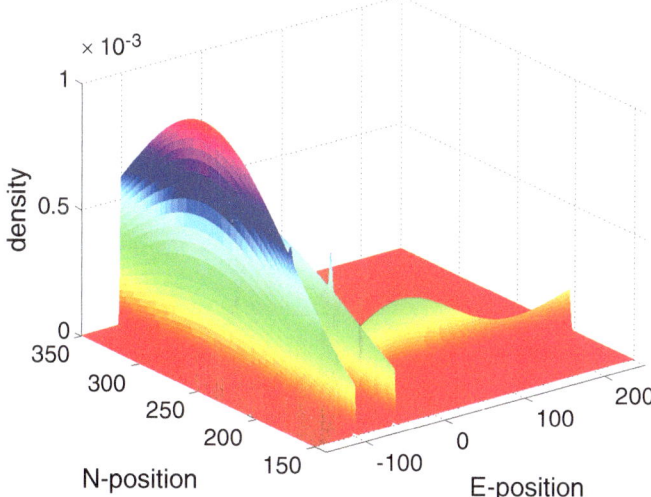

Fig. 8.22 The $\mathscr{Y}(29.5)^+$-notional location density has four distinct local maxima. The west-exit is favored

To frame $\boldsymbol{H}^+$ within the context of the **GWE** algorithm, we will view target reacquisition as an antecedent placement at a plausible kinematic state within $\mathbb{R}^O$. Specifically, we will generate $\boldsymbol{H}^+$ from a pseudo-measurement at $t = 29.5$ that puts the target near egress from $\mathbb{R}^O$.

To illustrate, let us look first at the west-exit hypothesis; $\iota^+ = 52 \ldots$. At $t = 30$, the GWE knows that the target must have been within $\mathbb{R}^O$ at $t = 29.5$ and be nearing exit. This would place the target slightly south of 251N on the road -100E. The mean speed would be half that at entry. The notional pre-exit longitudinal covariance is: $P_{y,Vy} = [5269\ 348; 348\ 29]$.

At $t = 29.5$ we know the target is on a road of width less than 9 m. And we know the target is northbound. We will generate $\mathbf{H}^+$ by postulating the existence of a location pseudo-measurement $y(t = 29.5) = (-100, 249)$ with 1 m standard error. That is, at $t = 29.5$ we assume we have a measurement that puts the target on the west road segment just south of $\mathbb{R}^C$; $\iota = 2 \ldots$. If we combine the pseudo-measurement with the extrapolated mean-covariance under $\iota = 2 \ldots$, we have small east-west speed and position errors. The north error is about a meter. The north speed is reduced since the GWE now recognizes that dead reckoning under $\iota = 2 \ldots$ took the target too far north. The adjusted ι-statistics are

$$\hat{x}(29.5) = \begin{bmatrix} -100 \\ 249 \\ 0.01 \\ 9.73 \end{bmatrix}$$

and

$$Pxx(29.5) = \begin{bmatrix} 1 & 0 & 0 & 0 \\ 0 & 1 & 0 & 0.066 \\ 0 & 0 & 0.01 & 0 \\ 0 & 0.066 & 0 & 5.60 \end{bmatrix}.$$

The antecedent placement of the target is slightly south of $\mathbb{R}^C$ within an EEP circle of radius less than a meter and a half. The velocity is strongly to the north. The east speed uncertainty is only about 0.1 m/s while the north speed uncertainty is 2.4 m/s.

Let us next look at the north-exit hypothesis at handoff; $\iota^+ = 63 \ldots$. As before, at $t = 30$, the GWE knows that the target is within $\mathbb{R}^O$ at $t = 29.5$ and nearing exit on the road -50E. The $t = 29.5$ state density is not Gaussian but is nearly so. We will use the projected approximation as the extrapolated longitudinal distribution:

$$\hat{x}_{y,Vy} = \begin{bmatrix} 279 \\ 11.6 \end{bmatrix}$$

and

$$P_{y,Vy} = \begin{bmatrix} 8940 & 341 \\ 341 & 24 \end{bmatrix}.$$

Again generate $\mathbf{H}^+$ by treating the antecedent event as a single pseudo-measurement of location with standard error 1 m: $y(t = 29.5) = 249N$. If we combine the pseudo-measurement with the extrapolated mean-covariance, we have

$$\hat{x}(29.5) = \begin{bmatrix} -50 \\ 249 \\ 0.01 \\ 10.5 \end{bmatrix}$$

and

$$Pxx(29.5) = \begin{bmatrix} 1 & 0 & 0 & 0 \\ 0 & 1 & 0 & 0.039 \\ 0 & 0 & 0.01 & 0 \\ 0 & 0.039 & 0 & 10.60 \end{bmatrix}.$$

The west-exit EEP is a circle of radius less than a meter and a half. The velocity is strongly to the north. The north-speed standard error is 3.3 m/s, an increase of 50 % over the west-exit mode.

The east-exit condition is more complicated than either of the two previous modal alternatives. As we pointed out earlier, the $\iota = 4\ldots$ extrapolation is so far from the $\iota = 8\ldots$ extrapolation that hypothesis merging does not appear reasonable. However at handoff, the east position error is resolved for both modes. We will assume the antecedent pseudo-measurement is 149E for both hypotheses. The measurement noise is again about 1 m and the pseudo-measurements are assumed to be independent. Then the corrected extrapolations are close enough to fuse yielding:

$$\hat{x}(29.5) = \begin{bmatrix} 149 \\ 205 \\ 12.79 \\ 0.01 \end{bmatrix}$$

and

$$Pxx(29.5) = \begin{bmatrix} 1 & 0 & 0.062 & 0 \\ 0 & 1 & 0 & 0 \\ 0.062 & 0 & 6.75 & 0 \\ 0 & 0 & 0 & 1 \end{bmatrix}.$$

We will call the fused pre-egress mode $\iota = 4\ldots$ despite the fact that we used this label singly earlier. The uncertainty is now primarily in east-west, and the variance is actually smaller than generated under the north hypothesis.

We now have the antecedent initial conditions. The modal state space is of dimension three. We will refer to the antecedent modes as: $\iota = 2 \dots$ west; $\iota = 3 \dots$ north; $\iota = 4 \dots$ east. The antecedent modal distribution is

$$\hat{\phi}(29.5)^+ = \begin{bmatrix} 0.44 \\ 0.34 \\ 0.22 \end{bmatrix}.$$

8.4.3.5 Motion Across $\mathbb{R}^C$

After handoff to the antecedent state, the GWE will cross the interrogation region. The target may pass along one of two northbound paths or the one eastbound path. They are labeled: $\iota^+ = 52 \dots$ west; $\iota^+ = 63 \dots$ north; $\iota^+ = 74 \dots$ east. The post-handoff probabilities are given above. At time $t = 30$, we receive the first post-egress range measurement.

The three extant modes require an extrapolation and an update. Extrapolation is made difficult by the fact that the precise exit time is not known. Is the target within $\mathbb{R}^O$ most of the $(29.5, 30)$ interval or is it there for only a small part of the interval? In the former case we would extrapolate using the $\hat{x}(29.5)$-speed estimate. In the latter, we would use roughly twice that speed. The choice is arbitrary without additional information. In what follows, we will take a centered value and say that κ^+-speed is one and a half times the antecedent speed. Of course, this leaves open the question of the forward speed at time $t = 30$: Should that be adjusted to bring the speed up to the entry speed? We will not make a further adjustment to speed. Rather, we will let the GWE adjust to the low exit speed.

We will not assume that the GWE has a map of $\mathbb{R}^C$. We assume only that the longitudinal speed uncertainty grows at $4.4\,\text{m/s}^2$ and the lateral speed uncertainty grows at $1.1\,\text{m/s}^2$: the target narrowly favors its exit direction.

The three values for $\hat{x}(30)$ can now be computed in the usual manner. The GWE-modal probabilities over $\kappa(30)$ are:

$$\hat{\phi}(30)^+ = \begin{bmatrix} 0.77 \\ 0.23 \\ 0.00 \end{bmatrix}.$$

The measurement, $y(30)$, is such as to reduce the probability of east motion to essentially zero. The $\mathscr{Y}(30)$-distribution, $\mathbb{P}(30)$, is thus a two-term Gaussian sum.

Figure 8.23 shows the notional $20\,\text{m}^2$ assurance region, $\mathbf{A}_{0.75}$. The region is the union of two disjoint ellipses; one centered on -100E and one centered on -50E. The target is actually found in the $\iota = 52 \dots$ ellipse. This is despite the fact that the mean north speed is high. Thus, the GWE correctly classifies the target during the first interrogation opportunity.

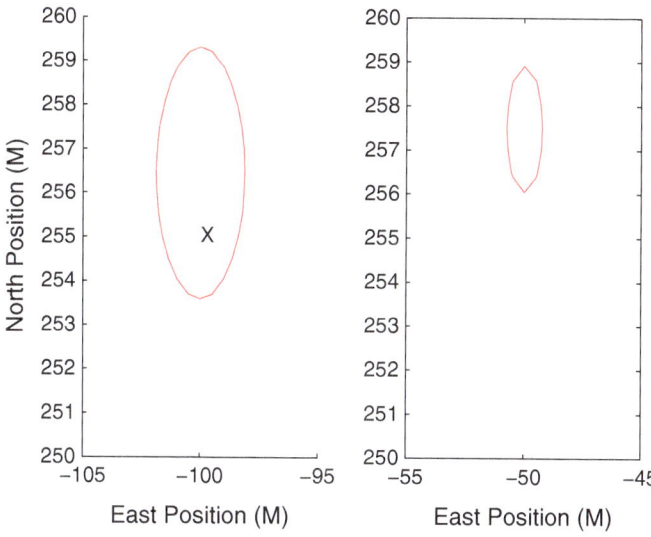

Fig. 8.23 At $t = 30$, the notional $20\,\mathrm{m}^2$ assurance region is the union of two disjoint ellipses. The east exit is no longer relevant. The mode $\iota = 52\ldots$ is favored over $\iota = 63\ldots$; i.e., the west ellipse is bigger than the north. The target is correctly classified at this time

As we move to $t = 30.5$, the **GWE** is better able to estimate the target's north speed. But the measurement is such as to favor $\iota = 63\ldots$. Indeed:

$$\hat{\phi}(30.5)^+ = \begin{bmatrix} 0.62 \\ 0.38 \\ 0.00 \end{bmatrix}.$$

Figure 8.24 shows the $\mathbf{A}_{0.75}$ region for $\mathscr{Y}(30.5)$. Again, the target is found in the $\iota = 5\ldots$ node.

At $t = 31$, the **GWE** resolves the modal uncertainty:

$$\hat{\phi}(31)^+ = \begin{bmatrix} 0.98 \\ 0.02 \\ 0.00 \end{bmatrix}.$$

The $\mathbf{A}_{0.75}$ region is a single ellipse. Figure 8.25 shows both $\mathbf{A}_{0.75}$ and the $\mathscr{Y}(31)$-density. The target is captured for the third consecutive time. The speed error remains and is a remnant of the speed error resulting from the poor egress estimate from $\mathbb{R}^O$. A more nuanced handoff formula would account for the actual speed increment on entry to $\mathbb{R}^C$.

Figure 8.26 shows the evolution of $\{\hat{\phi}[k]\}$ as the target moves from the interior of $\mathbb{R}^O$ to the exit from $\mathbb{R}^C$. Only the first two components of $\{\hat{\phi}[k]\}$ are shown so

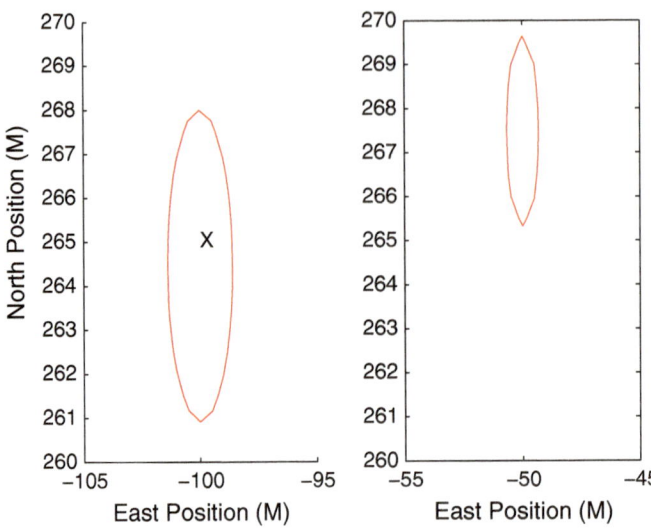

Fig. 8.24 At $t = 30.5$, $\mathbf{A}_{0.75}$ region is again the union of two disjoint ellipses. Mode $\iota = 52\ldots$ is favored over $\iota = 63\ldots$ but not by as much. The target is correctly classified again

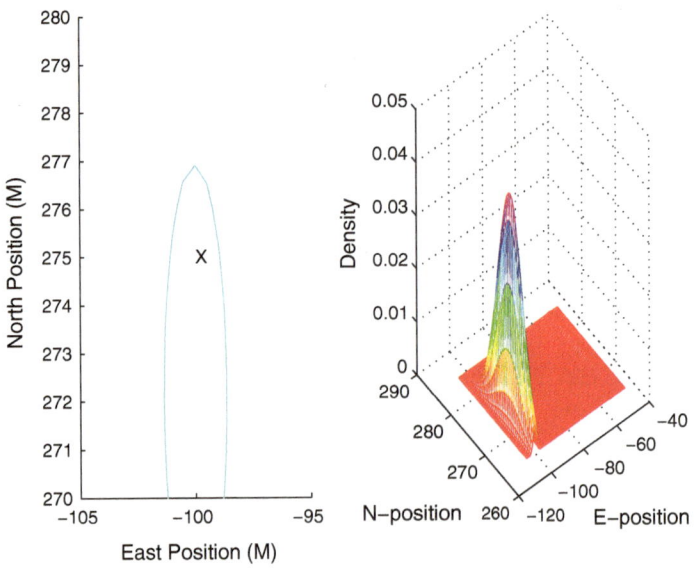

Fig. 8.25 The *left panel* shows the target within the $\mathbf{A}_{0.75}$ region. The notional density of location at $t = 31$ is shown in the *right panel*

the sum of the probabilities is less than one. All modes are equally likely at $t = 29$. At handoff, they separate. By the time of exit from $\mathbb{R}^C$, $t = 31.5$, the **GWE** has identified the correct motion mode with probability essentially one.

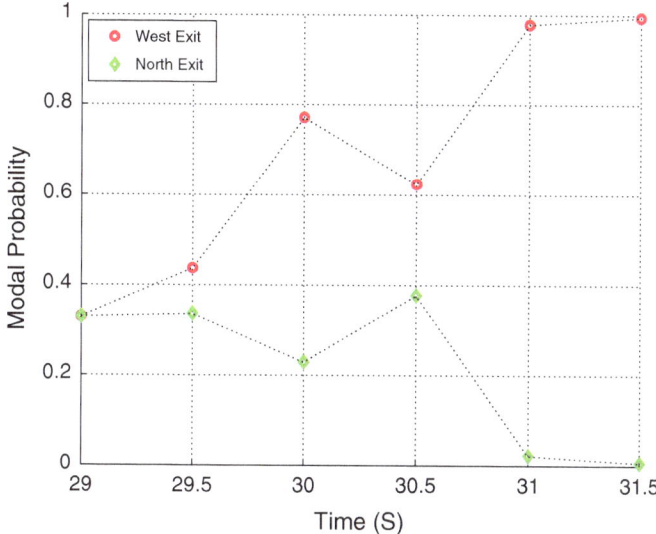

Fig. 8.26 As time progresses, the probability that the **GWE** identifies the correct mode; i.e., $\iota = e_5$, grows toward one. But the evolution is not monotonic

The range data set found deficient in the **EKF**-classifier is more than adequate with the **GWE**-classifier. The notional probability of correct target identification is 93 %, and this is conservative as we have three successive illuminations: the notional probability of three captures in $\mathbb{R}^C$ is only 19 %. Actually, the system architect could reduce the size of the assurance regions for this engagement.

8.5 Conclusions

This chapter explores the integration of serendipitous events into a difficult classificational engagement. A target is following a multi-regime path with obscuration. Target range is measured from three fixed locations. Within an interrogation region, a classification window is placed about the target and target intent determined. Once intent is known, appropriate countermeasures can be selected.

The simplest classification algorithm is based upon a CV motion model. An **EKF** tracks the target until it enters the interrogation region. An EEP-ellipse is placed about the target. With 83 % confidence, intent will be determined. If the areas of the EEP ellipses are small enough, the classification architecture is simple to implement and meets the broad resource allocation constraints.

The **EKF**-architecture is adequate for the task if the range noise is small and the $\mathscr{Y}[k]$-data set complete. But if the target enters a veiled region, the **EKF** loses its way. Eventually, this classifier can place the target, but it fails to do so in an expeditious manner.

A multi-model algorithm combined with an annotated map of the veiled region has much better performance. The GWE-classifier meets all of the performance specifications on the sample function used to test the algorithms. The GWE is able to integrate the map data with an abridged $\mathscr{Y}[k]$-data set. The GWE architecture is clearly the proper choice in this low-noise engagement.

When the sensor noise increases and/or the map annotation becomes less precise, the classification problem becomes more difficult. The sensor measurements are converted to a Euclidian reference frame, and this transformation is sensitive to the very location errors that the classifier is trying to minimize. The map information in the first encounter is sufficiently precise to enable the GWE to properly identify the mode upon egress from the veiled region. The second engagement is such that the mean time to traverse the veiled region is equal for all modes, and the exit time is less informative.

In the second engagement, the EKF architecture is unequal to the classification task. The abridged measurement set leads to instability in the coordinate transformation, and large errors in location and velocity result. No attempt was made to study the adequacy of the EEP-ellipses. They are clearly deficient.

With the same high-noise $\mathscr{Y}[k]$-data set, the GWE-architecture proves to be both adequate for the task assigned, and even conservative. The target is located and captured within the specified assurance region. Performance exceeds specifications. The $A_{0.75}$ region is not simply connected after ingress to the interrogation region. Thus, the GWE-architecture requires an ability to split the capture region. With this caveat, the GWE-classifier is certainly the architecture of choice.

Chapter 9
Deceiving a Classifier

Abstract A hostile target may do more than execute evasive maneuvers to avoid being tracked. Electronic countermeasures may be used to avoid intercept. In this chapter, we investigate range spoofing—the intentional transmission of a false range sequence to the command processor. The EKF has difficulty in recognizing a nuanced spoofing signal. In a poor GDOP environment it tends to balance all available kinematic measurements. We explore the advantage of a hybrid command algorithm in an environment that might encounter spoofed measurements. Even the GWE can be fooled by sophisticated spoofing, but the GWE is better able to stabilize the command-and-intercept algorithm than is the EKF. The engagement chosen illustrates the advantage of the hybrid approach. The adjustments to the algorithm are encounter-specific, but they provide a general guide to engineering synthesis.

9.1 Introduction

Target classification is hard enough in the presence of exogenous accelerations and sensor noise. The common reductionist kinematic models ignore the sequencing of motion modes: slowing, speeding up, turning at road junctions. Without some kind of complementary information or measurements, identifying uncooperative targets becomes problematic. Further, if the target has the ability to mislead the tracking algorithm, performance will quickly degrade.

The most direct approach to target classification is to use a model-based tracker to generate an estimate of target position along with a reliable measure of uncertainty. The classification protocol places an illumination window about the mean target location, and the threat is determined. If the target is not within the window, it is categorized according to some a priori convention. The larger the window, the more likely the target is to be labeled. However, the size of the window is constrained by the command resources.

For example, the command architect could use an EKF-tracker to locate the target. The EEP ellipses would then delineate the illumination windows. This architecture was used in the previous chapter. It meets the system specifications in the engagement studied as long as the sensor noise is small, and the sensor coverage is complete.

© Springer International Publishing Switzerland 2016
D.D. Sworder, J.E. Boyd, *Locating, Classifying and Countering Agile Land Vehicles*, DOI 10.1007/978-3-319-19431-8_9

In Chap 8 the primary engagement was made difficult by the fact that the road segment preceding illumination was obscured by some geological feature. With an abridged measurement set, the EKF-classifier no longer met the specifications. A successful alternative was one that used a hybrid tracker to place the illumination regions. The hybrid algorithm, based upon the GWE, used the same set of kinematic measurements found deficient in the EKF-classifier. The GWE used an annotated map $\mathcal{M}$ of the motion space to advantage.

The Chap. 8 engagement uses a measurement architecture based upon a set of range-only sensors. Range-only sensors are frequently chosen because of their simplicity. The range measurements are simply distances from a collection of fixed nodes. In planar tracking, the node locations are displayed in a 2-space registration table.

The tracking algorithm is based upon an established node/target geometry. In the classifier algorithms studied here, the raw ranges are first rotated into a Euclidian east-north coordinate frame to produce the kinematic data set, $\{\mathcal{Y}[k]\}$. This polar-to-Euclidian transformation is only approximate. Further, the pivot point is the estimated position of the target. This is itself a $\mathcal{Y}[k]$-random variable, and as a consequence, the data conversion is uncertain.

A discussion of this coordinate transformation is presented in Chap. 2. It is pointed out there that a range-based tracker is highly sensitive to errors in the registration table if GDOP is unfavorable. We found that an EKF is not well able to isolate registration errors because it tends to average faulty range measurements with those that are valid. Only with complementary information on target placement are we able to isolate faux-ranges.

The sensitivity of range-tracking algorithms to a faulty notional geometry suggests that a sophisticated antagonist could take effective measures to avoid classification. If the tracker can be drawn from the target path, or if the assurance regions can be made unacceptably large, a hostile target can avoid defensive action. For example, suppose the antagonist overrides one or more of the true range sequences with a mock range signal that initially deviates only slightly from truth. If the tracker accepts the signal, the faux range can be programmed to lead the tracker away from the true path. This will introduce error into the estimated target position, secondarily leading to instability in the polar-Euclidean coordinate transformation.

We say that an antagonist is *spoofing* the tracker-classifier when he causes one or more of the range measurement sequences to be overridden with a fictitious process. If done in a nuanced manner, spoofing is hard to detect and eliminate. In Chap. 2, spoofing was achieved by generating a pseudo-range signal from a node displaced from that listed in the registration table. Even this simple range bias was hard to isolate. A protocol in which an antagonist lets the bias slowly increase from zero is even more difficult. In this chapter, we will study such an engagement, and we will test alternative algorithms. We are particularly interested in the utility of a single-beacon update on a hybrid classifier.

9.2 The Engagement

As in the previous chapter, a target of undetermined intent is moving north at -50 E at speed 20 m/s. It is first detected at $(-50, -60)$, and we will set $t = 0$ at detection. After 12.5 s, the target slows and the speed drops to 10 m/s. After another 2.5 s, the target turns west and continues at a speed of 5 m/s. After a westward interval of 10 s, a turn north is made, and the speed returns to 10 m/s. Finally, 30 s after detection the target returns to a north velocity of 20 m/s, Fig. 8.1 of Chap. 8 shows the path and the regions of reduced speed.

The most commonly used tracker model uses the CV representation between speed and direction changes. Even if detection placement is accurate, and if the exogenous accelerations are small, the extrapolated target position can drift far from the road. Figure 9.1 shows a sample path of the target using the reductionist model adjusted for speed and direction as listed above. The wideband exogenous 1σ-acceleration is 0.63 m/s^2 longitudinally and 0.44 m/s^2 laterally. Despite the conservative temporal model and fortuitous north coordinate at the west turn, the sample motion drifts far from truth.

To counter this possibly hostile target, we need kinematic measurements. We will suppose that there are four synchronized transmitter-receivers that provide noisy range measurements every $T = 0.5$ s with a standard range error of 2 m. The sensor node locations are listed in order in a registration table:

$$(X_r, Y_r) = [(-100, 0); (-20, 0); (-20, 50); (-50, 500)].$$

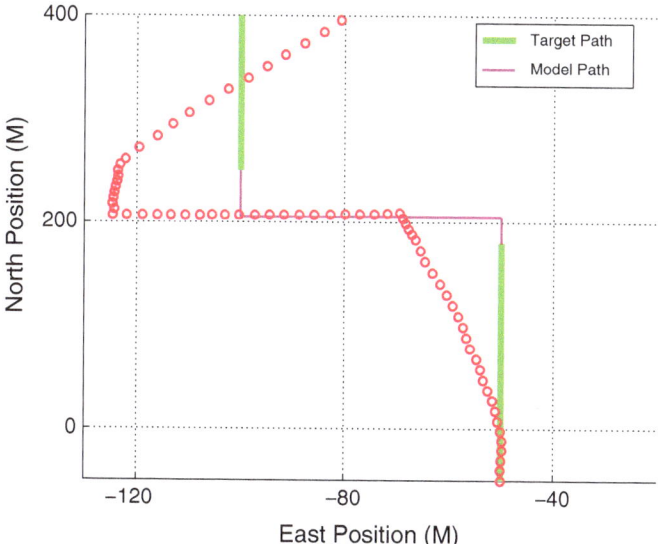

Fig. 9.1 A sample extrapolation of a *target path* may drift far from truth. (Note: In this figure and certain others, axes use unequal scales to enhance visibility of certain details)

This suite gives more comprehensive coverage than that prescribed in Chap. 8. The first three sensors are the south array of that chapter. Here, we provide an additional north sensor. Clearly, GDOP is improved in the north region with this augmented array. This architecture should protect against the instability in the north noted in Chap. 8. In this engagement we will assume the range measurements are available throughout; i.e., no blinding.

Later in the discussion, we shall introduce another sensor, a beacon or unattended ground sensor, that detects and reports the time a target passes. This detector, though very simple, will significantly enhance target classification in some engagements.

Our system specifications are similar to those in Chap. 8. The classifier has no grid map, but is aware that after the west turn at $t = 15$, the target will slow considerably. We will interrogate the target as it moves from $t = 15.5$ to $t = 17.0$. This captures the initial low speed section of the path and labels the target early in the engagement. The spatial accessibility interval will be called $\mathbb{R}^C$. The interrogation schedule provides four neighboring opportunities to label the target. We require that the threat be classified with probability of Type 1 and Type 2 errors of 12 % or less. The illumination window must be no more than 20 m^2. We assume that the classification is without error within an illumination region. If a target is missed, it is declared benign.

If the target is determined to be hostile, it will be countered at $t = 17.5$. The defensive resources are transferred instantaneously upon command (a DEW) and can cover an area of no more than 20 m^2. The intercept region need not be simply connected. If not classified as hostile, no action is taken.

This defensive system is subject to electronic contravention. As the target moves north of the detection point, it enters a region of possible hostile jamming. An adversary may override the range with a faux range purportedly from one of the range nodes; i.e., spoofing. Specifically, the defensive system is aware that node 1 or node 3 may be compromised as the target moves from detection through $\mathbb{R}^C$. Of course, many possible combinations of location and timing of spoofing are possible. We will suppose for this engagement that at most one node is compromised, and that node remains such until the target traverses the illumination region. The target is aware of the defensive protocol. If illuminated, it will turn off the spoofer after passing $\mathbb{R}^C$.

9.3 An EKF Classifier

9.3.1 No Spoofing

Let us first look at the classification problem in the absence of spoofing. The defensive system detects the target in the south region with good accuracy. First, we will use an EKF-tracker to place the target and generate the EEP ellipses. We will base this upon a time-variable CV representation of the kinematic motion.

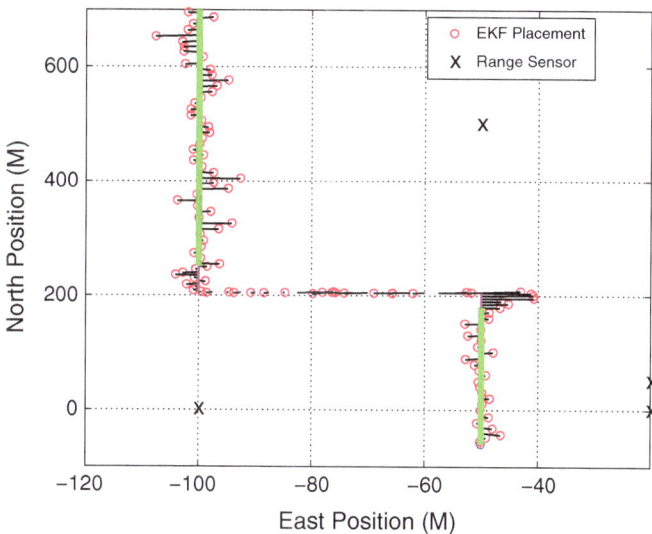

Fig. 9.2 With 2 m ranges, an EKF tracks the target rather well except for difficulty near the *west turn*. Note that the *north axis* is compressed

Even though the nominal times of slowing and speeding are incorporated into the CV equations, the intensity of the exogenous accelerations is conservative because, lacking a grid map, the location of the turns is not known precisely: the kinematic model used a 1σ-acceleration of 6.3 m/s^2 to the north-south, and 4.4 m/s^2 to the east-west.

Figure 9.2 shows the sample response of the EKF along the road path along with the four range sensors. Even though the CV model is aware of the speed reduction on the initial north leg, target placement drifts east for a time. But the GDOP is good in the neighborhood of $\mathbb{R}^C$, and the target is captured quickly on the west leg. On the second north leg, sensor 4 improves GDOP enough that the measurement instability beyond 400 N observed in Chap. 8 does not reoccur.

Figure 9.3 shows the EEP ellipses of the EKF-tracker during the accessibility interval. The east bias at $t = 15$ draws the first ellipse to the east. But a target capture occurs at $t = 17$. The areas of the ellipses range from under 13 m^2 to a little above 14 m^2. If we expanded the ellipses to their specified maximum, we would have two or three illuminations. On this sample, an EKF-based architecture is adequate for the labeling phase of the engagement: the target is successfully classified within an illumination region of area of 20 m^2.

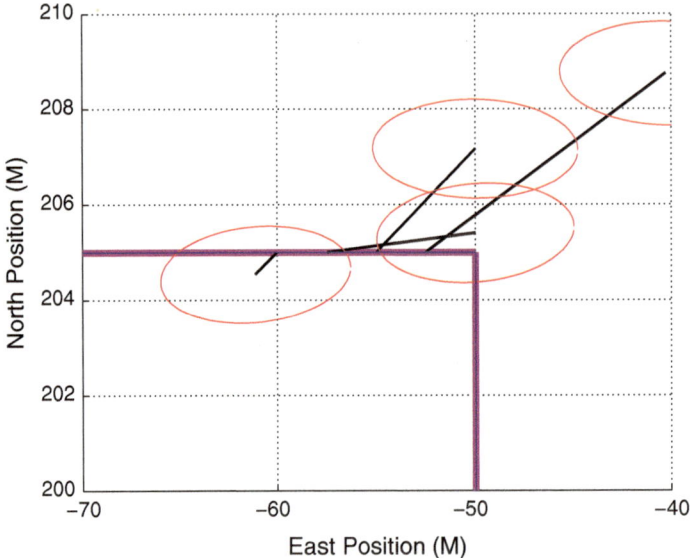

Fig. 9.3 Under nominal conditions, the EEP *ellipses* of the EKF capture the target once

9.3.2 Spoofing

Let us now suppose that the adversary recognizes detection and introduces a faux
range signal purportedly from one or more of the sensor nodes. Specifically, we will
suppose at most one of the nodes is compromised. The policy of the adversary is to
mimic a slow deviation from the registration table in the amount δS where

$$\delta S = P \otimes (r * t * d). \tag{9.1}$$

P is a 1×4 array that points to the counter-factual sensor (P is the null array in the
absence of spoofing); r is the pull-off rate; t is the time since spoofing began (the
time since target first noted its detection); d is the direction of displacement error.
For example, if we order the nodes in the registration table, then node 1 is $(-100, 0)$.
Suppose the signal from this node is counter-factual, $P = [1\ 0\ 0\ 0]$. If an antagonist
creates a pseudo-range from a displacement node that moves at the rate 2.1 m/s to
the north-east then $r = 1.5$ and $d = [1; 1]$.

As the encounter evolves, the faux node 1 moves closer to the north road segment,
and the range grows smaller than truth. This foreshortening draws the target estimate
to the west. Of course, node 1 is only one part of the sensor array, and the other
sensors will help keep the target placement from moving so far off the road as to
destabilize the tracker.

Figure 9.4 shows the response of the EKF tracker with node 1 spoofing. Both the
spoofed and the unspoofed EKF are shown. The primitive range measurements are

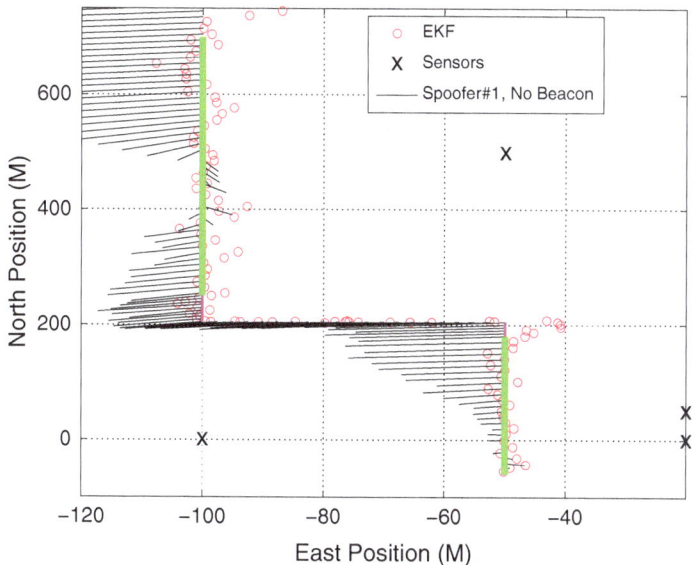

Fig. 9.4 A node 1, *north-east* spoofer pulls the tracker to the west during the accessibility interval. The unspoofed **EKF** is shown as a baseline

identical in this figure to those of Fig. 9.2. The ersatz range from node 1 is obtained by shifting the node 1 range by the spoofing offset; i.e., the wideband range noise is identical in Figs. 9.2 and 9.4. The performance of the tracker is that expected: a pull to the west where nominal GDOP is good; instability to the north where GDOP is poor.

It is clear from the figure that an **EKF** will not suffice in a spoofed engagement. Figure 9.5 shows the EEP ellipses in the accessibility interval. They are essentially the same size as were those in the basic **EKF**. But the tracker errors are several sigma from truth. The target will not be countered using this algorithm.

9.3.3 A Beacon Assist

The basic **EKF**-classifier is not suitable in the spoofed application. The CV model is aware of the temporal events, but the **EKF** is not sure of the path constraints. To ameliorate this lack, the **EKF** uses large pseudo-noise accelerations. The **EKF** drifts 40 m to the west of the north leg before the west turn. GDOP is better in the north region and the **EKF** returns to the road near 400 N. But the location estimate moves west after that, and the **EKF** is unstable.

To assist the classifier, suppose that an unattended ground sensor (a UGS) placed on the road at -50 E reports the target as it passes 190 N. In this engagement, the detection event occurs at $t = 13.5$ while the target is well south of $\mathbb{R}^C$. The UGS

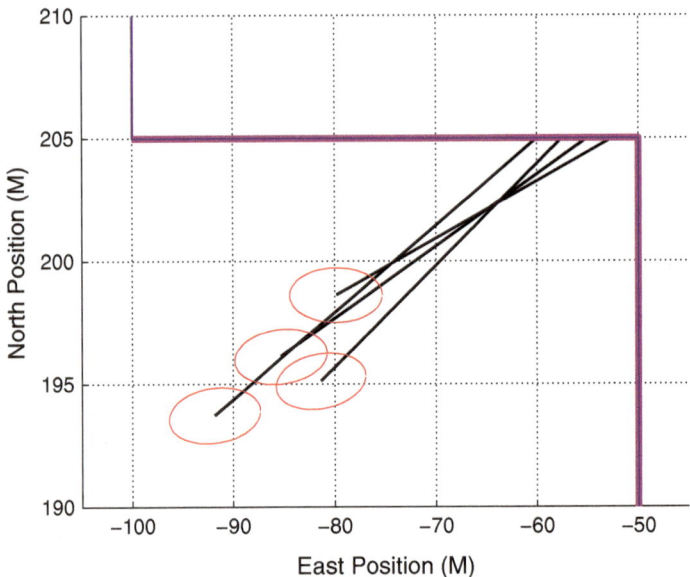

Fig. 9.5 The EEP *ellipses* of the spoofed EKF are not close to the target

makes and reports a single detection and provides no speed information. But the tracker is aware of the mainly north-south road path. The tracker accepts the beacon-reported passing event and sets the east velocity to a small value. The longitudinal velocity estimate from the EKF is, however, unchanged.

The beacon update is clearly advantageous. The west drift is eliminated along with the lateral velocity error. The beacon is 15 m south of $\mathbb{R}^C$, and the EEP ellipses should smaller and more representative of the actual target uncertainty.

Figure 9.6 shows the response of the beacon-aided EKF-tracker with node 1 spoofing. The range measurements are those of the previous section. Surprisingly, the beacon conveys little advantage outside a small nearby neighborhood. The aided EKF is indeed better at identifying the west turn. But within $\mathbb{R}^C$, the EEP ellipses (not shown) are still several standard deviations from truth. The areas of the ellipses in the illumination region are perhaps 20 % smaller than those of the un-aided EKF. On the -100 E leg, the aided EKF differs little from the unaided EKF.

9.4   A Hybrid Classifier

9.4.1   Spoofing

To illustrate the utility of the hybrid algorithm, let us again consider the classification problem without the beacon. The tracker begins the engagement with a good estimate of the kinematic state of the target. At $t = 0$, the classifier recognizes

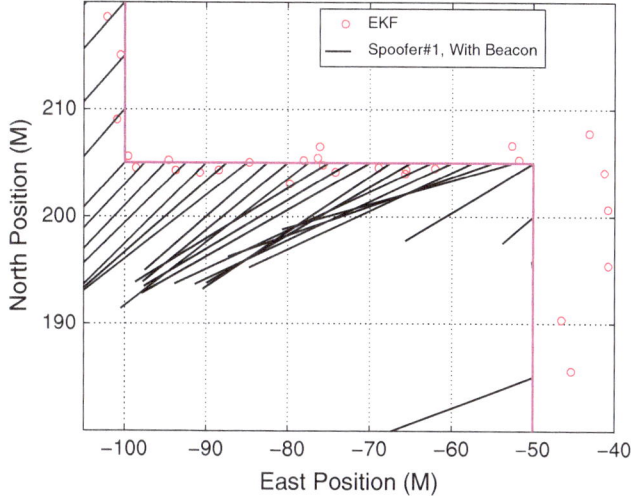

Fig. 9.6 A beacon conveys little advantage to the spoofed EKF

that the range measurements may be compromised as the target moves forward. There are three alternative modes: no spoofing; node 1 is compromised; node 3 is compromised. But neither the direction nor the pull-off rate is known. The classifier initially weights each equally and assumes that whatever the regime, it is constant from detection through illumination.

If the temporal spoofing policy were known, even the compromised sensor node would provide useful tracking information. The classifier could then attempt to cancel the spoofing component of the range signal. But lacking such sophistication, the algorithm proposed here simply ignores the compromised node; e.g., if node 3 is compromised, the classifier would use the registration table:

$$(X_r, Y_r) = [(-100, 0); (-20, 0); (-50, 500)]$$

Within $\mathbb{R}^C$, node 1 influences GDOP more than does node 3. Nodes 2 and 3 have a similar geometry when viewed from the road. So we would suppose that loss of node 1 would be more serious than loss of node 3. However, node 3 does confirm node 2 range, and its loss is a handicap. But we will ignore this subtlety and contrast three modal trackers: all four sensors are reliable ($\iota = 1$); sensors two through four are reliable ($\iota = 2$); sensors one, two, and four are reliable ($\iota = 3$). The spoofing condition exists until the target passes $\mathbb{R}^C$. Spoofing may cease immediately after illumination ceases or spoofing may continue for a half second.

We will again look at the engagement in which node 1 range is misleading; i.e.; $\iota = 2$. We will use that range data set used in the previous section. But recall that the $\{\mathscr{Y}[k]\}$ data set is random and depends upon the specific tracking algorithm; i.e., different algorithms use the same range data in different ways.

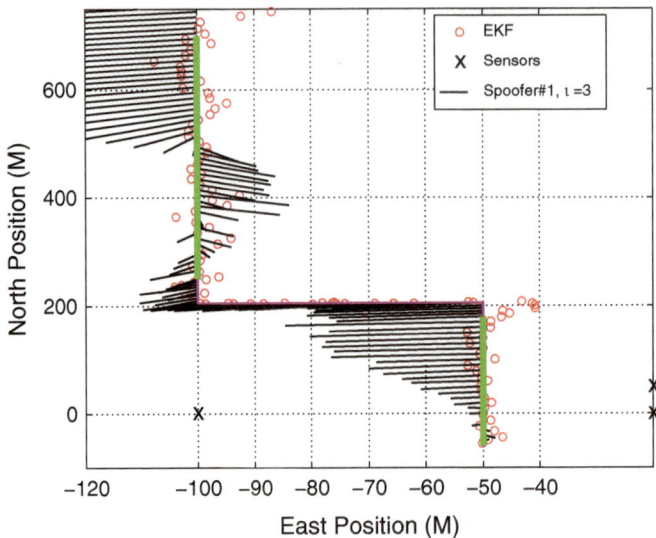

Fig. 9.7 If node 1 is compromised but we suspect node 3, performance is terrible

We already know that under $\iota = 2$, the basic **EKF**-classifier does not meet the system specifications (see Fig. 9.4). The situation is fundamentally worse under hypothesis $\iota = 3$ since the **EKF** would ignore a good sensor 3 and use a compromised sensor 1. We see this in Fig. 9.7. The location estimate under $\iota = 3$ differs little from the $\iota = 1$ tracker because the GDOP is little changed in the south with or without node 3. Both estimates are unstable in the north.

The algorithm that matches the engagement is that of $\iota = 2$. The $\iota = 2$ algorithm ignores node 1, and the spoofer is isolated: see Fig. 9.8. Unfortunately, rejecting the node 1 range increases GDOP significantly in the accessibility interval. This also increases the area of the EEP ellipses. Figure 9.9 shows the assurance regions. Though they cover the target three times, in only one case is the coverage well into the interior of the EEP ellipse. And the areas of the assurance regions are about $25\,\mathrm{m}^2$. This exceeds the area limit.

But the classification architecture proposed here fails for more elemental reasons. The tracker does not know the measurement mode and weights each hypothesis equally. Consider the location density at $t = 16$. The target is within $\mathbb{R}^C$ and the true position is $(-55, 205)$. Figure 9.10 shows the location density under the equi-likely hypothesis. Not only is the dominant lobe of the density well off the road (at approximately $(-82, 198)$), but the component associated with $\iota = 2$ hardly shows in the figure.

The dominant mode in Fig. 9.10 is a sum of the Gaussian densities associated with $\iota = 1$ and $\iota = 3$. While the densities differ, they blend when summed. Mode $\iota = 2$ is correct in this engagement, but the probability mass is spread thinly about

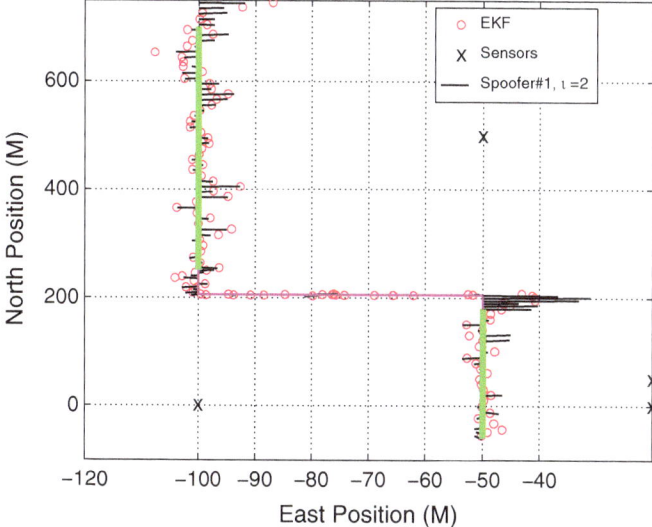

Fig. 9.8 If node 1 spoofing is correctly identified, the tracker performance is better but still unacceptable

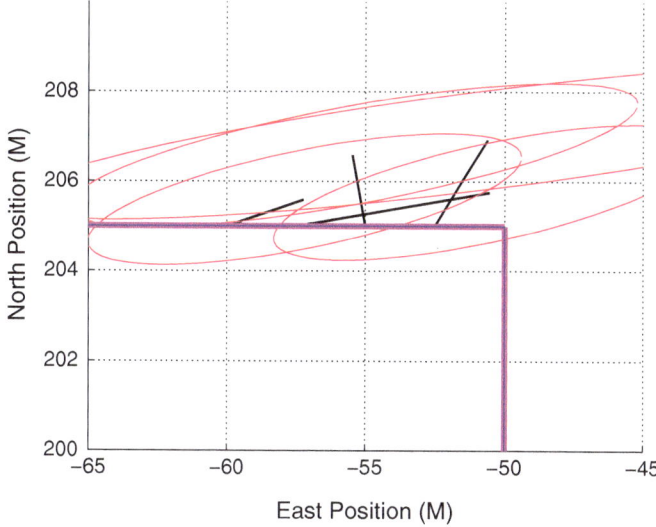

Fig. 9.9 Even under the node 1 spoofer hypothesis, the EEP *ellipses* scarcely meet the specification

the true target location. The diffuse nature of the ($\iota = 2$)-density is due to the large conditional covariance accruing to this mode. This is due in turn to the large ($\iota = 2$) GDOP. The sensor suite lacking node 1 does not have the compact coverage of the illumination region required in this application.

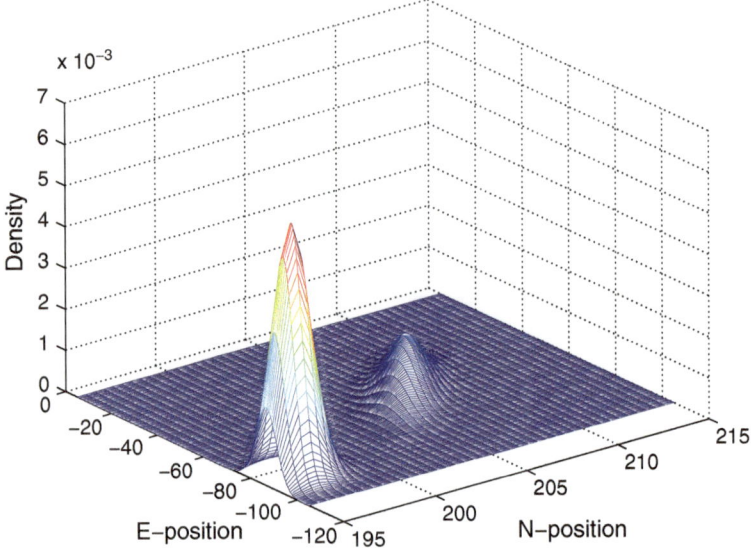

Fig. 9.10 The location density at $t = 16$ if node 1 is spoofed

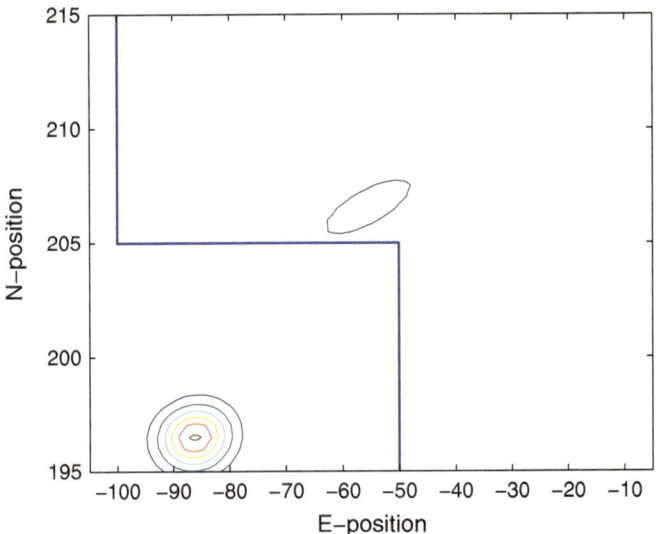

Fig. 9.11 A contour plot of the $t = 16$ conditional density shows that even if node 1 is correctly identified as being spoofed, there is little chance of correct classification

The tightest illumination regions are found by locating those points in the plane where the conditional density is greatest. Figure 9.11 shows a contour plot if the $t = 16$ density. The main illumination energy is focused about 50 m from the target.

9.4.2   A Beacon Assist

The basic hybrid classifier also does not meet the system specifications on this engagement. While acknowledging the possible modes, it has no way of distinguishing them with the available measurements. So the classifier spreads its interrogation over a large area.

Now let us augment our data set with a UGS beacon update at $t = 13.5$. In our study of the EKF-tracker we found that only in a neighborhood of the beacon is the location estimation improved to any significant degree. We concluded that the beacon is of marginal utility. However, in the hybrid architecture, the beacon signal allows us to separate the modal alternatives with high confidence.

The modal alternatives are equally likely before the beacon update. Each of the three modal trackers generates a normalized estimate of its location error at the beacon update. As previously noted, these errors are notionally Rayleigh distributed with mean 1.3. On this basis, the probability of $\iota = 1$ or $\iota = 3$ is essentially zero and the mode at $t = 13.5$ is almost surely $\iota = 2$. Once the mode $\iota = 2$ hypothesis is ratified, tracking is much improved. Figure 9.12 shows the EEP ellipses in $\mathbb{R}^C$. They are similar in size to those lacking the beacon. But the target is well into the interior of each ellipse. Thus, the hybrid classifier can be said to meet the specifications.

Figure 9.13 shows the GWE-location density at $t = 16$. The density is concentrated near the road with a north bias. The contour plot in Fig. 9.14 shows that the hybrid tracker meets the classification specifications.

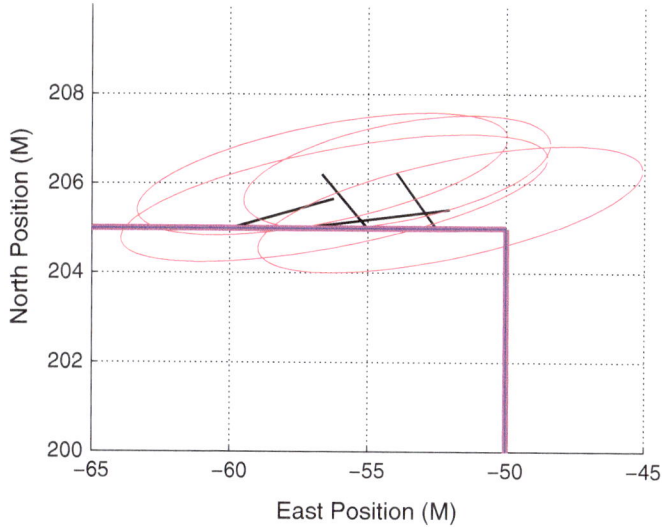

Fig. 9.12 Under the node 1 spoofing hypothesis, the EEP *ellipses* capture the target several times

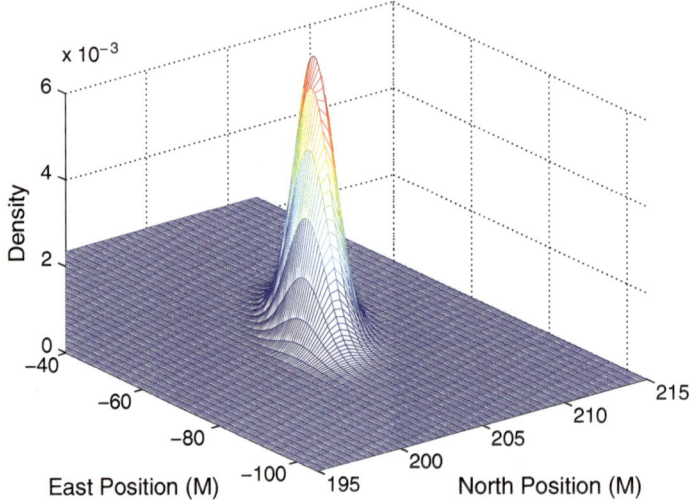

Fig. 9.13 The location density at $t = 16$ places the target near the 205 N road

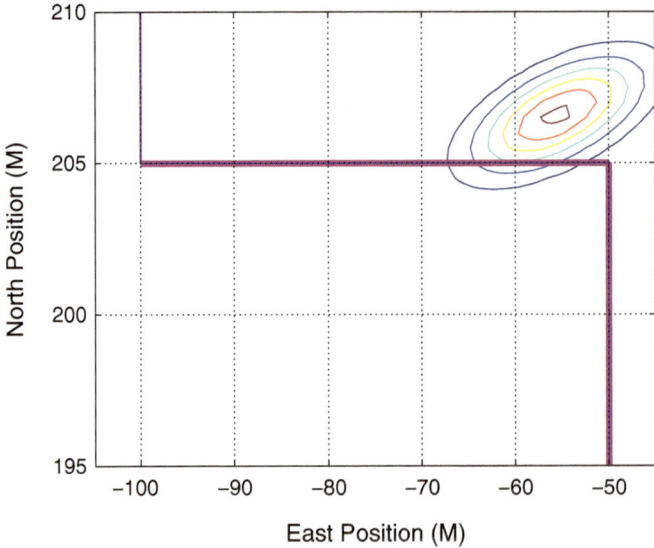

Fig. 9.14 A contour plot of the $t = 16$ conditional density shows that the hybrid classifier with a beacon meets the system specifications

9.5 Countermeasure Effectiveness

The GWE-classifier appears to be adequate for identifying the spoofer at $t = 17$ as $\iota = 2$. The illumination ellipses cover the target more than once, and as such, it is classified without error. Suppose the target is hostile. The defensive systems must now counter the threat by placing a lethal ellipse of specified size about the target at $t = 17.5$. We will suppose the transit time of the action elements is zero.

It is known that the adversary will cease spoofing after the target leaves $\mathbb{R}^C$ to avoid action against its jamming resources. But it is not known if spoofing will halt at $t = 17$ (exit from $\mathbb{R}^C$) or $t = 17.5$ (there is latency is recognizing exit from $\mathbb{R}^C$). The defensive system recognizes that the $\iota = 2 \rightarrow \iota = 1$ transition can take place at $t = 17$ (call this $\iota^+ = 122$) or at $t = 17.5$ (call this $\iota^+ = 112$).

Recognition of the temporal extent of spoofing is advantageous. The ($\iota = 1$) GDOP is significantly smaller than the ($\iota = 2$) GDOP. Figure 9.15 shows the $20\,\mathrm{m}^2$ removal regions for $\iota^+ = 122$. The target is successfully countered, but it is on the boundary of the ellipse. It actually would be preferable to wait until 19 to remove the target. The tracker would have had a chance to accommodate to the improved GDOP.

Unfortunately, the performance of Fig. 9.15 depends upon the cessation of spoofing at exit from the accessibility interval. Should the spoofing be extended for a half second ($\iota^+ = 112$), the elimination regions are moved far from the road. Figure 9.16 shows the removal ellipses under this condition. The tracker corrects

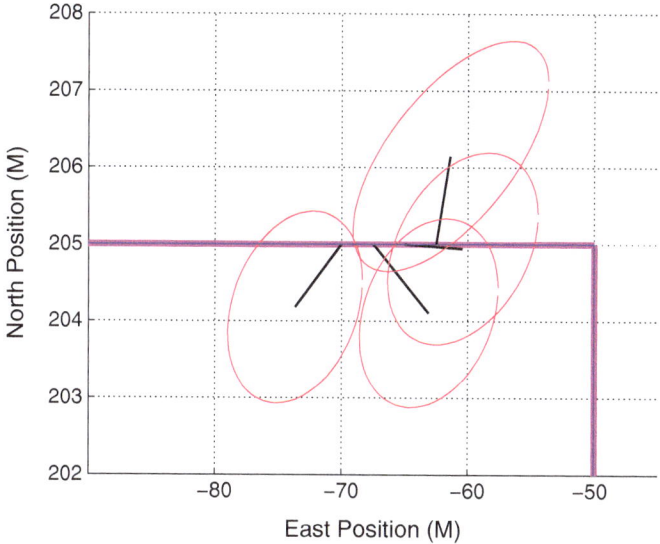

Fig. 9.15 The removal regions for $\iota^+ = 122$ on the interval $t = 17.5$ though $t = 19$

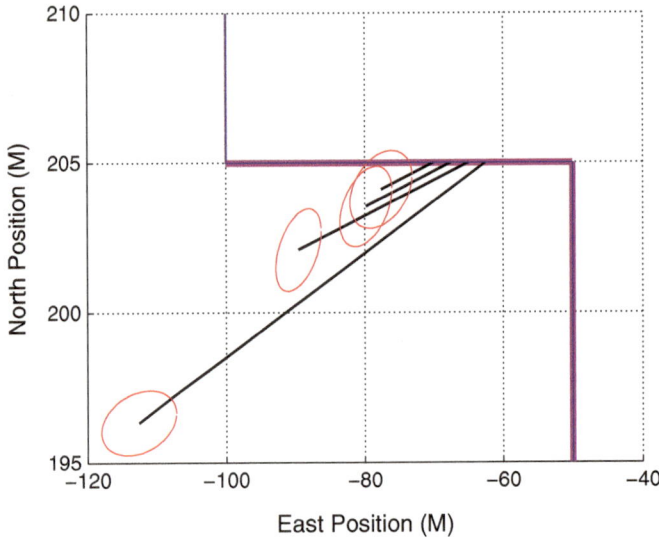

Fig. 9.16 The removal regions for $\iota^+ = 112$ on the interval $t = 17.5$ though $t = 19$

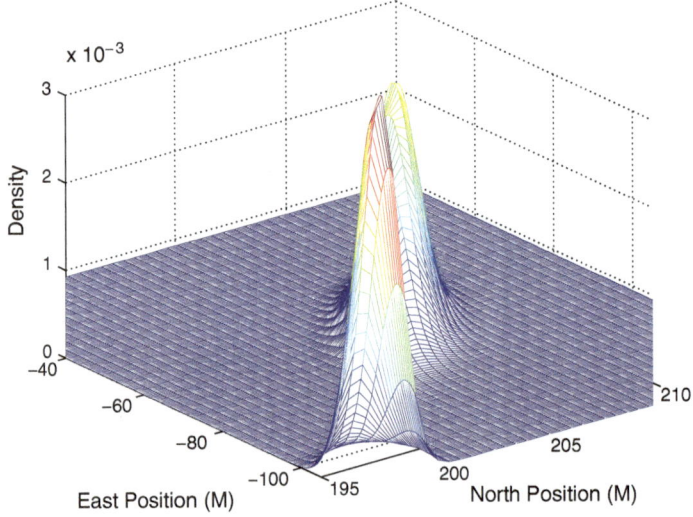

Fig. 9.17 The location density at $t = 17.5$ before the GWE update

the location estimates as time goes on. But at the decision time, the target is not captured by the countermeasures.

If the alternatives at $t = 17.5$ are equally likely, the location density is that shown in Fig. 9.17. The density is bi-modal with the sharper peak associated with $\iota^+ = 112$. This drains the defensive resources away from the target to a greater extent than the simple probabilities would suggest.

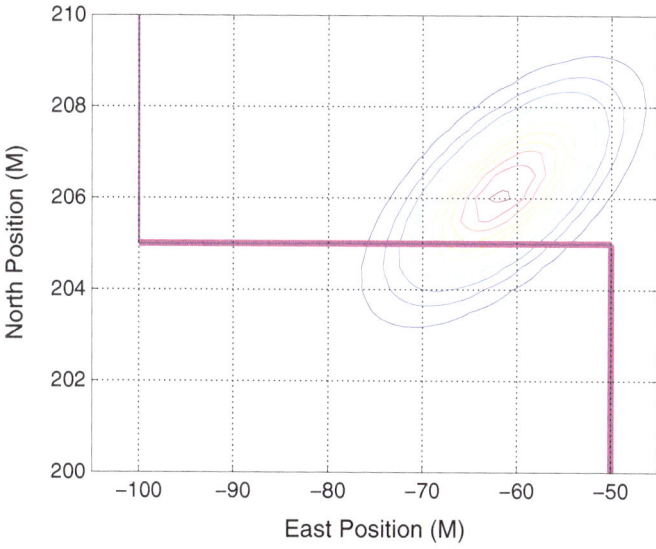

Fig. 9.18 A density contour at $t = 17.5$ after the GWE update

Now adjust the modal weights in accordance with the **GWE**: see (10.20) and (10.21):

$$L_t + [k + 1] = \Pi_{pi} z[k + 1]' \mathbf{D}_{.p} |F_t^y| \exp -\tfrac{1}{2} (\|y[k + 1]\|_{D_t^n}^2 - \Delta \|D_t^+\|_{P_t^+}^2)$$

Figure 9.18 shows the removal region after the **GWE** update. The hybrid classifier now meets all of the system specifications

9.6   Conclusions

This chapter explores the implications of target avoidance of classification using sensor spoofing. A target of undetermined intent is following a road that is not well delineated within the knowledge base of the defensive system. Target range is measured from four fixed locations. Adversary action could cause one or more of the range sequences to be false. This action is intended to cause the defender to lose the target and to fail to generate a threat alert. A defensive algorithm must interrogate the target within a specified time interval and classify it as hostile or benign. In the former case, the target must be neutralized at a specified time.

This engagement involves tracking, classifying, and countering the target with limited defensive resources. The simplest algorithm ignores spoofing and is based upon a time-variable CV motion model. The EEP ellipses generated by an **EKF** are illuminated and intent determined. The areas of the illumination ellipses are constrained. The architecture is simple to implement and meets the broad resource allocation limits.

The **EKF**-architecture is adequate for the task in the absence of spoofing. But a sophisticated spoofer can lead the EEP windows far from the target. Indeed, the **EKF**, despite its temporal model match, is unstable if but one of the range sensors is spoofed.

We propose a hybrid algorithm that identifies and isolates the spoofed sensor(s). We consider only a single spoofer in this chapter. The hybrid algorithm requires a complementary measurement to avoid the limitations of the conventional **EKF**. We suppose that there is a beacon at a known point on the road that gives a position update when the target passes. Despite having no Doppler information, this update resolves the location errors that accumulate from the time of detection to the time of beacon detection.

The **EKF**-classifier uses the beacon in a rational manner, but the benefit is lost soon after the target leaves a neighborhood of the beacon. In the interrogation interval $\mathbb{R}^C$, the **EKF** loses the target. Classification is impossible.

A hybrid tracker is superior. After the beacon update, the tracker prunes the incompatible modal hypotheses. This architecture makes classification possible within the specified windows. The hybrid classifier meets all system specifications.

The classification algorithm operates by isolating the suspect range sequence. While helpful in tracking, deleting a range measurement increases GDOP. This makes the EEP ellipses bigger and the conditional location estimate more diffuse. To counter the target, we want a tight assurance region. The EEP ellipses are smaller if we recognize the time when spoofing ceases. The full sensor suite has a smaller GDOP than does the abridged sensor suite.

But a premature decision to expand the range sequences is fraught with problems. In the chapter we show that a single location update in the **GWE** can isolate the spoofer, and with great confidence, neutralize of a hostile target.

This chapter introduces some of the issues of spoofing within a delimited context: range sensors, single spoofer, no spoofer cancellation, simple spoofer cessation, etc. Any of these restrictions on the engagement can be weakened, and indeed would necessarily be weakened in an application. But the engagement presented illustrates the utility of the hybrid formulation of the problem and the advantage of the beacon update in separating modal hypotheses.

Chapter 10
The Gaussian Wavelet Estimator

Abstract This appendix gives the details of the development of the GWE. These details are rather tedious and only the *very* interested reader will wish to explore this appendix. After some basic analytical development, we illustrate the GWE by isolating single time increments in the algorithm and providing a rationale for the behavior of the algorithm. Some of this work expands on issues discussed in some of the earlier chapters. It may indeed repeat the same comments used in earlier sections. One thing we emphasize here is the need for engineering judgement in modifying the underlying algorithm to suit the constraints of an actual engagement. We propose specific adjustments that we find plausible without any claim that they are optimal in a global sense. Again we note that the notation in the Appendix is very demanding; please refer to the Notation Conventions and Details section in Chap. 1.

10.1 GWE Details

10.1.1 Engagement Structure: Kinematics

Subject to the various approximations described in the earlier chapters, there is a probability space $(\Omega, \mathscr{F}, \mathbb{P}; \{\mathscr{F}[k]\})$ under which the comprehensive time-discrete dynamic model for the target motion is characterized by:

1. **Regime dynamics**: The regime proxy (or regime state) is a finite state Markov chain satisfying

$$\phi[k+1] = \Pi\phi[k] + m[k+1]; \qquad (10.1)$$

2. **Target dynamics**: The family of kinematic models satisfy

$$x[k] = x[k]^- + \Delta_t x[k]$$
$$x[k+1]^- = A_t x[k] + \sqrt{P_t^w}\, w[k+1]. \qquad (10.2)$$

The adjustment at time $t = kT$, $\Delta_t x[k]$, is called the handoff transformation, $\mathbf{H}^+$.

© Springer International Publishing Switzerland 2016
D.D. Sworder, J.E. Boyd, *Locating, Classifying and Countering Agile Land Vehicles*, DOI 10.1007/978-3-319-19431-8_10

The regime state is number selected from the alphabet $\mathbf{S} = \{1, \ldots S\}$. The Kleene closure of the alphabet, $\mathbf{S}^*$, is the set of all finite strings from this alphabet; i.e., the set of all finite $\mathbf{S}$-radix numbers. Denote by $\iota^{\oplus}[k]$ a string of length $k + 1$ in $\mathbf{S}^*$. At time $t = kT$, $\iota^{\oplus}$ represents a retrograde regime sequence beginning at time $t = 0$.

As long as $k \geq L - 1$, $\iota^{\oplus}$ has a prefix of length L; $\iota^{\oplus} = (\iota, \iota^{\oplus}/\iota)$. The set of all L-digit numbers is the modal state space, κ, and we write $\iota \in \kappa$. The probability structure on κ is inherited from (10.1).

An alternative notation for the regime state is created if we replace the number $i \in S$ with a canonical unit vector in S-space: $i \in \mathbf{S} \mapsto \mathbf{e}_i$. It is this convention that underlies (10.1): if $\phi[k] = \mathbf{e}_i$, the regime proxy for the $k$thinterval is the ith. With this notation, $\iota = (\mathbf{e}_i, \mathbf{e}_j, \ldots, \mathbf{e}_k, \mathbf{e}_l)$ or $\iota = ij \ldots kl$: the modal state is an L-string of unit vectors in S-space or an L-digit number. If we wish to append the forward regime to ι, we write $\iota^+ = pij \ldots kl$. Then $\iota^+ \in \kappa^+$, the set of all S-strings of length $L + 1$.

We will use ϕ as a pointer toward both unitary and composite events. For example, if $\phi[k] = \mathbf{e}_i$, $\phi[k]$ is an S-vector and points to the regime event i at time $t = kT$: $\phi_m[k] = 1$ if $m = i$ and 0 otherwise. In this case, $\phi[k]$ distinguishes a unitary event. A pointer to a composite event is illustrated by $\phi[k] = \mathbf{e}_\iota$: then $\phi[k] = 1$ if the event sequence is $(\mathbf{e}_i, \mathbf{e}_j, \ldots, \mathbf{e}_k, \mathbf{e}_l)$ and $\phi[k] = 0$ otherwise. In the latter case, $\phi[k]$ could be displayed as a vector, but we will view it as a pointer. The modal state, $\phi[k]$, is alternatively: a letter, a word, a number, a vector, an indicator function. These various usages will be employed without comment.

As indicated above, a subscript has many uses. It may identify time or the component of a vector (e.g., $\{v_1\}$ is a scalar process that is the first component of the vector process $\{v_t\}$); a matrix subscript can distinguish a matrix from others of the same class or can point to a matrix element. If a time-continuous process is sampled every T seconds, the discrete sequence so generated is written $\{y[k]\}$ where the index denotes sample number rather than time. A discontinuity in a time-continuous process $\{x_t\}$ is labeled $\Delta x_t = x_t - x_{t-}$. However, in the case of a time-discrete process, $\Delta x[k]$ is interpreted more broadly as the increment after some event at $t = kT$: $\Delta x[k] = x[k]^+ - x[k]^-$. The event could be exogenous; e.g., a regime change, or the event could be endogenous; e.g., a step of processing.

A Gaussian random variable with mean $\hat{x}$ and positive covariance P_{xx} is labeled $x \sim \mathbf{N}(\hat{x}, P_{xx})$. All of the covariance matrices encountered here are positive with a positive inverse and with a positive symmetric square root: $P_{xx}^{-1} = D_{xx} = F_{xx}^2$. The matrix D_{xx} is called the *information matrix*. Denote a unit Gaussian density with mean m by $\Phi_x(m)$. If A is a positive symmetric matrix and x a compatible vector, $x'Ax$ is denoted $\|x\|_A^2$.

The initial distribution of the kinematic state is $x[0] \sim \mathbf{N}(\hat{x}[0], P_{xx}[0])$. The initial regime proxy is independent of $x[0]$ and has distribution $\phi[0] \sim \hat{\phi}[0]$. We will define the initial modal distribution as follows: $\hat{\phi}_\iota = \hat{\phi}_i$ if ι is an i-string. Denote the filtration generated by $\{w[k]\}$ and $\{m[k]\}$ along with the initial conditions on the target by $\{\mathcal{O}[k]\}$: $\{\mathcal{O}[k]\}$ is the filtration underlying by the target motion. Then $w[k + 1]$ and $m[k + 1]$ are $\mathcal{O}[k]$-martingale increments, and the former is Gaussian.

10.1.2 Engagement Structure: Sensor Architecture

There are two primary data streams that give information about the target motion: a regime measurement and a kinematic measurement. On the probability space $(\Omega, \mathscr{F}, \mathbb{P}; \{\mathscr{F}[k]\})$, we have:

1. **Kinematic measurement**: The kinematic state is measured with gain and channel noise matrices (H_t, P_t^n). We will suppose that there is latency in the kinematic measurement; the measurement occurs before handoff:

$$y[k+1] = H_t x[k+1]^- + \sqrt{P_t^n} n[k+1] \qquad (10.3)$$

2. **Regime measurement**: The regime state is measured with quality matrix **D**. We will suppose that there is no latency; the regime measurement is a function of the forward regime:

$$z[k+1] = \mathbf{D}\phi[k+1] + \eta[k+1] \qquad (10.4)$$

The exogenous processes in (10.3)–(10.4) are centered. The processes $\{n[k]\}$ and $\{\eta[k]\}$ are $\mathscr{F}[k]$-martingale differences with the former a unit Gaussian white sequence. The collection of exogenous processes are mutually independent.

Denote the filtration generated by the kinematic measurements, $\{y[k]\}$, by $\{\mathscr{Y}[k]\}$ (respectively, that generated by the regime measurements and the para-measurements by $\{\mathscr{Z}[k]\}$). The observation filtration is $\mathscr{G}[k] = \mathscr{Y}[k] \vee \mathscr{Z}[k]$. In the absence of the regime measurements, $\mathscr{G}[k] = \mathscr{Y}[k]$. The exact nature of $\{\mathscr{G}[k]\}$ is clear within the context of the engagement. Conditional expectation is denoted with a circumflex with the relevant σ-field apparent from context.

Note that the construction of the observation filtration depends upon the mode of the system. Sometimes this is made explicit and sometimes not.

10.1.3 Hybrid Estimation

10.1.3.1 An Uninformative Sensor Suite

On the original sample space, let us consider an alternative sensor architecture, $(\Omega, \mathscr{F}, \mathbb{P}^\sim; \{\mathscr{F}[k]\})$, under which

1. **Regime measurement**: The regime measurement is an independent, identically distributed (iid) sequence that is uniformly distributed across $\{\mathbf{e}_1, \dots, \mathbf{e}_S\}$.

$$z[k+1] = \eta[k+1] \qquad (10.5)$$

2. **Kinematic measurement**: The kinematic measurement is a unit white Gaussian process:

$$y[k + 1] = n[k + 1] \tag{10.6}$$

The measure $\mathbb{P}^\sim$ is identical to $\mathbb{P}$ as regards the state dynamics. But the character of the observation is considerably simpler under $\mathbb{P}^\sim$. Under $\mathbb{P}^\sim$ the observation is uninformative; i.e., $E(x[k]\,|\,\mathscr{G}[k]) = E(x[k])$.

10.1.3.2   The Bayes Recurrence Formula

We wish to determine the target location under $\mathbb{P}$; $E(x[k+1]\,|\,\mathscr{G}[k+1])$. Define

$$\bar{\lambda}[l] = \frac{S}{\Phi(y[l])} \sum_l \phi_l (z[l]'\mathbf{D}\phi[l]\,|F_l^n|\Phi(F_l^n(y[l] - H_l x[l]^-))) \tag{10.7}$$

and let $\bar{\Lambda}[k]$ be the continuing product of the $\bar{\lambda}[l]$: $\bar{\Lambda}[k] = \prod_0^k \bar{\lambda}[l]$. It is pointed out in [19] that $\mathbb{P}$ and $\mathbb{P}^\sim$ are related by

$$\left.\frac{\partial \mathbb{P}}{\partial \mathbb{P}^\sim}\right|_{\mathscr{F}[k+1]} = \bar{\Lambda}[k+1](z[k+1], y[k+1]) \tag{10.8}$$

We seek a recursive approximation to the $\mathscr{G}[k+1]$-distribution of the target state of the form

$$p[k + 1] = \sum_l \hat{\phi}_l[k+1]p_l[k+1]. \tag{10.9}$$

where

$$\hat{\phi}_l = \mathbb{P}(\phi[k+1] = \mathbf{e}_l\,|\,\mathscr{G}[k+1])$$
$$p_l\,dz = \mathbb{P}(x[k+1] \in [z, z+dz]\,|\,\phi[k+1] = \mathbf{e}_l, \mathscr{G}[k+1])$$
$$= N(m_l[k+1], P_l[k+1])\,dz. \tag{10.10}$$

Equation (10.9) is written as a normalized Gaussian sum: $\sum_l \hat{\phi}_l[k] = 1$. Alternatively, we might describe the target state with an unnormalized distribution:

$$q[k + 1] = \sum_l q_l[k+1],$$

where

$$q_l[k + 1] = \alpha_l[k+1]p_l[k+1]; \alpha_l \geq 0.$$

The map

$$\alpha_\iota \mapsto \frac{\alpha_\iota}{\sum_\iota \alpha_\iota} = \hat{\phi}_\iota$$

converts $q[k+1]$ to $p[k+1]$, and we will use the distributions interchangeably.

We seek an implementable mapping from $(p[k], y[k+1], z[k+1])$ to $p[k+1]$. Denote expectation with respect to $\bar{\mathbb{P}}$ by $\bar{E}$. The conditional Bayes Theorem [6, Theorem 3.2, Chap. 2] relates an expectation with respect to $\mathbb{P}$ to an expectation with respect to $\mathbb{P}^\sim$. Let f be a scalar test function. The $\mathbb{P}$-expectation of $\phi_\iota + f(x[k+1])$ can be computed using $\mathbb{P}^\sim$ as follows.

$$E[\phi_\iota + f(x[k+1]) \,|\, \mathscr{G}[k+1]] = \frac{\bar{E}[\phi_\iota + \bar{\Lambda}[k+1] f(x[k+1]) \,|\, \mathscr{G}[k+1]]}{\bar{E}[\bar{\Lambda}[k+1] \,|\, \mathscr{G}[k+1]]} \qquad (10.11)$$

The denominator is independent of the specific f selected. The numerator is thought of as the expectation of $f(x[k+1])$ with respect to the *unnormalized* density $q_\iota + [k+1]$.

$$\bar{E}[\phi_\iota + \bar{\Lambda}[k+1] f(x[k+1]) \,|\, \mathscr{G}[k+1]] = \int_\Omega f(z) q_\iota + [k+1](z) dz. \qquad (10.12)$$

Suppose $f(x[k+1]) \equiv 1$. Equation (10.12) becomes

$$E[\phi_\iota + \,|\, \mathscr{G}[k+1]] = \frac{\bar{E}[\phi_\iota + \bar{\Lambda}[k+1] \,|\, \mathscr{G}[k+1]]}{\bar{E}[\bar{\Lambda}[k+1] \,|\, \mathscr{G}[k+1]]}. \qquad (10.13)$$

Therefore

$$\bar{E}[\bar{\Lambda}[k+1] \,|\, \mathscr{G}[k+1]] = \sum_{\iota +} \bar{E}[\phi_\iota + \bar{\Lambda}[k+1] \,|\, \mathscr{G}[k+1]]. \qquad (10.14)$$

To find $\{q_\iota + [k+1]\}$, expand the numerator (10.13) by replacing $\phi[k+1]$, $\bar{\Lambda}[k+1]$ and $x[k+1]$ with their values:

$$\bar{E}[\phi_\iota + \bar{\Lambda}[k+1] f(x[k+1]) \,|\, \mathscr{G}[k+1]] = \frac{S}{\Phi(y[k+1])} \cdot$$
$$\bar{E}[e_p'(\Pi \phi[k] + m[k+1]) \bar{\Lambda}[k] z[k+1]' \mathbf{De}_p | F_\iota^n | \cdot$$
$$\Phi_y(F_\iota^n(y[k+1] - H_\iota x[k+1]^-)) \cdot \qquad (10.15)$$
$$f\left(\sum_\iota (A_\iota x[k] + \sqrt{P_\iota^n} w[k+1]) \phi_\iota[k]\right) \,|\, \mathscr{G}[k+1]].$$

It is true that $\sum_\iota \phi[k]' e_\iota \equiv 1$. Substituting this into (10.15),

$$\bar{E}[\phi_\iota + \bar{A}[k+1]f(x[k+1]) \,|\, \mathscr{G}[k+1]] = \frac{S}{\Phi(y[k+1])}\cdot$$

$$\bar{E}\Big[\sum_\iota (\phi[k]'\mathbf{e}_\iota)(\mathbf{e}_p'\varPi\mathbf{e}_\iota)\bar{A}[k]z[k+1]'\mathbf{De}_p|F_\iota^n|\cdot$$

$$\Phi_y(F_\iota^n(y[k+1] - H_\iota x[k+1]^-))\cdot$$

$$f(A_\iota x[k] + \sqrt{P_\iota^n}w[k+1])\phi_\iota[k]) \,|\, \mathscr{G}[k+1]\Big].$$

Under $\mathbb{P}^\sim$, $\{\mathscr{G}[k]\}$ is uninformative with respect to the hybrid-state. In the unnormalized distribution, factors common to all regimes can and will be ignored in what follows:

$$\bar{E}[\phi_\iota + \bar{A}[k+1]f(x[k+1]) \,|\, \mathscr{G}[k+1]]$$

$$= \int_\Omega \sum_\iota \varPi_{pi} z[k+1]'\mathbf{D}_{\cdot p}\Phi_y(F_\iota^n(y[k+1] - (H_\iota(A_\iota x + \sqrt{P_\iota^n}w))))\cdot$$

$$|F_\iota^n|f(A_\iota x + \sqrt{P_\iota^n}w)q_\iota[k]\Phi_w(w)\,dx\,dw.$$

To simplify this, make the change of variable $z = A_\iota x + \sqrt{P_\iota^n}w$. Then $dx\,dw = |F_\iota^w|dx\,dz$ and

$$\int_\Omega \sum_\iota \varPi_{pi} z[k+1]'\mathbf{D}_{\cdot p}|F_\iota^n||F_\iota^w|\Phi_y(F_\iota^n(y[k+1] - H_\iota z))\cdot$$

$$f(z)q_\iota[k](x)\Phi_w(F_\iota^w(z - A_\iota x))\,dx\,dz = \int_\Omega f(z)q_\iota + [k+1](z)\,dz.$$

So we have the following set of updates for the progeny of the ιth filter

---

Observation update

$$q_\iota + [k+1](x) = \varPi_{pi}|F_\iota^n||F_\iota^w|\Phi_y(F_\iota^n(y[k+1] - H_\iota x))\cdot$$

$$z[k+1]'\mathbf{D}_{\cdot p}\int_\Omega \Phi_w(F_\iota^w(z - A_\iota\zeta)q_\iota[k])\,d\zeta \qquad (10.16)$$

Equation (10.16) is the recurrence formula for the ι^+-term in the unnormalized density of the hybrid-state. It is based upon a sensor suite in which the modal sensor is instantaneous, and there is an incremental delay in the kinematic sensor. In applications, the latency conditions will differ, and (10.16) will be modified as appropriate.

10.1.3.3 The **GWE** Approximation

The **GWE** approximates the $\mathscr{G}[k]$-distribution of the target state with a Gaussian sum:

$$q[k] = \sum_{\iota \in \kappa} \alpha_\iota[k] N(m_\iota[k], P_\iota[k])$$

$$= \sum_{\iota \in \kappa} q_\iota[k]. \tag{10.17}$$

We will not usually make explicit the distinction between the normalized density and an unnormalized version. It will be assumed the densities are normalized when appropriate.

Let us focus on kth-time interval, $t \in [kT, (k+1)T)$ with regime proxy $\mathbf{e}_i$. The predecessor regime is $\mathbf{e}_j$, and the successor is $\mathbf{e}_p$. A specific regime sequence of length L moving back from $t = (k+1)T^-$ is written $\iota = [\mathbf{e}_i, \mathbf{e}_j, \ldots, \mathbf{e}_l]$. For example, if $L = 4$, then $\iota = [\mathbf{e}_1, \mathbf{e}_1, \mathbf{e}_1, \mathbf{e}_3]$ identifies the retrograde path $[\phi[k] = \mathbf{e}_1, \phi[k-1] = \mathbf{e}_1, \phi[k-2] = \mathbf{e}_1, \phi[k-3] = \mathbf{e}_3$: the system was in regime e_3 three samples ago but then changed to regime one and continues therein. Using our shorthand, $\iota = [\mathbf{e}_1, \mathbf{e}_1, \mathbf{e}_1, \mathbf{e}_3] \Leftrightarrow 1113$. If we refer to the (1113)-local filter, we point to the preceding regime sequence.

If we append a forward primitive to ι, $\phi[k+1] = \mathbf{e}_p$, we lengthen the modal index. For example, $\iota^+ = 31113$ evidences a return to mode three at $t = (k+1)T$. Alternatively, if we wanted to distinguish the modal past, the foreshortened index $\iota^- = [\mathbf{e}_j, \ldots, \mathbf{e}_l]$ will suffice; e.g., $\iota^- = 113 \in \kappa^-$. From ι^- we can construct elements in κ^+; e.g., $[\mathbf{e}_p, \mathbf{e}_i, \iota^-] \in \kappa^+$.

Sometimes it is important to isolate a subset of κ with the same digit (or digits). Denote the collection of elements in κ with prefix i by $\kappa(i)$. So $\kappa(i)$ is a partition of κ into S subsets with differing first digit. We can extend this to a common second digit: $\kappa(j)$ is interpreted as the subset of κ with common second digit.

This notation can be extended to sets of regime strings in which all but the suffix l is fixed. For example, $\kappa^+(\iota, l)$ is the set of all $L+1$ digit numbers of the form (ι, l) for fixed ι: there are exactly S such subsets of κ^+. Another partition of κ^+ is written $\kappa^+(p, \iota)$ which is the set of all $(L+1)$ digit numbers of the form (p, ι). A sum over the elements of such a partition is ambiguous in the abstract; i.e., $\sum_{\iota^+ \in (p,\iota)}$ could be a function of p or ι. Context will be determinative here.

The internal processing within the **GWE** balances the influence of sequences of the S regime primitives. Each term in (10.17) is identified with a specific string. There is a separate update for each of the filters at time $(k+1)T$. We will maintain local trackers for all strings of length L.

Consider the ιth local filter. This filter carries the kinematic state forward from $x[k]^+$ to $x[k+1]^-$. At $t = [k+1]T$, the tracker measures $z[k+1]$—a function of ϕ_p—and $y[k+1]$—a function of $x[k+1]^-$. With the value of q_ι from (10.10):

$$q_{\iota^+}[k+1](z) = \alpha_\iota[k]\Pi_{pi}z[k+1]'\boldsymbol{D}_{\cdot p}|F_\iota^n||F_\iota^w||F_\iota|\Phi_y(F_\iota^n(y[k+1]-H_\iota z)\cdot$$
$$\int_\Omega \Phi_z(F_\iota^w(z-A_\iota+\zeta))\Phi_\zeta(F_\iota m_\iota)d\zeta,$$

where we ignore factors common to all $\iota^+ \in \kappa^+$. Look first at the product of exponential pattern functions:

$$\Phi_y(F_\iota^n(y[k+1]-H_\iota z))\Phi_z(F_\iota^w(z-A_\iota\zeta))\Phi_\zeta(F_\iota m_\iota) = \exp-\tfrac{1}{2}J_1$$

where

$$J_1 = \|\zeta - m_\iota\|_{D_\iota}^2 + \|z - A_\iota\zeta\|_{D_\iota^w}^2 + \|y[k+1] - H_\iota z\|_{D_\iota^n}^2.$$

It is clear that J_1 is a quadratic form in ζ and z. It is an exercise to complete the squares in both variables. Some useful notations are:

$$\boldsymbol{D}_\iota^- = D_\iota + A_\iota'D_\iota^w A_\iota$$
$$\boldsymbol{P}_\iota^- = P_\iota - P_\iota A_\iota'(P_\iota^w + A_\iota P_\iota A_\iota')^{-1}A_\iota P_\iota$$
$$\boldsymbol{D}_\iota^+ = D_\iota^w - D_\iota^w A_\iota \boldsymbol{P}_\iota^- A_\iota'D_\iota^w + H_\iota'D_\iota^n H_\iota$$
$$m_\iota^-[k+1] = A_\iota m_\iota$$
$$P_\iota^-[k+1] = A_\iota P_\iota A_\iota' + P_\iota^w$$
$$\Delta d_\iota[k+1] = H_\iota'D_\iota^n y[k+1]$$
$$\Delta D_\iota = H_\iota'D_\iota^n H_\iota$$
$$P_\iota^y[k+1] = P_\iota^n + H_\iota[k+1]P_\iota^- H_\iota[k+1]'.$$

Consider the first two terms in J_1:

$$\|\zeta - m_\iota\|_{D_\iota}^2 + \|z - A_\iota\zeta\|_{D_\iota^w}^2 = \zeta'(D_\iota + A_\iota'D_\iota^w A_\iota)\zeta$$
$$-2\zeta'(D_\iota m_\iota + A_\iota'D_\iota^w z) + \|m_\iota\|_{D_\iota}^2 + \|z\|_{D_\iota^w}^2,$$

or

$$\|\zeta - m_\iota\|_{D_\iota}^2 + \|z - A_\iota\zeta\|_{D_\iota^w}^2 = \|\zeta - \boldsymbol{P}_\iota^-(D_\iota m_\iota + A_\iota'D_\iota^w z)\|_{D_\iota^-}^2$$
$$-\|D_\iota m_\iota + A_\iota'D_\iota^w z\|_{\boldsymbol{P}_\iota^-}^2 + \|m_\iota\|_{D_\iota}^2 + \|z\|_{D_\iota^w}^2.$$

Next collect the terms involving z:

$$\|z\|_{D_\iota^w}^2 - \|d_\iota + A_\iota'D_\iota^w z\|_{P_\iota^-}^2 + \|y[k+1] - H_\iota z\|_{D_\iota^n}^2 = z'D_\iota^+ z$$
$$- 2z'(H_\iota'D_\iota^n y + D_\iota^w A_\iota P_\iota^- d_\iota) - \|d_\iota\|_{P_\iota^-}^2 + \|y\|_{D_\iota^n}^2,$$

or

$$\|z\|_{D_\iota}^2 - \|d_\iota + A_\iota'D_\iota^w z\|_{P_\iota^-}^2 + \|y[k+1] - H_\iota z\|_{D_\iota^n}^2$$
$$= \|z - P_\iota^+ (H_\iota'D_\iota^n y + D_\iota^w A_\iota P_\iota^- d_\iota)\|_{D_\iota^+}^2$$
$$- \|H_\iota'D_\iota^n y + D_\iota^w A_\iota P_\iota^- d_\iota\|_{P_\iota^+}^2 - \|d_\iota\|_{P_\iota^-}^2 + \|y\|_{D_\iota^n}^2.$$

Then note that

$$D_\iota^+ = D_\iota^w - D_\iota^w A_\iota P_\iota^- A_\iota' D_\iota^w + H_\iota'D_\iota^n H_\iota$$
$$= (P_\iota^w + A_\iota P_\iota A_\iota')^{-1} + H_\iota'D_\iota^n H_\iota$$
$$= D_\iota^- + \Delta D_\iota$$
$$= D_\iota^+.$$

Further

$$P_\iota^+ (H_\iota'D_\iota^n y + D_\iota^w A_\iota P_\iota^- d_\iota) = P_\iota^+ (\Delta D_\iota + D_\iota^w A_\iota (D_\iota + A_\iota'D_\iota^w A_\iota)^{-1} d_\iota)$$
$$= P_\iota^+ (\Delta D_\iota + (P_\iota^w + A_\iota P_\iota A_\iota')^{-1} A_\iota' P_\iota d_\iota)$$
$$= P_\iota^+ (\Delta D_\iota + D_\iota^- m_\iota^-)$$
$$= m_\iota^+.$$

Therefore

$$\|z\|_{D_\iota}^2 - \|d_\iota + A_\iota'D_\iota^w z\|_{P_\iota^-}^2 + \|y[k+1] - H_\iota z\|_{D_\iota^n}^2$$
$$= \|z - m_\iota^+\|_{D_\iota^+}^2 - \|m_\iota^+\|_{D_\iota^+}^2 - \|d_\iota\|_{P_\iota^-}^2 + \|y\|_{D_\iota^n}^2.$$

Hence

$$J_1 = \|\zeta - P_\iota^- (D_\iota m_\iota + A_\iota'D_\iota^w z)\|_{D_\iota^-}^2 + \|m_\iota\|_{D_\iota}^2$$
$$\|z - m_\iota^+\|_{D_\iota^+}^2 - \|m_\iota^+\|_{D_\iota^+}^2 - \|d_\iota\|_{P_\iota^-}^2 + \|y\|_{D_\iota^n}^2.$$

But

$$\|m_\iota\|_{D_\iota}^2 - \|d_\iota\|_{P_\iota^-}^2 = d_\iota'(P_\iota - P_\iota^-)d_\iota$$
$$= d_\iota' P_\iota A_\iota'(P_\iota^w + A_\iota P_\iota A_\iota')^{-1} A_\iota P_\iota d_\iota$$
$$= m_\iota' A_\iota'(P_\iota^w + A_\iota P_\iota A_\iota')^{-1} A_\iota m_\iota$$
$$= m_\iota^{-'} D_\iota^- P_\iota^- D_\iota^- m_\iota$$
$$= \|D_\iota^-\|_{P_\iota^-}^2,$$

so

$$\|m_\iota\|_{D_\iota}^2 - \|d_\iota\|_{P_\iota^-}^2 - \|m_\iota^+\|_{D_\iota}^2 = -\Delta\|D_\iota^+\|_{P_\iota^+}^2$$

In these terms, J_1 can be written:

$$J_1 = \|\zeta - P_\iota^-(d_\iota + A_\iota' D_\iota^w z)\|_{D_\iota^-}^2 + \|z - m_\iota^+\|_{D_\iota^-}^2 + \|y[k+1]\|_{D_\iota^n}^2 - \Delta\|D_\iota^+\|_{P_\iota^+}^2.$$

Substituting,

$$q_{\iota^+}[k+1](z) = \alpha_\iota[k]\Pi_{pi} z[k+1]' D_{.p} |F_\iota^n| |F_\iota^w| |F_\iota| (|F_\iota^-| |F_\iota^+|)^{-1} \cdot$$
$$\exp -\tfrac{1}{2}(\|y[k+1]\|_{D_\iota^n}^2 - \Delta\|D_\iota^+\|_{P_\iota^+}^2) N_z(m_\iota^+, P_\iota^+) \cdot$$
$$\int_\Omega N_\zeta(P_\iota^-(d_\iota + A_\iota' D_\iota^w z), P_\iota^-) d\zeta.$$

The density associated with this local filter is thus

$$q_{\iota^+}[k+1](z) = \alpha_{\iota^+}[k+1] N_z(m_\iota^+, P_\iota^+) \qquad (10.18)$$

where

$$\alpha_{\iota^+}[k+1] = \alpha_\iota[k]\Pi_{pi} z[k+1]' D_{.p} |F_\iota^n| |F_\iota^w| |F_\iota| (|F_\iota^-| |F_\iota^+|)^{-1} \cdot$$
$$\exp -\tfrac{1}{2}(\|y[k+1]\|_{D_\iota^n}^2 - \Delta\|D_\iota^+\|_{P_\iota^+}^2).$$

It is interesting to note that the factor $\|y[k+1]\|_{D_\iota^n}^2 - \Delta\|D_\iota^+\|_{P_\iota^+}^2$ can be written in another way:

$$\|y[k+1]\|_{D_\iota^n}^2 - \Delta\|D_\iota^+\|_{P_\iota^+}^2 = -(D_\iota^- + \Delta D_\iota)' P_\iota^+ (D_\iota^- + \Delta D_\iota)$$
$$+ D_\iota^{-'} P_\iota^- D_\iota^- + y[k+1]' D_\iota^n y[k+1]$$
$$= -D_\iota^{-'}(P_\iota^+ - P_\iota^-)D_\iota^- - 2D_\iota^- P_\iota^+ H_\iota D_\iota^n y[k+1]$$
$$- y[k+1]'(D_\iota^n H_\iota P_\iota^+ H_\iota' D_\iota^n - D_\iota^n)y[k+1].$$

But

$$
\begin{aligned}
P_t^+ H_t' D_t^n &= (D_t^- + H_t' D_t^n H_t)^{-1} H_t' D_t^n \\
&= P_t^- H_t' (H_t P_t^- H_t' + P_t^n)^{-1} \\
&= P_t^- H_t' D_t^y,
\end{aligned}
$$

and

$$
\begin{aligned}
D_t^n H_t P_t^+ H_t' D_t^n - D_t^n &= D_t^n H_t (P_t^- + H_t' D_t H_t)^{-1} H_t' D_t^n - D_t^n \\
&= -(P_t^n + H_t P_t^- H_t')^{-1} \\
&= -D_t^y,
\end{aligned}
$$

so

$$
\|y[k+1]\|_{D_t^n}^2 - \Delta \|D_t^+\|_{P_t^+}^2 = \|y[k+1] - H_t m_t^-\|_{D_t^y}^2. \tag{10.19}
$$

The product of determinants in (10.18) can be simplified:

$$
\begin{aligned}
|D_t||\boldsymbol{P}_t^-| &= |I - A_t'(P_t^w + A_t P_t A_t')^{-1} A_t P_t| \\
&= |I - A_t' D_t^- A_t P_t| \\
&= |I - A_t P_t A_t' D_t^-| \\
&= |I - (P_t^- - P_t^w) D_t^-| \\
&= |P_t^w||D_t^-|.
\end{aligned}
$$

Thus

$$
|D_t^n||D_t^w||D_t||\boldsymbol{P}_t^-||P_t^+| = |D_t^n||D_t^-||P_t^+|,
$$

but

$$
\begin{aligned}
|D_t^+||P_t^-| &= |(D_t^- + H_t' D_t^n H_t) P_t^-| \\
&= |I + H_t' D_t^n H_t P_t^-| \\
&= |I + H_t P_t^- H_t' D_t^n|.
\end{aligned}
$$

Therefore

$$
\begin{aligned}
|P_t^n||D_t^+||P_t^-| &= |P_t^w + H_t P_t^- H_t'| \\
&= |P_t^y|.
\end{aligned}
$$

Let

$$L_{\iota+}[k+1] = \Pi_{pi} z[k+1]' \boldsymbol{D}_{,p} |F_\iota^y| \exp -\tfrac{1}{2}(\|y[k+1]\|_{D_\iota^n}^2 - \Delta \|D_\iota^+\|_{P_\iota^+}^2). \qquad (10.20)$$

Then

$$\alpha_{\iota+}[k+1] = \alpha_\iota[k] L_{\iota+}[k+1]. \qquad (10.21)$$

If there are conformity conditions on the estimate, they can be introduced at this point:

$$(m_\iota^+[k+1], P_\iota^+[k+1]) \xrightarrow{\mathcal{CM}} (m_\iota[k+1], P_\iota[k+1]).$$

Then (10.16) can be rewritten.

> **GWE estimate**
>
> $$q_{\iota+}[k+1](z) = \alpha_{\iota+}[k] L_{\iota+}[k+1] N_z(m_\iota, P_\iota) \qquad (10.22)$$

10.1.3.4   Broad moments

From (10.22), we can find the broad moments of the state distribution. First normalize $\{\alpha_{\iota+}[k+1]\}$. Then

$$\hat{\phi}_p[k+1] = \sum_{\kappa^+(p\iota)} \hat{\phi}_{\iota+}[k+1] \quad \text{forward mode}$$

$$\hat{\phi}_i[k+1] = \sum_{\kappa^+(pi\iota^-)} \hat{\phi}_{\iota+}[k+1] \quad \text{smoothed regime proxy}$$

$$\hat{\phi}_\iota[k+1]^+ = \sum_{\kappa^+(p,\iota)} \hat{\phi}_{\iota+}[k+1] \quad \text{smoothed mode}$$

$$\hat{x}[k+1] = \sum_\iota \hat{\phi}_\iota[k+1] m_\iota[k+1]$$

$$P[k+1] = \sum_\iota \hat{\phi}_\iota[k+1](P_\iota[k+1]$$

$$+ (m_\iota[k+1] - \hat{x}[k+1])(m_\iota[k+1] - \hat{x}[k+1])').$$

Unfortunately, the number of filters in (10.22) extends the size of the index variable by a factor S at every update. To maintain a bound on complexity, pruning and/or merging must be used to reduce the number of terms. For the purposes of this investigation, we will use the conventional Gaussian sum merging formula given in [1]. Let

$$m_\iota\text{-}[k+1] = \sum_{\kappa(\iota^-,l)} \hat{\phi}_\iota[k+1]m_\iota[k+1] \tag{10.23}$$

$$P_\iota\text{-}[k+1] = \sum_{\kappa(\iota^-,l)} \hat{\phi}_\iota[k+1](P_\iota[k+1]$$

$$+(m_\iota[k+1]-m_\iota\text{-}[k+1])(\dots)'). \tag{10.24}$$

The next iteration of the **GWE**-estimator begins with a set of κ terms:

---

updated distribution

$$q[k+1]^- = \sum_\iota \hat{\phi}_\iota[k+1]N(m_\iota\text{-}[k+1],P_\iota\text{-}[k+1]) \tag{10.25}$$

10.1.3.5 Variants on the GWE

Variants on the **GWE** as presented here have been derived for kindred plant representations. They differ in the timing of the handoff and the base state measurement: see, for example, [7, Eq. (8)], [5, Eq. (10)], [8, Eq. (4)]. The simplest situation is that in which the channel noise matrix is independent of the mode; there are neither regime measurements nor handoff increments. If $\{\mathscr{G}[k]\} = \{\mathscr{Y}[k]\}$,

$$L_\iota[k+1] = |F_\iota^y| \exp \tfrac{1}{2}\Delta \|d_\iota^+\|_{P_\iota^+}^2 \tag{10.26}$$

and the processing is

- Initialize filter at $t = kT$: $(\alpha_\iota[k], m_\iota\text{-}[k], P_\iota\text{-}[k]) \Rightarrow (\alpha_\iota[k], m_\iota[k], P_\iota[k])$
- Extrapolate : $(\alpha_\iota[k], A_i m_\iota, A_i' P_\iota A_i)) \xrightarrow{\text{EKF}} (\alpha_\iota[k], m_\iota[k+1]^-, P_\iota[k+1]^-)$
- Update $y[k+1]$: $(\alpha_\iota[k]L_\iota, d_\iota[k+1]^- + H'D^n y[k+1], D_\iota[k+1]^- + H'D^n H) \xrightarrow{\text{EKF}}$ $(\alpha_\iota[k+1]^+, d_\iota[k+1]^+, D_\iota[k+1]^+)$

Of course, the simplest **GWE** is associated with a unitary system. The model index is redundant, and we have the conventional **EKF** algorithm:

$$m[k] \Rightarrow Am[k]$$

$$P[k] \Rightarrow AP[k]A' + P^w$$

$$\Delta d[k+1] \Rightarrow H'D^n y[k+1]$$

$$\Delta D[k+1] \Rightarrow H'D^n H.$$

The updated distribution for the **EKF** is simply

$$q[k+1] = N(m[k+1],P[k+1]). \tag{10.27}$$

10.2 A Single-Step Example of the GWE

10.2.1 Introduction

To illustrate the GWE-tracker, let us use an adjusted version of the uncooperative engagement as described in Chap. 2. The engagement has three regime states given earlier: $S = 3$. The time-continuous kinematic model is presented in (2.1), and the path of the vehicle is shown in Fig. 2.1—see Fig. 10.1.

In contrast with the centered EKF, the GWE has the flexibility to use a mode-dependent acceleration. However to make this example as simple as possible, we will use a constant action matrix. In this engagement, there are neither handoff discontinuities nor para-measurements. Hence, the kinematic state is continuous at sample times: $x[k + 1] = x[k + 1]^- = x[k + 1]^+$.

Let us look at local estimator, $\iota^- = j \ldots l$ at time kT. The GWE begins with local state: $(m_\iota{-}[k], P_\iota{-}[k])$. More specifically, let us pick a time before the first turn: $t = 20\,\mathrm{s}$ ($k = 10$). The target has been moving in the CV regime since the beginning of the engagement. The $\iota^- = 22$ GWE-estimator should be fairly close to truth. The centered EKF-tracker actually computed the $22 \ldots$ estimate of the kinematic state at $t = 20$: $(\hat{x}[10], P_{xx}[10])$. Despite the fact that the regime string is correct, $\hat{x}[10]$ does not match $x[10]$; the initial error and the kinematic measurements draw the estimate from truth.

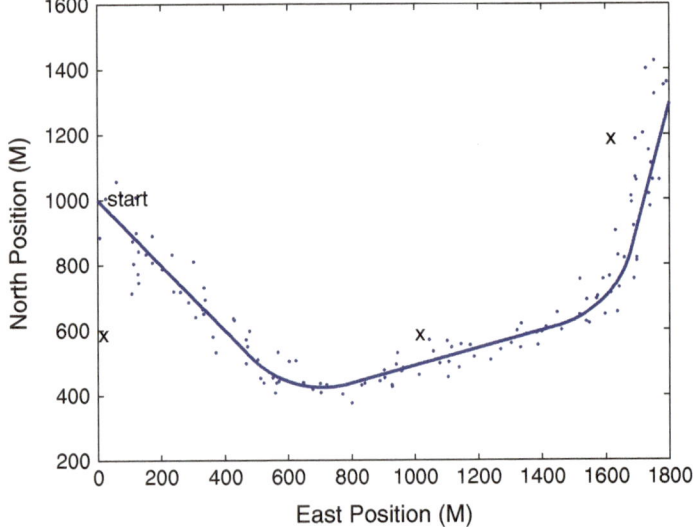

Fig. 10.1 A ground target is moving at 28 m/s. Three range-bearing sensors are shown with an *times* along with the cloud of measurements generated by the sensors

10.2.2   *Kinematic Extrapolation*

To make the GWE algorithm easier to visualize, let us initialize the $\iota^- = 22$-local filter at the true kinematic state of the target: $(m_{\iota^-} = x[10], P_{\iota^-} = P_{xx}[10])$. The covariance is that computed by the cooperative EKF. Even the cooperative EKF-tracker is not capable of such good target placement. This perfect initialization permits us to see the sources that draw the GWE-tracker away from the path.

The mode ι^- has three progeny at $t = 20$: $\iota = 122$ or turn-up; $\iota = 222$ or CV; $\iota = 322$ or turn-down. The local extrapolation models for the GWE flow from (2.1):

$$m_\iota[11] = A_i m_{\iota^-}[10]$$

$$P_\iota[11] = A_i' P_{\iota^-}[10]A_i + P^w. \tag{10.28}$$

Figure 10.2 shows the kinematic extrapolation of the ι-local filter. The initial placement and velocity of the target is exact. This is shown with the icon $\times$ in the figure. From the initial placement, three extrapolations are possible as we move to $t = 22$. The target is on a CV phase of its trajectory, and $\iota = 222$ gives the true extrapolation: since the velocity at $t = 20$ is correct, $\hat{\chi}_{222}[11]$ lies on the true path. This is shown with the icon $\diamond$ in the figure.

But there are two alternative extrapolations in the 22-filter. The target could turn-up: $\iota = 122$. The icon for this is $\triangle$. Or the target could turn-down, $\iota = 322$; the icon for this is $\triangledown$. The 122-filter extrapolates to a position above the path, and the 322-filter extrapolates to a position below the path.

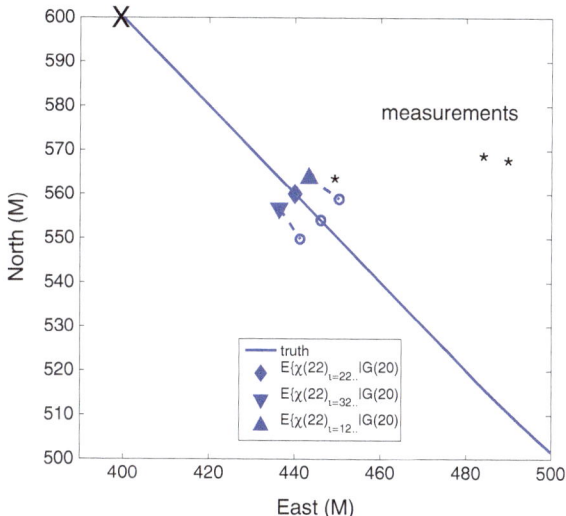

Fig. 10.2 The target has been moving at constant velocity for 20 s: $\iota^- = 22$. The ι^--local filter has been initialized at the true position and velocity labeled *times symbol*. The ι^--filter spawns three extrapolations toward $t = 22$: truth *diamond*; turn-up *triangle*, turn-down *inverted triangle*. Each extrapolation is shown along with an affixed velocity vector. The three kinematic measurements at $t = 22$ s are also shown

The velocity estimate for each of the progeny is affixed to the estimate in Fig. 10.2. The initial velocity is correct, and $\hat{v}_{222}[11]$ must lie along the path, However, the velocities of the other two filters move the estimates away from truth: e.g., with $\iota = 122$, $\hat{\chi}_{122}[11]$ moves the target in a counterclockwise direction.

10.2.3   The Measurements

The measurement processes conform to the state partition. In this engagement, there are three sensors and each provides range and bearing to the target. Specifically, each sensor measures the target location every 2 s with standard error: 56 m in range and 28 mr in bearing. The sensor noise is independent temporally and spatially.

The location of the sensors is given, but necessarily, the location of the target is not. We must first convert the polar coordinate system of the kinematic measurements into an east-north coordinate system using the geometry of the sensors and the extrapolated location of the target. Each extrapolation-sensor combination has a different geometry and hence a different coordinate transformation. In applications that follow, the coefficients of measurement model are indexed by ι. But in this simple illustration, there are only three extrapolations and we will just use the prefix i to label the sub-models.

The kinematic measurement is written

$$y_i[11] = H_i x[11] + \sqrt{P_i^n} n[11])F, \tag{10.29}$$

with additive white unit-Gaussian noise, $\{n[k]\}$. The channel noise matrix, P_i^n, is positive: $(P_i^n)^{-1} = D_i^n = (F_i^n)^2 > 0$. The coefficients in (10.29) depend on the regime index only because the $m_i[11]$ are different for different $i \in S$.

Figure 10.2 shows the three range-bearing measurements taken at $t = 22$ s with the icon $*$. The vector $y[11]$ is that in the $\{\mathscr{Y}[k]\}$ data set used to exercise the EKF-trackers in Chap. 2. Each measurement is associated with a specific sensor, but the sensor label is not displayed in the figure. As it happens, all of the measurements are northeast of the extrapolated positions. This reinforces the *false* conclusion that the target is turning up—$\iota = 122$.

We did not avail ourselves of a regime measurement in the uncooperative EKF scenario. With the GWE-tracker, we will suppose that there is also a direct measurement of the regime, ϕ_p. At $t = 22$ (or $k = 11$) let us assume that there is differential doppler sensor that gives the rotation rate of the spatially extended target. If the ostensible rotation is below a threshold, the motion is declared to be CV. Else, the target is declared to be turning in the compatible direction. The threshold is high and a turn regime is classified as CV 40 % of the time even when it turns. If the motion is CV, the $z[11]$ classifies it correctly only 60 % of the time:

$$z[11] = D\phi_p[11] + \eta[11], \tag{10.30}$$

where

$$\mathbf{D} = \begin{bmatrix} 0.4\ 0.2\ 0.2 \\ 0.4\ 0.6\ 0.4 \\ 0.2\ 0.2\ 0.4 \end{bmatrix}. \tag{10.31}$$

The sequence $\{z[k]\}$ is not particularly informative, but it does give a hint of a change in regime.

10.2.4   The GWE Update

10.2.4.1   Kinematic

Let us update the ιth estimator on the time interval $t \in (22, 22)$. After extrapolation, we have the covariance statistic, $(m_i[11], P_i[11]$ for $i \in \mathbf{S}$, and i is the prefix of $\iota = i22$. The information statistic is $(d_i[11], D_i[11])$,

$$D_i[11] = P[11]^{-1} \tag{10.32}$$

$$d_i[11] = D_i[11]m_\iota[11]. \tag{10.33}$$

Upon receiving the measurement $y[11]$, we make the replacement

$$d_i[11] \mapsto d_i[11] + H_i'D_i^n y[11] \tag{10.34}$$

$$D_i[11] \mapsto D_i[11] + H_i'D_i^n H_i. \tag{10.35}$$

In (10.34) we have written the three range-bearing observations as a single vector array.

Figure 10.3 shows the result of the update. Because of the initialization, the forward extrapolation of the $\iota = 222$ tracker is exact. The actual increment in the information state is $H_i'D_i^n y[11]$ where $y[11]$ is a three-vector. The change is proportional to the channel gain H_i (related to the measurement geometry), inversely proportional to the channel noise P_i^n, and proportional to the observation $y[11]$. The gain and channel geometry are both evaluated using ι as the modal state and $\hat{\chi}_i$ as the imputed location. Hence, $\Delta d_i[11]$ will be tend larger when the geometry/noise covariance is advantageous; e.g., the location increment is large when the SNR is large.

It is evident from the figure that the measurements pull the state estimates away from the path. One of the measurements was actually quite close to the counterclockwise extrapolation. The updated location estimates congregate to the northeast of the path. The best of the three estimates is actually $\iota = 322$. A counterclockwise turn best fits $y[11]$ even though $\iota = 222$ is true. The updated velocities are also rotated counterclockwise to some degree—and actually are slightly bigger.

Fig. 10.3 After receiving the kinematic measurement at $t = 22$ s, the GWE updates the three local filters: truth *diamond*; up *triangle*, down *inverted triangle*. The measurements are shown. The pre- and post-update one sigma error ellipses of the $\iota = 222$-local estimator are shown centered on $\hat{\chi}_i$

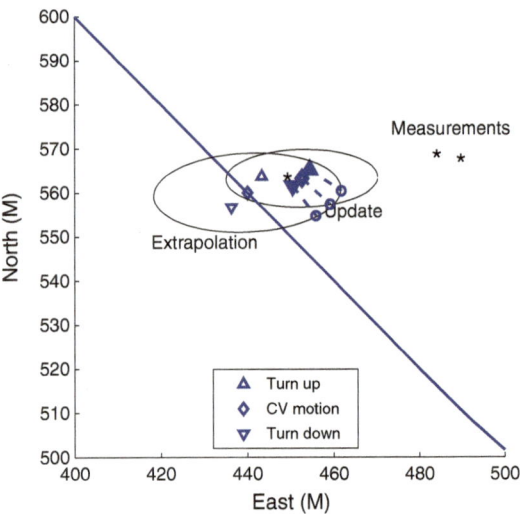

The increment in the information matrix is also proportional to a term involving the interaction of the target/sensor geometry with the primary noise directions: $\Delta D_i[11] = H_i' D_i^n H_i$. To illustrate the change in the uncertainty in the **GWE**-estimate, Fig. 10.3 shows the one sigma $P_{\chi\chi}$ error ellipse for the $i = 2$ local filter before and after the kinematic update. Before the update, the error ellipse is centered on the true location. The acceleration uncertainty is greater in the east direction. The error ellipse after extrapolation still favors east-west with about a 30 m span.

The observation is such to compress the error ellipse. The shape of the P_{222} matrix gives an indication of the direction of the uncertainty in the estimate. Since only the position uncertainty is shown, we should not over interpret the result. With that caveat, we would say that the CV-tracker is more confident of its north coordinate than it is of east coordinate after the kinematic measurement. The error ellipse is smaller after the measurement as the three range-bearing observations convey information on the location (directly) and the velocity indirectly.

The reference path is actually a simple one: the target moves with constant speed and bearing for $t \in [0, 24)$. If the path map, $\mathbf{C}^{\mathcal{M}}$, were known, we would see that none of the $E\{\chi_i | \mathcal{Y}[11]\}$ fall on the target track. With a map para-measurement, we could improve the estimate by translating all of the $\{m_i\}$ back to the path and adjusting the covariances:

$$(m_i[11], P_i[11]) \xrightarrow{\mathbf{C}^{\mathcal{M}}} (m_i[11], P_i[11])$$

10.2.4.2   Modal

The regime dynamics utilize a Markov model:

$$\phi[k + 1] = \Pi \phi[k] + m[k + 1], \qquad (10.36)$$

where $\{m[k]\}$ is a time-discrete, martingale difference sequence and Π is as presented in Chap. 3,

$$\Pi = \begin{bmatrix} 0.86 & 0.04 & 0.05 \\ 0.09 & 0.93 & 0.09 \\ 0.05 & 0.04 & 0.86 \end{bmatrix}. \qquad (10.37)$$

At the beginning of the sample interval, the **GWE** assigns a probability mass function to the local filters in κ. Call these $\mathscr{G}[k]$-probabilities $\{\alpha_\iota[k]; \iota \in \kappa\}$. The probabilities could be normalized or unnormalized. If we want to emphasize the former, we may write the mass function as $\{\hat{\phi}_\iota[k]; \iota \in \kappa\}$. The mass function is updated in two steps: first with the kinematic measurement $y[k + 1]$, and second with the regime measurement $z[k + 1]$.

Let us look at the $\iota^- = 22$-tracker with statistic $(\hat{\phi}_\iota[10], m_\iota-[10], P_\iota-[10])$. At time $t = 20$, the filter splits and generates under three distinct progeny labeled by $i \in \mathbf{S}$. At $t = 22^-$, these progeny filters have the statistic $(\hat{\phi}_\iota[10], m_\iota^-[11], P_\iota^-[11])$.

The tracker receives the kinematic measurement, $y[11] = H_\iota x[11] + \sqrt{P_\iota^n} n[11]$; the measurement geometry and the SNR depend upon the mode. Denote the covariance of the kinematic measurement by P_ι^y:

$$P_\iota^y[11] = H_\iota[11] P_\iota[11] H_\iota[11]' + P_\iota^n[11], \qquad (10.38)$$

where $P_\iota^y > 0; \iota \in \kappa$. Note that both $P_\iota^n[11]$ and $P_\iota[11]$, convey information on the modal state.

Define

$$E_\iota^y = \Delta \|m_\iota\|_{D_\iota}^2 - \|y[11]\|_{D_\iota^y}^2. \qquad (10.39)$$

E_ι^y is a unit-free quantity that is a function of the kinematic measurement and the mode. When $y[11]$ is observed, both m_ι and D_ι change. To the degree that $y[11]$ corresponds to the extrapolated state, the corresponding mode is favored. For a given mode, the bigger $\Delta \|m_\iota\|_{D_\iota}^2$ is, the more likely ι is the correct mode: a filter moves a larger multiple of the standard units of m_ι when the kinematic measurement is matched to the filter.

To adjust for the raw size of the measurement another unit-free term, $\|y[11]\|_{D_\iota^n}^2$, is included in E_ι^y. For each local filter, the kinematic update of the unnormalized modal mass function is

$$\alpha_\iota[11] = \alpha_\iota[10] |F_\iota^y| \exp(\tfrac{1}{2} E_\iota^y). \qquad (10.40)$$

Equation (10.40) updates the $\mathscr{G}[11]$-mass function with the $y[11]$ data. As with the kinematic state, the local filters are updated in parallel. At this step we see that

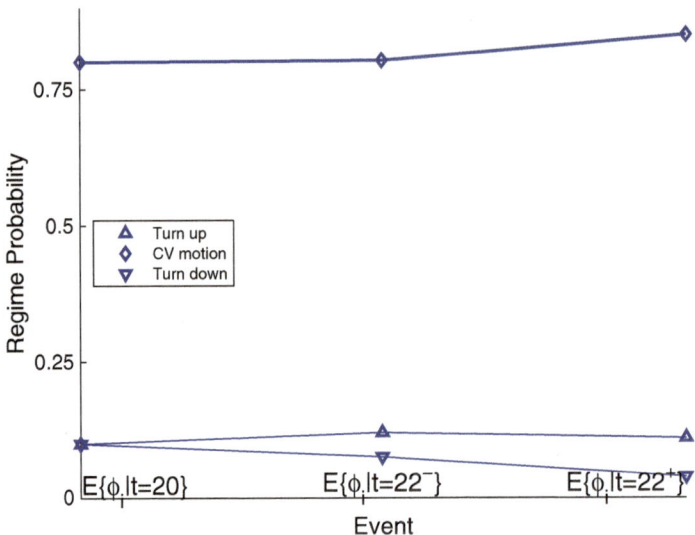

Fig. 10.4 From the initial distribution favoring $\iota = 222$, the GWE receives a kinematic measurement at $t = 22^-$: the event is labeled $E\{\phi_i|t = 22^-\}$. The measurement increases $\hat{\phi}_1$ (or turn-up) primarily at the expense of $\hat{\phi}_3$ (or turn-down). The regime measurement at $t = 22^+$, actually favors the CV-mode despite the fact that $z[11] = e_1$

any common factors in E_ι^y can be ignored in computing $\alpha_\iota[11]$. For example, if $D_\iota^n \equiv D^n$, the term $\|y[11]\|^2_{D_\iota^n}$ can be ignored.

To illustrate the influence of (10.40), let us return to the example displayed in Fig. 10.3. Let us use the kinematic measurements shown in the figure, and append a modal probability at $t = 20$ s. Since the target has been following a CV path, let us arbitrarily initialize the local filters: $\hat{\phi}_{222} = 0.8$; $\hat{\phi}_{122} = 0.1$; and $\hat{\phi}_{322} = 0.1$. At time $t = 20$, the GWE is confident at the 80 % level that CV will continue and a probability of 10 % that a turn in either direction will occur.

With the kinematic observations displayed in the figure, $|F_\iota^y|\exp(\frac{1}{2}E_\iota^y)$ is quite small. But by normalizing with $|F_{222}^y|\exp(\frac{1}{2}E_{222}^y)$, more manageable numbers are derived: $|F_\iota^y|\exp(\frac{1}{2}E_\iota^y) = [1.200; 1.0; 0.758]$. The range-bearing measurements shown in the figure clearly favor an *up* regime.

Figure 10.4 shows the evolution of $\hat{\phi}_\iota$. Initially, $\iota = 222$ is strongly favored. The kinematic measurements increase $\hat{\phi}_1$ mainly at the expense of $\hat{\phi}_3$. The confidence in the CV continuation is changed only slightly. The kinematic measurement actually increases $\hat{\phi}_2$ a bit. The *up* regime is favored primarily at the expense of the *down* regime.

The next step in the modal update is the integration of the regime measurement. To understand the structure of the regime update equation, observe that the element Π_{pi} is the probability that the mode makes the transition $(\iota = ijk) \mapsto (\iota^+ = pijk)$. This probability can be expressed simply as a function of the regimes; i.e., $\mathbb{P}(i \mapsto p)$.

As the GWE looks to the next sample interval, a cue to the forward regime proxy is supplied by the regime measurement $z[11]$. The discernibility matrix, $\mathbf{D}$, has columns that are probability vectors. The regime measurement picks one of them. The scalar $z[11]'\mathbf{D}_{.p}$ tends to be higher in the forward direction of travel.

In the engagement we studied using the centered EKF, there were no regime measurements. Let us arbitrarily add a regime measurement to the observation mix. Suppose $z[11] = \mathbf{e}_1$. This measurement is intentionally chosen to be a bad one. The target is in a CV regime. The kinematic measurements favor the *up* regime, and the modal measurement does as well.

The increment in the modal probability mass function is

$$\alpha_{\iota+}[11] = \alpha_{\iota}^{-}[11]\Pi_{pi}z[11]'\mathbf{D}_{.p}. \tag{10.41}$$

Equation (10.41) generates a mass function on κ^{+}—a set S-times the size of κ. The mass function $\hat{\phi}_{\iota+}$ generated by (10.41) can be used to find $\hat{\phi}_{\iota}$:

$$\hat{\phi}_{\iota}[11] = \sum_{p} \hat{\phi}_{\iota+}[11]. \tag{10.42}$$

Because it depends on the forward regime, $\hat{\phi}_{\iota}[11]$ can be thought of as a smoothed estimate of the mode. Figure 10.4 shows $\hat{\phi}_i$ after the regime update. The probability of a counterclockwise turn, $\hat{\phi}_1$, which increased by 20 % after the range bearing measurement, actually decreases to $\hat{\phi}_1 = 0.11$ after the confirming measurement. Over the interval, the likelihood of CV motion rises to 85 %. Despite the fact that the kinematic and the modal measurement favor the up turn, the CV hypothesis has strengthened by 6 %. Of course the *up* hypothesis has been strengthened even more, by 10 %. The clear loser in all of this is the *down* regime.

10.2.5 The Location Estimate

10.2.5.1 Gaussian Sum or the Projected Distribution

At $t = 22$, we finalize the tracker with the $\mathcal{G}[11]$-distribution for the kinematic state. The distribution is a Gaussian sum with three terms:

$$\mathbb{P} = \sum_{\iota \in \mathbf{S}} \hat{\phi}_{\iota}[11]\mathbf{N}(m_{\iota}[11], P_{\iota}[11]) \tag{10.43}$$

For the single interval under study, Fig. 10.5 shows the location density. The string $\iota = 222$ is true and it has the largest weight in $\mathbb{P}$. The density is unimodal and the mean differs from the true location primarily in the north coordinate: $\chi[11] = (440, 560)$ and $\hat{\chi}[11] = (453, 563)$. This is due to the fact that the

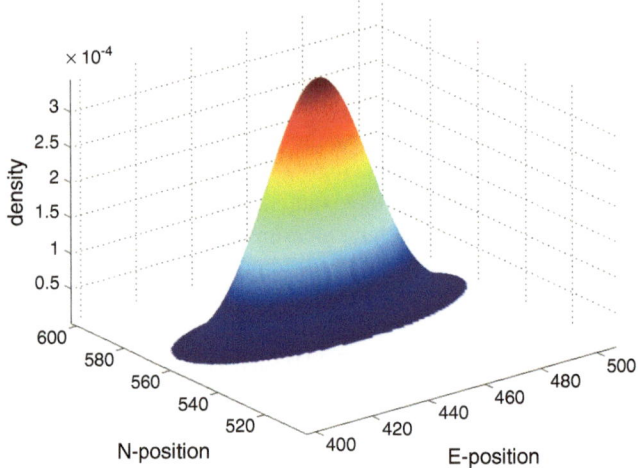

Fig. 10.5 The $\mathscr{G}[11]$-density of location is shown

kinetic measurements were north-east of the targets. The measurements pulled the velocity estimates toward the north-east as well: $\hat{v}[11] = (21.6, -19.3)$ and $v[11] = (20, -20)$.

The projected distribution, $\hat{\mathbb{P}}$, is derived from $\mathbb{P}$ by matching the first two moments. The $\mathscr{G}[11]$-position distribution is:

$$\hat{\mathbb{P}}_{\chi} = \mathbf{N}(\hat{\chi}[11], P_{\chi\chi}[11]) \tag{10.44}$$

The projected distribution differs little from the notional distribution shown in the figure; both $\mathbb{P}$ and $\hat{\mathbb{P}}$ are unimodal and closely centered on the conditional mean. Figure 10.6 shows the $\mathbb{P}$ and $\hat{\mathbb{P}}$ densities in the lateral direction and centered on $\hat{\chi}[11] = (453, 563)$. The actual 6.5 m mean lateral error appears on the flank of both densities. The two densities differ little in the scale of the figure. For purposes of delineating assurance regions, either $\mathbb{P}$ or $\hat{\mathbb{P}}$ would suffice with the latter far easier to work with.

The putative advantage of $\mathbb{P}$ is that it distributes probability over a larger region when the modal uncertainty is greater. This is sometimes referred to by saying the $\mathbb{P}$ has a *thicker tail* than does $\hat{\mathbb{P}}$. The tail comparison is not apparent in Fig. 10.5 but can be seen in Fig. 10.6. In the lateral direction, $\mathbb{P}_{\chi}$ is thicker than $\hat{\mathbb{P}}_{\chi}$ by a factor nearing 2 at a distance of 20 m (Fig. 10.7).

10.2.5.2 $\mathscr{G}[11]$ statistics

The $\mathscr{G}[k+1]$ distribution provides a number of important tracking statistics. They include

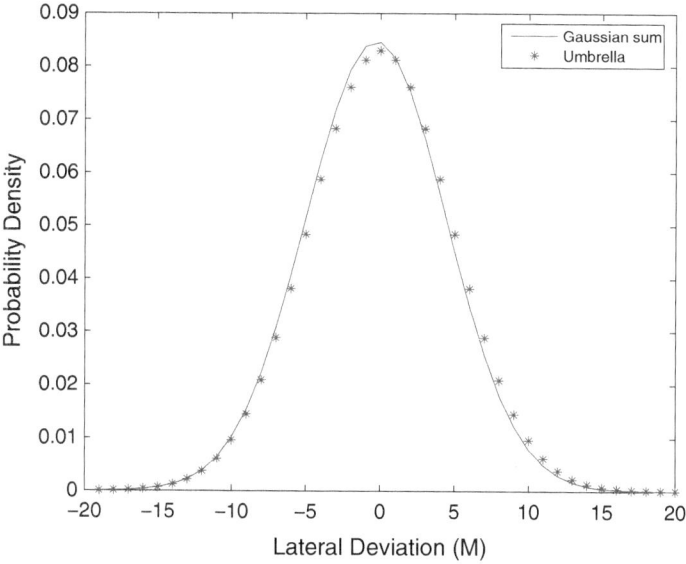

Fig. 10.6 The $\mathscr{G}[11]$-density of lateral position error. Both $\mathbb{P}$ and $\hat{\mathbb{P}}$ are shown centered on $\hat{\chi}$, and the lateral direction is perpendicular to $\hat{\nu}[11]$

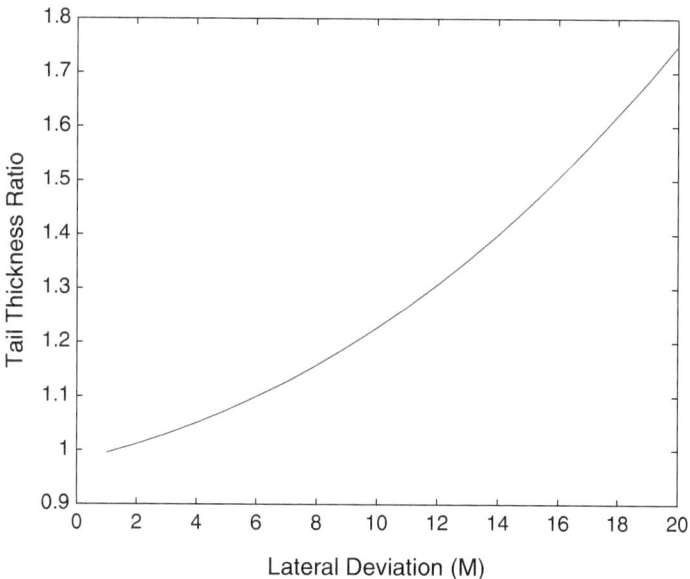

Fig. 10.7 Both $\mathbb{P}$ and $\hat{\mathbb{P}}$ have thin tails. But the tails of $\mathbb{P}$ are significantly thicker

$$\hat{\phi}_p[11] = \sum_\iota \hat{\phi}_{\iota^+}[11] \quad \text{forward regime} \tag{10.45}$$

$$\hat{\phi}_\iota[11] = \sum_p \hat{\phi}_{\iota^+}[11] \quad \text{modal state} \tag{10.46}$$

$$\hat{\phi}_i[11] = \sum_{jl} \hat{\phi}_\iota[11] \quad \text{primary regime} \tag{10.47}$$

$$\hat{x}[11] = \sum_i \hat{\phi}_\iota[10]m_\iota[11] \tag{10.48}$$

$$P_{xx}[11] = \sum_i \alpha_\iota[10](P_\iota[11]$$
$$+ (m_\iota[11] - \hat{x}[11])(m_\iota[11] - \hat{x}[11])'). \tag{10.49}$$

The global statistics given above are generated outside the local filters, and they have no effect on the accuracy of the tracker. Neither handoff transformations nor para-measurement compliance arise in this illustrative example. We will discuss such issues as they present themselves in later applications.

10.2.6   Reinitialization

To prepare the GWE for the interval $t \in [22, 24]$, we must initialize the state of the local trackers. First observe that $\iota[10] = ijl$ and $\iota^+[11] = pijl$. Let us shift the modal state up one step: $\iota[11] = pij$. The initial modal probabilities at $t = 22$ are

$$\hat{\phi}_\iota[11] = \sum_l \hat{\phi}_{pijl}[11]. \tag{10.50}$$

Now produce the $\mathscr{G}[11]$-kinematic distribution by averaging the statics of the local filter over the most remote event:

$$m_{ij}[11] = \sum_l m_{ijl}[11]\hat{\phi}_{ijl}[11] \tag{10.51}$$

$$P_{ij}[11] = \sum_l \hat{\phi}_{ijl}[11](P_{ijl}[11] + (m_{ij}[11] - m_{ijl}[11])(\ldots)'. \tag{10.52}$$

This is a *soft* reduction in the complexity of the distribution by the factor S; the number of (mean, covariance) pairs is reduced by a factor of S as we re-initialize. Alternatively, a *hard* reduction would be achieved by pruning: Map the S smallest $\hat{\phi}_{ijl}[11]$ to zero and ignore them in the update. In the latter case, our unique identification of the modal state with a regime sequence would be lost.

10.2.6.1 Handoff and Measurement at the Beginning of the Interval.

Suppose that handoff occurs at the beginning of the sample interval, and there is no latency in the state observation.

- Initialize filter at $t = kT$: $(\alpha_t[k], m_t[k], P_t[k]) \Rightarrow (\alpha_t[k], m_t[k], P_t[k])$
- Extrapolate : $(\alpha_t[k], A_i m_t^+, A_i P_t^+ A_i') \xrightarrow{\text{EKF}} (\alpha_t[k], m_t[k+1]^-, P_t[k+1]^-)$
- Handoff to $\mathbf{e}_p$: $(\alpha_t[k], A_{ip} m_t^- + B_{ip}, A_{ip} P_t^- A_{ip}') \xrightarrow{\mathbf{H}^+} (\alpha_t, m_t[k+1]^-, P_t[k]^-)$
- Update $y[k+1]$: $(\alpha_t[k], D_t^- + H_t' D_t^n(y[k+1] - H_t B_{ip} - \Upsilon_p^m), D_t[k+1]^- + H_t' D_t^n H_t) \xrightarrow{\text{EKF}} (\alpha_t[k], D_t[k+1]^+, D_t[k+1]^+)$
- Update $z[k+1]$:$(\alpha_t[k] L_t +, m_t[k+1]^+, P_t[k+1]^+) \xrightarrow{\text{GWE}} (\alpha_t + [k], m_t[k+1]^+, P_t[k+1]^+)$
- Conform to external conditions: $(\alpha_t + [k], m_t[k+1]^+, P_t[k+1]^+) \xrightarrow{\mathbf{C}^{\mathcal{M}}} (\alpha_t + [k+1], m_t[k+1], P_t[k+1])$

10.2.6.2 Handoff at the Beginning of the Interval but Latency in the Base-State Measurement.

Because of latency, the measurement of the base state occurs at $t = (k+1)T^-$. This means that the estimate so generated is (m_t, P_t). Hence, we begin the interval with a step delay. Suppose that handoff occurs at the beginning of the sample interval. The initialization map is then $(m_t, P_t) = (m_t-, P_t-)$

- Initialize filter at $t = kT$: $(\alpha_t[k], m_t-[k], P_t-[k]) \Rightarrow (\alpha_t[k], m_t[k], P_t[k])$
- Handoff: $(\alpha_t[k], A_{ji} m_t + B_{ji}, A_{ji}' P_t[k] A_{ji}) \xrightarrow{\mathbf{H}^+} (\alpha_t[k], m_t[k]^+, P_t[k]^+)$
- Extrapolate : $(\alpha_t[k], A_i m_t^+, A_i' P_t^+ A_i)) \xrightarrow{\text{EKF}} (\alpha_t[k], m_t[k+1]^-, P_t[k+1]^-)$
- Update $y[k+1]$: $(\alpha_t[k], d_t[k+1]^- + (H_i' D_i^n(y[k+1] - H_i B_{ji} - \Upsilon_i^m)), D_t[k+1]^- + H_i' D_i^n H_i) \xrightarrow{\text{EKF}} (\alpha_t[k], d_t[k+1]^+, D_t[k+1]^+)$
- Update $z[k+1]$:$(\alpha_t[k] L_t +, m_t[k+1]^+, P_t[k+1]^+) \xrightarrow{\text{GWE}} (\alpha_t + [k], m_t[k+1]^+, P_t[k+1]^+)$
- Conform to external conditions: $(\alpha_t + [k], m_t[k+1]^+, P_t[k+1]^+) \xrightarrow{\mathbf{C}^{\mathcal{M}}} (\alpha_t + [k+1], m_t[k+1], P_t[k+1])$

10.3 The Map-Enhanced GWE

The performance of a map-enhanced GWE tracker/illuminator is presented in Chap. 6. In this section, we will give the rationale for the choices we have made to accommodate the basic algorithm to the data set appropriate to that engagement. We will do this by focusing on a specific time and specific local estimator.

10.3.1   The (i44)-Local Estimator

10.3.1.1   Kinematic Extrapolation

The GWE is composed of $\mathbf{S}^L$ local filters operating in parallel on a common data sequence. In this example the regime set will be the cardinal directions, and we will choose $L = 3$: there are 64 local filters. With a sample interval of five seconds, the explicit memory in the GWE is 15 s.

To extrapolate the kinematic states of the target, the GWE must first identify where it is on the map and determine the closest junction accordant with estimated location and velocity. Suppose the target is located at $(-800, -800)$ and is moving west; $\phi[k] = \mathbf{e}_4$. The nearest junction accepting a westbound vehicle is $\chi_5 = (-950, -800)$. Access to the junctions at $\chi_4 = (-750, -800)$ and $\chi_3 = (-350, -800)$ is precluded since U-turns are not permitted. Since we are prescient, we know that the vehicle will turn north at χ_5. The GWE-tracker only knows that a turn will be made, but the direction is not known.

To illustrate the alternatives within the GWE in more detail, let us focus on the ιth local filter. Suppose it begins the kth interval with mean state $m_\iota- \approx (-800, -800, 0, -10)'$, covariance $P_\iota-$, and modal suffix, $\iota^- = (44)$: the tracker has been initialized near the true location and with the true direction. The following are possible on the interval $t \in (kT, (k + 1)T)$:

The target could continue west (i = 4) and not project beyond the χ_5 junction.
In this case, $\iota = (444)$.

$$(m_\iota[k], P_\iota[k]) \quad = \quad (m_\iota-[k], P_\iota-[k])$$

$$(m_\iota[k], P_\iota[k]) \xrightarrow{\mathrm{EKF}_\iota} (m_\iota^-[k + 1], P_\iota^-[k + 1]) \qquad (10.53)$$

This is the conventional extrapolation used in the EKF. But in the GWE, the action matrix P_i^w depends on the regime. Because the initial condition is diffuse, the extrapolated mean position is used to test $\mathcal{M}$-constraint satisfaction. Some error is introduced thereby.

The target could continue west (i = 4), but the extrapolation project beyond χ_5.
Here, $\iota = (444)$. In this case, the GWE stops the vehicle at χ_5.

$$(m_\iota[k], P_\iota[k]) \quad = \quad (m_\iota-[k], P_\iota-[k])$$

$$(m_\iota[k], P_\iota[k]) \xrightarrow{\mathcal{M}} ([\chi_5; 0; 0]', P_{\chi_5}) \qquad (10.54)$$

The modal state is still $\iota = (444)$. But the west extrapolation stops (the speed goes to zero) when $\hat{\chi}$ reaches the west terminus of the road segment.

The target could turn north (i = 1). Now $\iota = (144)$. The nearest junction accordant with a westbound target is χ_5. We will reinitialize the state and extrapolation follows using the northbound local model.

$$(m_\iota[k], P_\iota[k]) \quad = \quad ([\chi_5; 20; 0], P_{\chi_5})$$

$$(m_\iota[k], P_\iota[k]) \xrightarrow{\text{EKF}_\iota} (m_\iota^-[k+1], P_\iota^-[k+1]) \tag{10.55}$$

We use the map to place the target at the junction with the regime-nominal velocity. We then extrapolate using the EKF. The covariance matrix P_{χ_5} describes the uncertainty in the precise junction location—$P_\iota-$ could even be the zero matrix without creating a singularity in the GWE.

The target could turn south (i = 2).   In this case, $\iota = (244)$. The nearest junction accordant with a W-bound target is again χ_5. We will reinitialize the state and extrapolation follows using the southbound local model.

$$(m_\iota[k], P_\iota[k]) \quad = \quad ([\chi_5; -20; 0], P_{\chi_5})$$

$$(m_\iota[k], P_\iota[k]) \xrightarrow{\text{EKF}_\iota} (m_\iota^-[k+1], P_\iota^-[k+1]) \tag{10.56}$$

The target cannot turn east (i = 3) $\iota = (344)$. If the local estimate is such that no accordant junctions exists; e.g., a U-turn is not permitted at χ_5, the extrapolation step is ignored.

$$(m_\iota[k], P_\iota[k]) \quad = \quad (m_\iota-[k], P_\iota-[k])$$

$$(m_\iota[k], P_\iota[k]) \xrightarrow{\text{I}} (m_\iota^-[k+1], P_\iota^-[k+1]) \tag{10.57}$$

Extrapolation yields a swarm of 64 target locations, $\{m_\iota^-[k+1]; \iota \in \kappa\}$ with their associated covariances. Figure 10.8 illustrates one event complex. The target is westbound at $(-850, -800)$, and is approaching a T-junction at $(-950, -800)$. Let us look at the modal suffix $\iota^- = (44)$: the target has been westbound for the past 10 s.

Looking forward, the GWE sees four alternatives. The first is the most likely and is in fact the truth: the target continues west $\iota = (444)$. The extrapolated position is shown at $(-890, -800)$ with a bold **x** and the label 444. A local 1σ-error ellipse is shown about the extrapolation point. The uncertainty is longer longitudinally than laterally because P_{444}^w emphasizes east-west acceleration.

Another regime prefix is $\phi[k] = e_1$: the target could have turned north. But the nearest (and only) west-admitting, north-exiting junction is that at $(-950, -800)$. If the target turned north, it would have had to do it from that junction, and it would arrive at the point $(-950, -700)$ after 5 seconds while travelling at the nominal north speed. This extrapolation is shown in the figure with **x** and the label 144. Note that the east-west error is excised, and that the error ellipse is much smaller than associated with $\iota = 444$. This is because the longitudinal uncertainty is reduced at the junction, and the extrapolation begins with the small covariance determined by the geometry of the junction.

Fig. 10.8 Extrapolation of
the westbound target yields a
swarm of four possible
locations. The large
times symbol is the true
location; labeled 444. The
times symbol labeled 144
indicates the target has turned
north. The range-bearing
measurements are indicated
with +

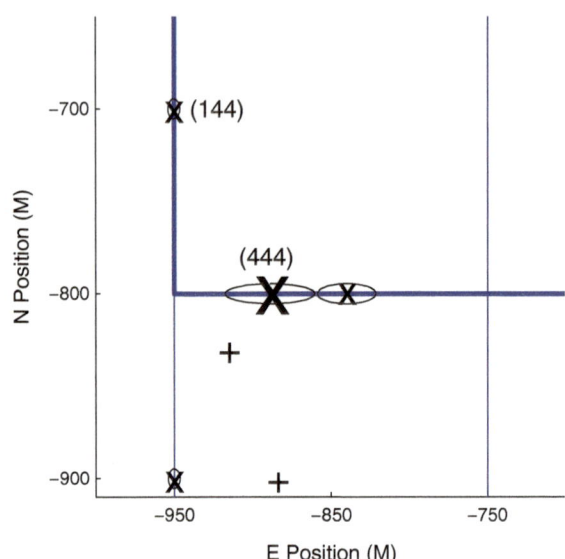

The target could have turned south; $\iota = 244$. All of the above comments apply symmetrically here. The point extrapolation moves to $(-950, -900)$ after 5 s. The error ellipse is like that for $\iota = 144$.

When we look at $\iota = 344$, we see that this corresponds to a U-turn to the east in mid-street. In this engagement, such a regime change is prohibited: $\iota^- = 44 \nrightarrow \iota = 344$ is not a permitted modal event. In this case, the **GWE** makes no extrapolation: the extrapolated state at $\iota = 344$ is that at $\iota^- = 44$. The error ellipse remains as it was.

Figure 10.8 shows the four progeny of $\iota^- = 44$. The true forward modal string is $44 \mapsto 444$. This subset of four extrapolations shows how the **GWE** separates its forward projections according to the regime possibilities. In the next section, we will see how the observations are weighted in determining the assurance regions.

10.3.1.2   Kinematic Update

The **GWE** fuses data from disparate sources. It first uses the location measurement $\{y[k + 1]\}$. The initial update is accomplished using the classical **EKF** algorithm. For each $\iota \in \kappa$,

$$(m_\iota^-[k + 1], P_\iota^-[k + 1]) \xrightarrow{\text{EKF}_\iota} (m_\iota^+[k + 1], P_\iota^+[k + 1]). \tag{10.58}$$

This $\mathcal{Y}[k + 1]$-increment in the kinematic state is written

$$\Delta d_\iota = H_\iota' D_\iota^n y[k + 1]$$

Fig. 10.9 The initial $\mathscr{Y}[k+1]$-update moves the extrapolated position (labeled *times symbol*) to the updated position (labeled *open circle*). The correct local filter, $\iota = 444$, is shown with larger icons

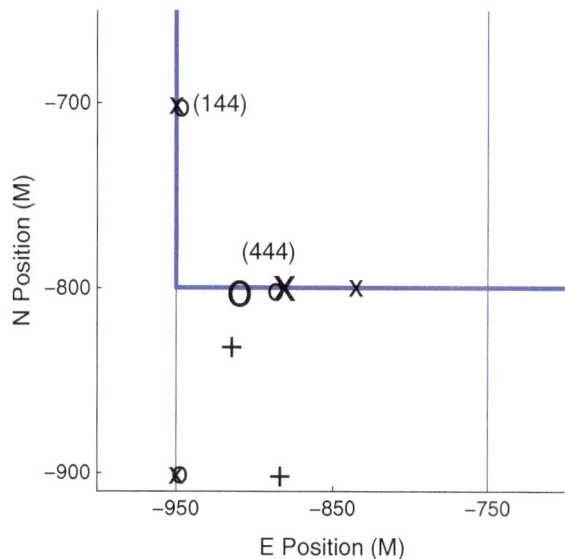

$$\Delta D_\iota = H_\iota' D_\iota^n H_\iota.$$

The information matrix, ΔD_ι, increases most when H_ι is large (good geometry) and D_ι^n is large (high SNR). The change in the information state, Δd_ι, is in the direction of $y[k+1]$ and is largest again when H_ι and D_ι^n are large. Figure 10.9 shows the two kinematic measurements with a +: The position measurements are $(-891, -902)$ and $(-922, -830)$. The latter favors $\iota = 244$ while the former favors the true mode, $\iota = 444$.

Figure 10.9 shows the adjustment position for the four progeny of $\iota^- = 44$. The adjustment for the $\iota = 444$ local filter is shown in the figure with the larger font. The adjustment is 30 m in an essentially west direction: even though $(y[k+1] - H_\iota m_\iota^-[k+1])$ is more north-south than east-west, the increment in $\hat{\chi}_\iota$ is primarily westward.

The adjustment for the 144-filter is tiny despite the large measurement residual. Again this is due to the small covariance associated with this modal string. This is true for the 244-filter as well. This same process is replicated for the 64 local filers that make up this **GWE**.

Unfortunately, the diffuse sensor noise will cause $\{m_\iota^+[k+1]\}$ to move from the grid. To restore map compliance, we use the $\mathbf{C}^{\mathcal{M}}$ transformation to find the point on the map, $\hat{\chi}_\iota^{+F}$ nearest $\hat{\chi}_\iota^+$. We then move the estimate to the map. This is the first phase of the $\mathbf{C}^{\mathcal{M}}$ transformation.

Map compliance also requires that the velocity be adjusted. To first order, the velocity must be longitudinal and correspond to the specific regime prefix in ι: $\hat{v}_1$ is small if $\iota = 144$ or $\iota = 244$; $\hat{v}_2$ is small if $\iota = 344$ or $\iota = 444$. And similarly the covariance $P_\iota^+[k+1]_{11}$ is small if $\iota = 144$ or $\iota = 244$; and so on.

The kinematic update is now complete, and each of the local filters is map compliant:

$$(m_\iota^+[k+1], P_\iota^+[k+1]) \xrightarrow{\mathcal{M}} (m_\iota^{+F}[k+1], P_\iota^{+F}[k+1]). \qquad (10.59)$$

Since (10.59) represents a simple replacement, we will write

$$(m_\iota^{+F}[k+1], P_\iota^{+F}[k+1]) \mapsto (m_\iota[k+1], P_\iota[k+1])$$

This completes the kinematic update. There are several engineering choices that are made to establish the local algorithm; e.g., the initialization at a turn, the intensity of the exogenous lateral accelerations. In the cases studied, performance has not been sensitive to these design selections.

10.3.1.3   Modal Update

The GWE delineates the likelihoods of the various modal hypotheses in terms of a (normalized or unnormalized) $\mathcal{G}[k+1]$-modal probability mass function which we write as $\{\alpha_\iota[k+1]; \iota \in \kappa\}$. The mass function is updated in two steps. First, with the kinematic adjustment based upon $y[k+1]$, and second, with the regime adjustment based upon $z[k+1]$.

Let us first look at the kinematic update. The $\iota = ijl$-filter begins with an initial condition $(\alpha_\iota[k], m_\iota[k], P_\iota[k])$. It extrapolates forward under the prefix $i \in \mathbf{S}$. At $t = [k+1]T^-$, it is now in state $(\alpha_\iota[k], m_\iota^-[k+1], P_\iota^-[k+1])$. The filter index remains ι, but the local moments move up one sample time.

At time $t = [k+1]T^-$, the tracker makes a measurement represented by $y[k+1] = H_\iota x[k+1] + \sqrt{P_\iota^n} n[k+1]$: the model relates the raw measurement $y[k+1]$ to the model state ι through both the geometry of the sensors (H_ι) and the peculiarities of the exogenous noise (P_ι^n). The covariance of the kinematic measurement will be labeled P^y:

$$P_\iota^y = H_\iota P_\iota^- H_\iota + P_\iota^n. \qquad (10.60)$$

The engagement model is such that $P_\iota^y > 0; \iota \in \kappa$. As such, $D_\iota^y > 0$ and $F_\iota^y > 0; \iota \in \kappa$.

The channel noise matrix requires a coordinate change from a polar systems to a rectilinear one. The transformation is based upon the local estimate of position; i.e., on $\hat{\chi}_\iota$. Hence, the intensity of the sensor noise, P_ι^n, conveys some information on the modal state; the size of the noise in $y[k+1]$ could favor a specific groups of modes.

The situation regarding P_ι^- is more subtle. The covariance is intimately linked to the path of the vehicle; e.g., is the target moving east or north? And the current velocity is a function of the full modal string—of which ι is a prefix.

As noted earlier.

$$E_\iota^y = \Delta \|m_\iota\|_{D_\iota}^2 - \|y[k+1]\|_{D_\iota^n}^2 \qquad (10.61)$$

is a unit-free quantity that is a dependent upon the kinematic measurement and the mode. To see more clearly what the ιth update is trying to accomplish, let us focus just on the position components of m_ι and on the upper block of D_ι. The units of the mean state are meters, and the units of the upper elements of D_ι are m^{-2}. So the vector $D_\iota m_\iota$ has units m^{-1}. The term $D_\iota m_\iota$ expands the mean state by emphasizing those components those value is the most sure. The direction in which we have more confidence (say lateral) is favored when multiplied by D_ι over the direction in which we have less confidence (say longitudinal).

When $y[k+1]$ is observed, both m_ι and D_ι change: The former in the direction of the measurement, and the latter gets bigger to the degree and in the direction that the measurement conveys useful information. This means that $\|m_\iota\|_{D_\iota}^2$ will change as well. When $y[k+1]$ is near the extrapolated state, the associated mode is favored: for a given mode, ι, the bigger $\Delta\|m_\iota\|_{D_\iota}^2$ is, the more likely the ιth mode is the correct one. Internally, the GWE computes both d_ι and P_ι and uses the equivalence $\Delta\|d_\iota\|_{P_\iota}^2 = \Delta\|m_\iota\|_{D_\iota}^2$ to compute this term in E_ι^y.

The above comment is for purpose of illustration. The units of m_ι and D_ι contain compounds of both meters and seconds. But the term $\|m_\iota\|_{D_\iota}^2$ emphasizes the kinematic states in which the ιth filter has most confidence. The increments in $\Delta\|m_\iota\|_{D_\iota}^2$ should favor the correct sub-filter: this filter moves a larger multiple of the standard units of information when the kinematic measurement is matched to the filter.

It is also true that the size of the state correction is partially due to the size of the measurement itself. To adjust for this, another unit-free term $-\|y[k+1]\|_{D_\iota^n}^2$ is included in E_ι^y. This term is bigger when the measurement is in a direction of lower SNR, and it acts to normalize for the size of the raw observation. Because E_ι^y is used to contrast the importance of alternative modes, when D_ι^n is independent of the mode, $\|y[k+1]\|_{D_\iota^n}^2$ is constant across modes, and it will be ignored in computing E_ι^y.

For each of the 64 filters in the GWE, the kinematic update of the modal mass function is

$$\alpha_\iota^-[k+1] = \alpha_\iota[k]|F_\iota^y|\exp(\tfrac{1}{2}E_\iota^y). \qquad (10.62)$$

Equation (10.62) updates the $\mathscr{G}[k+1]$-mass function with the $y[k+1]$ data. As with the kinematic state, the local filters are updated in parallel. As written, $\alpha_\iota^-[k+1]$; $\iota \in \kappa$ is unnormalized.

The next step in generating the $\mathscr{G}[k+1]$-modal mass function is the integration of the regime measurement. As we discussed earlier, we must choose the preceding regime, $z[k]$ or the successor, $z[k+1]$. The answer to this depends upon the delay in reducing a raw observation data to a classificational statement: at time $t = (k+1)T^-$ the regime is actually p while the current regime proxy is i. With the simple acoustic measurement, we will assume that the regime state is measured with quality matrix $\mathbf{D}$ and without latency. For $i \in \mathbf{S}$:

$$z[k + 1] = \mathbf{D}\phi_p[k + 1] + \eta[k + 1], \tag{10.63}$$

where $\mathbf{D}$ is the discernibility matrix. For $\iota^+ \in \kappa^+$, the update for a regime measurement is

$$\alpha_{\iota^+}[k + 1] = \alpha_{\iota}^-[k + 1]\Pi_{pi}z[k + 1]'\mathbf{D}_{\cdot p}. \tag{10.64}$$

Equation (10.64) generates a mass function on κ^+—a set S-times the size of κ.

To understand the structure of the regime update equation, observe that the factor Π_{pi} is the probability that the mode makes the transition $(ijl) \mapsto (pijl)$. Since the GWE formalism is based upon the (compromised) premise that $\{\phi[k]\}$ is a Markov process, this probability can be expressed as a function of the regimes; i.e., $\mathbb{P}(i \mapsto p)$.

The proposed transition is then confirmed by a measurement: $z[k + 1]'\mathbf{D}_{\cdot p}$. This factor tends to be higher in the actual direction of travel. For example, at the beginning of the interval at time $t = 23\,T = 115\,$s, the target is westbound: $\iota = 444$. It is also westbound at the end of the interval: $t = 120\,$s $: \iota^+ = 4444$. The acoustic sensor will classify the motion as westbound with probability 38 %, eastbound with probability 38 %, northbound with probability 12 %, southbound with probability 12 %. The GWE will weight the elements of $\alpha_{\iota}^-[k + 1]$ that correspond to east-west at three times the level of the north-south elements.

The nature of the forward regime transition probability matrix is somewhat contrary in this engagement. A premise in the development of the GWE is that $\{\phi[k]\}$ is exogenous: $\Pi_{ip} = \mathbb{P}(\phi[k + 2] = \mathbf{e}_p|\phi[k + 1] = \mathbf{e}_i)$ is a function of the ordered pair (pi) alone—though it could be a function of k with little change. In this case, the regime sequence is precursory to the kinematic state sequence. But in the tracking example, the regime transitions and the kinematic state are entwined. To illustrate, consider a westbound target at $\chi[k] = (-800, -800)$. The vehicle cannot turn north over the next interval: $444 \not\mapsto 1444$. We would have to say that $\Pi_{41} = 0$. Alternatively, if $\chi[k] = (-950, -800)$, a turn must be made: $444 \not\mapsto 4444$ and perhaps $\Pi_{41} = 0.5$. Unfortunately at $t = 115\,$s, both $\chi[k] = (-800, -800)$ and $\chi[k] = (-950, -800)$ are possible since the support of the $\mathscr{G}[k+1]$-conditional distribution of location covers both alternatives—though the former is far more likely.

To complete the synthesis of the GWE, we will approximate the forward transition probability with a Π matrix that takes into account the peculiarities of the path. Specifically, we will compose Π from a pair of primitive rate matrices: I and $\Pi_{\vdash}$. The former captures the situation when the target is between junctions, and the latter captures the situation when the vehicle is at a junction.

Between junctions, the regime transitions are not permitted; the target maintains its current direction. This simply means that $\mathbb{P}(p = i) = 1$. At a junction, the situation is not as clearly delineated. If the target encounters a north-south T-junction from the east; e.g., $(-950, -800)$, $\mathbb{P}(p = 1 \text{ or } p = 2) = 0.5$. But the 4-way junction at $(-750, -800)$ would be captured with $\mathbb{P}(p = 1) = \mathbb{P}(p = 2) = \mathbb{P}(p = 4) = 0.33$.

The transition behavior at a junction depends upon the junction type: does a westbound target encounter a 4-way or a T junction; can the target make a U-turn? In this engagement, we will simplify things and delineate junction behavior with the single matrix $\Pi_\vdash$ corresponding to a 4-way without U-turns. This is not a fully prescriptive choice: it makes every junction accessible from every point on the road with compatible east-north coordinates. But it simplifies internal structure of the algorithm. The **GWE** is not required to find those junctions accepting a given direction of motion. However, the way *closeness* is defined in the **GWE** has the effect of achieving directional agreement.

In principle, the tracker uses the appropriate rate matrix, Π, at every observation time. Unfortunately, the **GWE** does not place the vehicle determinately on $\mathcal{M}$; the probability that the tracker is in a neighborhood of every junction is positive, albeit small in most cases. Hence, the choice between $\Pi = \Pi_\vdash$ and $\Pi = \mathbf{I}$ cannot be made with certainty. For example, the westbound, 444-tracker at $(-850, -800)$, projects ahead to a neighborhood $(-900, -800)$. Even though this is well short of $(-950, -800)$, the junction is within the support of the $\mathscr{G}[k]$-distribution of the local filter.

More specifically, consider the 444-filter at time $t = 23\,T^-$. The target is moving west and $\hat{\chi} = (-920, -800)$. The west-variance is $281\,m^2$— $\sigma = 17$ m. Looking west, the filter sees the junction at $\hat{\chi} = (-950, -800)$: the junction is $1.7\,\sigma$ away in a west direction. The normative position distribution for this local filter is Gaussian. On this basis, the probability that the target lies west of the road terminus is given by the complementary error function, **efrc**: But $0.5\mathrm{erfc}(1.7/\sqrt{(2)}) = 0.04$: there is a 4 % chance that the target lies in the prohibited zone west of the junction. A plot of $0.5\mathrm{erfc}0.7\sigma)$ is shown in Fig. 10.10. If the **GWE** places the target at the junction: $\hat{\chi}_\iota = (-950, -800)$, there is still a 50 % probability that the target has not reached the junction. On the other hand, if m_ι is one σ east of the junction, the probability the target lies at the junction is 15 %.

To capture the ambiguity in location and the concomitant ambiguity in transition matrix, we will use the complementary error function, **erfc** to blend the primitives in the **GWE**, smoothly changing from I to $\Pi_\vdash$ as the target nears a junction.

To quantify the notion of approach to a junction, we use the Mahalanobis distance to measure the target/junction separation. At time $t = (k + 1)T$, the local position statistics of target position are $(\hat{\chi}, P_{\chi\chi})$. Suppose that the rth junction is compatible with ι: if $\iota = 444$, and χ_r is west of $\hat{\chi}_\iota$ with the same north coordinate, then χ_r is a possible regime transition point for the target.

The Mahalanobis distance, ϱ, from the target to the rth junction is defined as: $\varrho[m_\iota[k + 1], \chi_r] = \sqrt{\|\hat{\chi}_\iota[k + 1] - \chi_r\|_{D_\chi}^2}$. After a $y[k + 1]$ update, let $\varrho(m_\iota[k + 1])$ be distance to the nearest congruent junction. For example, a westbound target, $(\iota = 444)$, would look for the nearest west accessible junction with the same north coordinate. A gauge of the likelihood of encountering the junction during the next sample time is $\mathrm{erfc}(0.7\varrho(m_\iota[k + 1])$.

This **GWE** adjusts the Π matrix as a function of target position. It replaces the true position with the local mean, $\hat{\chi}_\iota$. It assumes the likelihood of making a regime

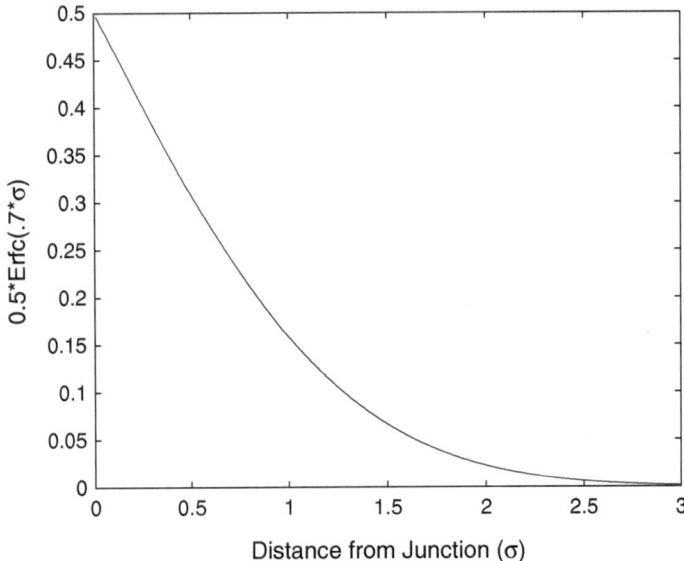

Fig. 10.10 The complementary error function decreases rapidly as a function of the standard deviation

change is dependent upon the longitudinal separation from the nearest junction measured in standard units. The modal transition matrix used in this engagement is:

$$\Pi_\iota = \Pi_\vdash \mathrm{erfc}(0.7\varrho(m_\iota[k+1]^+)) + I\mathrm{erf}(0.7\varrho(m_\iota[k+1]^+)) \tag{10.65}$$

If there are no compatible junctions, we will set $\Pi = \mathbf{I}$.

Equation (10.65) provides a smooth flow from $\mathbf{I}$ to $\Pi_\vdash$ as the target approaches a junction. Properly $\Pi_\vdash$ should be matched to the junction type. This refinement has not been incorporated into the **GWE** displayed here: $\Pi_\vdash$ corresponds to 4-way junctions without U-turns. This means that χ_ι is the distance to the *nearest* junction even if that junction is inaccessible. This will create anomalous estimates from time to time.

10.3.1.4 The $\mathscr{G}$[k+1]-Estimate of Target State

We have produced an algorithm that updates each of the 64 parallel filters that make up the **GWE**. Several engineering judgments were made to simplify the architecture particularly regarding the table of junctions. Each local tracker is map compliant, and the trackers interact weakly. Specifically we begin the ιth sub-filter at $t = kT$ with the condition $(\alpha_\iota, m_\iota-, P_\iota-); \iota \in \kappa$. Under $(y[k+1], z[k+1])$, we then update the sub-tracker state:

$$\alpha_\iota \mapsto \alpha_\iota + \tag{10.66}$$

$$m_\iota- \mapsto m_\iota[k+1]$$

$$P_\iota- \mapsto P_\iota[k+1].$$

To obtain the $\mathscr{G}[k+1]$-distribution of the kinematic state, we will first normalize the modal distribution: $\alpha_\iota + \mapsto \hat{\phi}_\iota+$. This links the modal filters, but normalization takes place outside of the internal estimates generated by the local algorithms that proceed in parallel.

From $\hat{\phi}_\iota+$, we can find several relevant marginal distributions:

$$\hat{\phi}_p[k+1] = \sum_\iota \hat{\phi}_{p\iota}[k+1] \quad \text{forward mode} \tag{10.67}$$

$$\hat{\phi}_i[k+1] = \sum_{p.\iota^-} \hat{\phi}_{p\iota}-[k+1] \quad \text{regime} \tag{10.68}$$

$$\hat{\phi}_\iota[k+1] = \sum_p \hat{\phi}_{p\iota}[k+1] \quad \text{mode} \tag{10.69}$$

$$\hat{\phi}_{\iota\odot}[k+1] = \hat{\phi}_\iota[k+1]/\sum_l \hat{\phi}_{\iota-l}[k+1] \quad \text{antecedent given } \iota. \tag{10.70}$$

From this we can write the $\mathscr{G}[k+1]$-distribution of the state:

$$\mathbb{P}[k+1] = \sum_{\iota\in\kappa} \hat{\phi}_\iota[k+1]\mathbf{N}(m_\iota[k+1], P_\iota[k+1]). \tag{10.71}$$

The distribution is a 64-term Gaussian sum. Equation (10.71) completes the internal cycle of the GWE.

The projected distribution, $\hat{\mathbb{P}}$ derives from $\mathbb{P}$:

$$\hat{\mathbb{P}}[k+1] = \mathbf{N}(\hat{x}[k+1], P[k+1]) \tag{10.72}$$

where

$$\hat{x}[k+1] = \sum_\iota \hat{\phi}_\iota[k+1]m_\iota[k+1] \tag{10.73}$$

$$P[k+1] = \sum_\iota \hat{\phi}_\iota[k+1](P_\iota[k+1]$$

$$+(m_\iota[k+1] - \hat{x}[k+1])(m_\iota[k+1] - \hat{x}[k+1])') \tag{10.74}$$

The former is a multi-term composite, and the latter is unitary Gaussian.

From $\hat{x}[k+1]$, the notional estimate of location, $\hat{\chi}[k+1]$, is extracted. However, because of its construction, $(\hat{x}[k+1], P[k+1])$ need not satisfy the external state space constraints. For example, $\hat{x}[k+1]$ may be located in a prohibited zone even though none of the $\{m_\iota; \iota \in \kappa\}$ are: a convex combination of local estimates may fail to meet the global constraints. In this event, we will adjust $(\hat{x}[k+1], P[k+1])$ in a suitable manner to force state map compliance. In this example, we will replace the $\mathcal{G}[k+1]$-target state with one that corresponds to the closest position on $\mathcal{M}$. In most cases, the distance from the map, $\|\hat{\chi}[k+1] - \mathcal{M}\|$, is slight: one element, $\hat{\phi}_\iota$, carries essentially all of the probability. In any case, we will make the following replacement without comment:

$$(\hat{x}[k+1], P[k+1]) \xrightarrow{\text{c}\mathcal{M}} (\hat{x}[k+1], P[k+1]). \qquad (10.75)$$

This suffices to create the projected distribution, $\hat{\mathbb{P}}$. The global estimates in $\hat{\mathbb{P}}$ take place outside the **GWE** inner loops, and they have no effect on the local filters.

Finally we must initialize the local filters for the next time step. For the $\iota[k+1]$th filter, let

$$m_\iota - [k+1] = \sum_l \hat{\phi}_\iota \odot [k+1] m_\iota[k+1]$$

$$P_\iota - [k+1] = \sum_l \hat{\phi}_\iota \odot [k+1] (P_\iota[k+1]$$

$$+ (m_\iota[k+1] - m_\iota - [k+1])(\ldots)').$$

These equations average the local statistics over the regime suffix in ι. It is only at this point that the local trackers are linked in the **GWE**. To the degree that the filter memory degrades quickly, this averaging is unobtrusive.

References

1. B.D.O Anderson, J.B. Moore, *Optimal Filtering* (Prentice-Hall, Englewood Cliffs, 1979)
2. C.G. Cassandras, S. Lafortune, *Introduction to Discrete Event Systems*, 2nd edn. (Springer, New York, 2008)
3. D.R. Cox, H.D. Miller, *The Theory of Stochastic Processes* (Wiley, New York, 1965)
4. R.J. Elliott, *Stochastic Calculus and Applications* (Springer, New York, 1982)
5. R.J. Elliott, J. van der Hoek, A finite dimensional filter for hybrid observations. IEEE Trans. Automat. Control **AC-43**, 736–739 (1998)
6. R.J. Elliott, L. Aggoun, J.B. Moore, *Hidden Markov Models: Estimation and Control* (Springer, New York, 1995)
7. R.J. Elliott, F. Dufour, D.D. Sworder, Exact hybrid filters in discrete time. IEEE Trans. Automat. Control **41**(12), 1807–1810 (1996)
8. J.S. Evans, R.J. Evans, Image-enhanced multiple model tracking. Automatica **35**, 1769–1786 (1999)
9. M.S. Grewal, A.P. Andrews, *Kalman Filtering: Theory and Practice* (Prentice-Hall, Englewood Cliffs, 1993)
10. A.H. Haddad, *Probabilistic Systems and Random Signals* (Prentice-Hall, New York, 2006)
11. W. Holzapfel, M. Sofsky, U. Neuschaefer-Rube, Road profile recognition for autonomous car navigation and navstar GPS support. IEEE Trans. Aerosp. Electron. Syst. **39**(1), 2–12 (2003)
12. A.H. Jazwinski, *Stochastic Processes and Filtering Theory* (Academic Press, New York, 1970)
13. X.R. Li, V.P. Jilkov, A survey of maneuvering target tracking—part III: measurement models, in *Proc. of the SPIE: Signal and Data Processing of Small Targets*, vol. 4473 (2001)
14. X.R. Li, V.P. Jilkov, Survey of maneuvering target tracking. Part 1: dynamic models. IEEE Trans. Aerosp. Electron. Syst. **39**(4), 1333–1364 (2003)
15. P.S. Maybeck, *Stochastic Models Estimation and Control*, vol. 2. (Academic Press, New York, 1982)
16. P. Misra, P. Enge, *Global Positioning System: Signals Measurements and Performance* (Ganga-Jamuna Press, Lincoln, 2001)
17. J.H. Painter, D.R. Kerstettet, S. Jowers, Reconciling steady-state kalman and alpha-beta filter design. IEEE Trans. Aerosp. Electron. Syst. **26**(6), 986–991 (1990)
18. B. Pannetier, K. Benameur, Ground moving target tracking with road constraint. in *Proc. of the SPIE: Signal and Data Processing of Small Targets*, vol. 5428 (2004)
19. D.D. Sworder, J.E. Boyd, *Estimation Problems in Hybrid Systems* (Cambridge University Press, Cambridge, 1999)
20. D.D. Sworder, J.E. Boyd, Target recognition from motion paths, in *Proc. of the SPIE: Signal and Data Processing of Small Targets*, vol. 3809 (1999), pp. 425–434

© Springer International Publishing Switzerland 2016
D.D. Sworder, J.E. Boyd, *Locating, Classifying and Countering Agile Land Vehicles*, DOI 10.1007/978-3-319-19431-8

21. D.D. Sworder, J.E. Boyd, Target recognition and tracking. Int. J. Syst. Sci. **32**(7), 937–945 (2001)
22. D.D. Sworder, J.E. Boyd, Maneuver sequence identification, in *Proc. of the SPIE: Signal and Data Processing of Small Targets*, vol. 5204 (2003), pp. 155–164
23. D.D. Sworder, J.E. Boyd, Motion models for tracking. IEEE Trans. Aerosp. Electron. Syst. **41**(3), 1052–1056 (2005)
24. D.D. Sworder, R. Vojak, Tracking mobile vehicles using a non-Markovian maneuver model. J. Guid. Control Dyn. **17**(4), 870–873 (1994)
25. D.D. Sworder, P.F. Singer, D. Doria, R.G. Hutchins, Image enhanced estimation methods. Proc. IEEE **81**(6), 797–814 (1993)
26. D.D. Sworder, M. Kent, R. Vojak, R.G. Hutchins, Renewal models for maneuvering targets. IEEE Trans. Aerosp. Electron. Syst. **31**(1), 138–150 (1995)
27. X. Wang, S Challa, R.J. Evans, Gating techniques for maneuvering target tracking in clutter. IEEE Trans. Aerosp. Electron. Syst. **38**(2), 1087–1096 (2002)